Synthesis Lectures on Mechanical Engineering

This series publishes short books in mechanical engineering (ME), the engineering branch that combines engineering, physics and mathematics principles with materials science to design, analyze, manufacture, and maintain mechanical systems. It involves the production and usage of heat and mechanical power for the design, production and operation of machines and tools. This series publishes within all areas of ME and follows the ASME technical division categories.

Khalid M. Mosalam · Yuqing Gao

Artificial Intelligence in Vision-Based Structural Health Monitoring

 Springer

Khalid M. Mosalam
Department of Civil and Environmental
Engineering
University of California
Berkeley, CA, USA

Yuqing Gao
Department of Civil and Environmental
Engineering
University of California
Berkeley, CA, USA

ISSN 2573-3168   ISSN 2573-3176 (electronic)
Synthesis Lectures on Mechanical Engineering
ISBN 978-3-031-52406-6   ISBN 978-3-031-52407-3 (eBook)
https://doi.org/10.1007/978-3-031-52407-3

To the engineers interested in pursuing transdisciplinary research.

Foreword

Structural health monitoring (SHM) is the process of developing global, continuous, online, automated, and quantified *in situ* damage detection capabilities for all types of engineered systems. The term SHM started to appear regularly in the aerospace, civil, and mechanical engineering literature around the later 1980s and early 1990s where it can be seen to evolve from more traditional local non-destructive evaluation (NDE) methods. The online aspect of SHM attempts to address the shortcoming of NDE that almost always require the system being assessed to be taken out of service and/or disassembled. Looking back still further reveals that both SHM and NDE have their origins in the *ad hoc* qualitative visual, audio, and tactile damage detection methods that man has employed, most likely, since tools have come into use. Today, researchers worldwide continue to develop improved SHM processes driven by the economic and life-safety societal benefits that this technology has the potential to provide, which motivates the need for additional textbooks on this subject.

Like many technology developments, advances in SHM have been made possible by the confluence of advances in different enabling technologies (hardware and software) that engineering practitioners and researchers have integrated in creative ways to produce higher-fidelity damage-detection capabilities. From a hardware perspective, SHM has benefited from reductions in cost and increases in resolution of various sensing modalities and associated data acquisition hardware (e.g., accelerometers, high-resolution cameras, Analog-to-Digital (A-to-D) converters) coupled with the continuous reduction in costs in computing hardware (processors and memory) and increased capabilities of these assets.

The revolution in data science over the last twenty years has had a tremendous impact on all aspects of the software used to implement the data analysis portion of SHM including sensor optimization, the identification of damage-sensitive features, the ability to separate changes in these features caused by damage from changes caused by varying operational and environmental conditions, and the statistical classification of the features into damaged and undamaged classes. During this time the SHM field has adopted many data-driven algorithms from disparate fields such as radar and sonar detection, artificial intelligence/machine learning, speech-pattern recognition, statistical decision theory, and econometrics.

In the future I anticipate that all engineers will be expected to understand and apply state-of-the-art data science tools to solve a wide variety of currently intractable engineering challenges. In this regard, the book *Artificial Intelligence in Vision-Based Structural Health Monitoring* is particularly timely because it gives the reader a detailed summary of many recently developed data science tools in an engineering context. The artificial intelligence/machine learning material presented in Part I of the book is generally applicable to all approaches to SHM and NDE as well as engineering problems beyond those focused on damage detection. It is refreshing to see these topics presented in an engineering context. For me personally, seeing some of the concepts from data science (e.g., support vector machines and the "kernel trick") explained by engineers has given me better insight than I have been able to glean from the data science literature.

Parts II to IV of this book apply the artificial intelligence/machine learning approaches from Part I to vision-based structural health monitoring where computers are trained to identify damage from images. Understanding these vision-based approaches to the damage detection problem should be of particular interest to the broader SHM community because of the potential to incorporate imagers onto robotic inspection vehicles (submerged, surface (land and sea), and aerial), the potential for large-area surveillance that such vehicles provide, and their ability to perform the surveillance in hazardous environments. In addition to giving the reader an in-depth overview of vision-based SHM, these chapters provide detailed examples showing engineering applications of artificial intelligence/machine learning from Part I.

Finally, it should be noted that the authors are considering some of the leading researchers in the field of vision-based SHM. I find their explanations to be well organized and they have organized the material in a very logical manner. The best compliment I can provide to Prof. Mosalam and Dr. Gao regarding this book is that I plan to incorporate some of the material contained in both Part I and Parts II–IV into the graduate SHM course I teach as an adjunct faculty member at the University of California, San Diego and in the SHM short course I have been teaching for 20+ years.

Los Alamos, New Mexico, USA   Charles R. Farrar
December 2023   Los Alamos National Laboratory Fellow

Preface

The genesis of this book is the doctoral dissertation of the second author under the supervision of the first author. Several journal publications, included in the list of references, by the two authors together with their collaborators contributed to the material organized in this book. Textbooks focusing on Structural Health Monitoring (SHM) coupled with recent advances in Computer Vision (CV) using machine and deep learning are very limited. The authors are interested in developing and offering courses that encompass the two topics of SHM and CV as part of the engineering curriculum. Moreover, short courses and training on such a dual topic can be offered as part of the continuing education for practicing engineers, especially those interested in rapid reconnaissance activities. These two reasons motivated us to embark on the project of writing this book, which builds on our previously published research.

The development of this book is timely because of the current activities related to the growing need for rapid and effective reconnaissance efforts. This is currently emphasized through the Structural Extreme Events Reconnaissance (StEER) Network. We hope this book will fill the gap in combining SHM with CV techniques to be used as a reference to students in courses related to modern SHM and to engineers interested in reconnaissance activities following extreme events.

The book is divided into four parts which can be taught over two semesters. The first semester, after the introduction in Chap. 1, can cover basic materials discussed in Chaps. 2–4 (representing Part I). Applications in Chaps. 5–7 (within Part II) can complement the basic materials of the first semester. Students will be able to carry out a course project at the conclusion of the first semester using the material covered in the first 7 chapters of the book. The second-semester offering can consider the remaining material of the book, i.e., Chap. 8 of Part II, Chaps. 9–11 (Part III), and Chaps. 12 and 13 (Part IV). Upon the completion of the second semester, students will be able to embark on graduate-level research, e.g., toward Ph.D. dissertations, related to using CV for many problem in SHM.

Berkeley, CA, USA   Khalid M. Mosalam
December 2023   Yuqing Gao

Acknowledgements

The authors would like to thank many individuals who participated in their research activities to provide the foundational material for this book. These include collaborators: Dr. Selim Günay, Dr. Sifat Muin, Dr. Chenglong Li, Dr. Yue Zheng, Dr. Cheng Fang, Dr. Jianfei Yang, Dr. Hanjie Qian, Mr. Boyuan Kong, and Mr. Shiyuan Lu, graduate and undergraduate students of the first author: Guanren Zhou, Fan Hu, Issac Kwok-Tai Pang, Chrystal Chern, Claudio Perez, Jiawei Chen, Gaofeng Su, Juan Meriles, and Pengyuan Zhai. Many funding agencies contributed resources used toward the different research projects that led to many developments and applications presented in this book. These are Taisei Chair of Civil Engineering, National Science Foundation (NSF) for the Project "Structural Extreme Events Reconnaissance (StEER) Network", Pacific Earthquake Engineering Research (PEER) Center for the project related to the development of PEER Hub ImageNet (ϕ-Net), California Department of Transportation (Caltrans) for the project "Bridge Rapid Assessment Center for Extreme Events (BRACE2)", California Geological Survey for several projects related to data interpretation from the "California Strong Motion Instrumentation Program", and NSF & United States Department of Agriculture (USDA) for the "AI Institute for Next Generation Food Systems (AIFS)" project. Opinions expressed in this book are those of the authors and they do not reflect the points of view of these sponsors. We wish to extend our special thanks to Guanren Zhou for his outstanding work in conducting case studies on structural damage localization and segmentation in Chapters 7 and 8. Special thanks are given to Fan Hu for her dedicated efforts in careful editing and meticulously checking the technical content of the entire manuscript.

Yuqing moved to Tongji University, Shanghai, China, on August 2023, and he would like to thank Profs. Xilin Lu, Ying Zhou, and Wensheng Lu for their instructive suggestions. The writing of this book started in summer of 2022 and the complete proofreading took place in late 2023 during Khalid's sabbatical leave at the Disaster Prevention Research Institute (DPRI), Kyoto University, Japan. He would like to express his gratitude to his host Prof. Masahiro Kurata of DPRI for all the arrangements made during his sabbatical time that enabled the completion of this book. Last but not least, the efforts of both authors would not have been possible without the continuous sacrifice and encouragement of their family members.

Contents

Acronyms

1D	One-Dimensional
2D	Two-Dimensional
3D	Three-Dimensional
AL	Active Learning
AE	Autoencoder
AI	Artificial Intelligence
ANN	Artificial NN
AP	Average Precision
AR	Average Recall
AUC	Area Under the Curve
Bbox	Bounding box
BHM	Bridge Health Monitoring
BMP	Bitmap
BN	Batch Normalization (or BatchNorm)
BSS-GAN	Balanced Semi-Supervised GAN
CAE	Convolutional AE
CAL	Convolutional Active Learning
CAM	Class Activation Map
CART	Classification and Regression Tree
CCD	Concrete Crack Detection
CCIC	Concrete Crack Images for Classification
CelebA	CelebFaces Attributes dataset
CIFAR	Canadian Institute for Advanced Research
CM	Confusion Matrix
C-MTL	Convolutional-MTL
CNN	Convolutional NN
Conv	Convolution
CPU	Central Processing Unit
CV	Computer Vision
DA	Data Augmentation

DCGAN	Deep Convolutional GAN
DE	Decoder
Deconv	Deconvolution
DeconvNet	Deconvolution Network
DL	Deep Learning
DT	Decision Tree
EERI	Earthquake Engineering Research Institute
ELU	Exponential Linear Unit
EN	Encoder
FC	Fully-Connected
FCN	Fully CNN
FHWA	Federal Highway Administration
FID	Fréchet Inception Distance
FM	Feature Matching
FN	False Negative
FP	False Positive
FPN	Feature Pyramid Network
GA	Generalization Ability
GAN	Generative Adversarial Network
GAP	Global Average Pooling
GBP	Guided Backpropagation
GPU	Graphics Processing Unit
Grad-CAM	Gradient-weighted CAM
GRU	Gated Recurrent Unit
H	Heuristic
HiL	Human-in-the-Loop
HTL	Hierarchical TL
ILSVRC	ImageNet Large Scale Visual Recognition Challenge
IoU	Intersection over Union
IR	Information Retrieval
IS	Inception Score
JPEG	Joint Photographic Experts Group
JSON	JavaScript Object Notation
KKT	Karush–Kuhn–Tucker
KL	Kullback–Leibler
LB	Leaf-Bootstrapping
LB-GAN	Leaf-Bootstrapping GAN
LFE	Learning From Earthquakes
LIME	Local Interpretable Model-Agnostic Explanation
LN	Layer Normalization
LR	Logistic Regression

LReLU	Leaky Rectified Linear Unit
LSTM	Long Short-Term Memory
LSUN	Large-scale Scene Understanding
LSV	Latent Space Vector
MAE	Mean Absolute Error
MAMT2	Multi-Attribute Multi-Task Transformer
mAP	mean AP
MBGD	Mini-Batch Gradient Descent
MD	Message-Digest
MHA	Multi-Head Attention
MINST	Modified National Institute of Standards and Technology
mIoU	mean IoU
ML	Machine Learning
MLP	Multi-Layer Perceptron
MSA	Multi-head SA
MS COCO	Microsoft Common Objects in Context
MSE	Mean Square Error
MSRA	Microsoft Research Asia
MTL	Multi-Task Learning
NA	Not Applicable
NLP	Natural Language Processing
NMS	Non-Maximum Suppression
NN	Neural Network
PANet	Path Aggregation Network
PE	Positional Encoding
PEER	Pacific Earthquake Engineering Research
PGGAN	Progressive Growing GAN
PHI-Net (or ϕ-Net)	PEER Hub ImageNet
PHI-NeXt (or ϕ-NeXt)	Next generation PHI-Net
PNG	Portable Network Graphics
PR	Precision-Recall
PSD	Positive Semi-Definite
PSP-Net	Pyramid Scene Sparse Network
RBF	Radial Basis Function
RC	Reinforced Concrete
R-CNN	Regional CNN
ReLU	Rectified Linear Unit
ResNet	Residual Network
RF	Random Forest
RGB	Red, Green & Blue
RMSE	Root Mean Square Error

RNN	Recurrent NN
ROC	Receiver Operating Characteristic
RoI	Regions of Interest
RPN	Region Proposal Network
SA	Self-Attention
SCAE	Stacked Convolutional AE
SDF	Synthetic Data Fine-tuning
SDNET	Structural Defects Network
Self-T	Self-training
SGD	Stochastic Gradient Descent
SHM	Structural Health Monitoring
SIGMA-Box	Structural Image Guided Map Analysis Box
SIM	Structural ImageNet Model
SIS	Self-IS
SPPNet	Spatial Pyramid Pooling Network
SSD	Single Shot Detection
SSL	Semi-Supervised Learning
Std	Standard deviation
StEER	Structural Extreme Events Reconnaissance
SV	Support Vector
SVM	SV Machine
Swin	Shifted window
SW-MSA	Shifted Window MSA
TIFF	Tagged Image File Format
TL	Transfer Learning
TN	True Negative
TNR	True Negative Rate
TP	True Positive
TPR	True Positive Rate
TRIMM	Tomorrow's Road Infrastructure Monitoring and Management
UAV	Unmanned Aerial Vehicle
VGGNet	Visual Geometry Group Net
ViT	Vision Transformer
WGAN	Wasserstein GAN
W-MSA	Window MSA
WTT	Wind Turbine Tower
XAI	eXplainable Artificial Intelligence
YOLO	You Only Look Once

List of Figures

List of Tables

List of Algorithms

Introduction

1.1   Background and Motivation

Structural health monitoring (SHM), as defined in [1], is a "paradigm" of rapidly identifying structural damage in an instrumented structural system and can be classified in terms of their scale–local or global damage detection methods. Whereas global methods employ numerical models that intake global characteristics of a structure (such as modal frequencies) that are indicative of possible damage, local methods attempt to detect damage by screening structural systems at the individual element scale.

During the life cycle of a structure, it typically goes through different types of loading stages, from the common daily loads to severe destructive external loads caused by earthquakes, tsunamis, hurricanes, etc. These external loads introduce different degrees of structural damage that may lead to losses, injuries, and even fatalities. Therefore, SHM in the form of rapid and automated damage assessment and reconnaissance efforts during daily service conditions or after natural hazards has become an important focus in the field of Structural Engineering, which aims for effective decision-making toward building more resilient infrastructural systems.

Traditionally, SHM and reconnaissance efforts require onsite human inspection and potential field testing. For example, bridge conditions are identified through regular bridge inspections performed by contractors on-site, as mandated by regulations enforced by governmental highway agencies, e.g., U.S. Federal Highway Administration (FHWA). However, such traditional inspections and testing are not conducted on a regular basis because of limited human resources, financial burdens, and time-consuming efforts. Given the large number of infrastructural systems and buildings, the total amount of necessary labor and resources would make this process impractical to be conducted on a regional scale and regular basis. Moreover, in some cases, it is difficult to make onsite inspections where some damage states are hard to notice, or it is unsafe and risky right after an extreme event to enter damaged buildings to make the necessary detailed inspection [2].

Regarding the history of SHM, besides traditional human inspection and specimen testing, a variety of approaches and damage criteria were developed through pattern analysis from data, which is known as data-driven SHM. Through analyzing the relationship between the input and output data, the current states and characteristics of the instrumented structures can be obtained, which are used for evaluating the health conditions of these structures and making decisions during the SHM process. Moreover, the boosting of computer hardware and the increasing capacity and decreasing cost of data storage make the data-driven SHM attract increasing attention from the Structural Engineering community. Based on the data type and source, current data-driven SHM can be mainly divided into two categories: (1) vision-based SHM and (2) vibration-based SHM. The vision-based SHM relies on visual data (mainly images and videos) collected from the field. On the other hand, the vibration-based SHM exploits the vibration signals from deployed sensors, usually accelerometers. The current focus of data-driven SHM aims to augment and eventually replace repetitive human engineering works with automated recognition systems using data based on artificial intelligence (AI) technology. The scope of this book is vision-based SHM. However, most of the theoretical background in Part I can be easily extended to be applicable to vibration-based SHM.

1.2   Vision-Based SHM

Nowadays, structural damage recognition using images and videos is one of the important topics in vision-based SHM and structural reconnaissance, which heavily relies on human visual inspection and experience. Take an example of a bridge damage[1] inspection in practice. The inspectors and engineers may utilize inspection vehicles, equipped with a variety of detection instruments for data recording, to inspect all parts of the bridge and recommend appropriate maintenance actions, Fig. 1.1. However, human inspection is typically laborious and subjective resulting in possible erroneous measurements and ultimately the wrong decisions.

The bridge inspection vehicle is a special transport vehicle equipped with various detection instruments and workbenches and can detect or repair various parts of the bridge. The commonly used bridge inspection vehicles include the folding arm type (Fig. 1.2a) and the truss type (Fig. 1.2b), which can be used for bridge damage detection at different scales. Using inspection vehicles often requires the inspectors and engineers to reach the intended position for inspection through platforms and end up working at high elevations above the ground, which brings about several safety concerns.

---

[1] This can be in the form of cracking due to accumulated environmental deterioration or due to an extreme event such as a major earthquake.

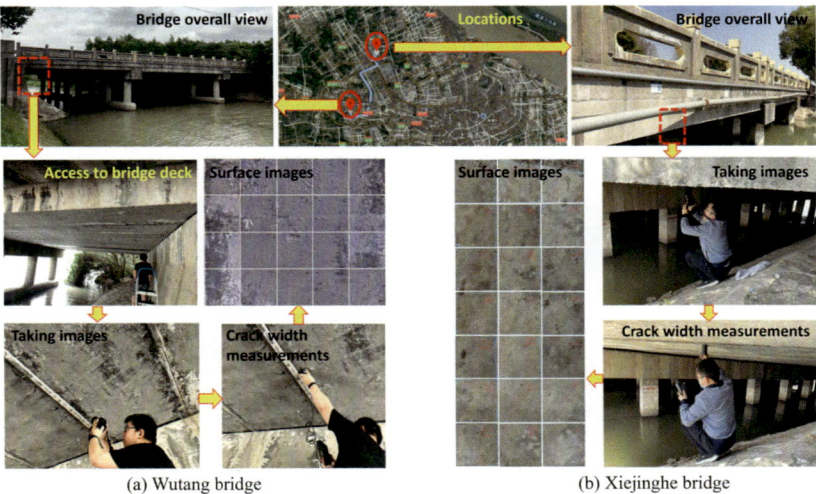

(a) Wutang bridge   (b) Xiejinghe bridge

Fig. 1.1 Bridge cracking manual inspection for two bridges in Shanghai, China [3]

In the past few decades, computer vision (CV) technologies have provided another direction for SHM. Many traditional CV methods, e.g., histogram transformation [4], threshold segmentation [4–7], edge detection [8–10], and seeded region growing [11–13], have been developed and explored in the vision-based crack detection applications of different structures.

In 1979, Nobuyuki Otsu [4] proposed an adaptive threshold determination method for image binarization, also known as the maximum inter-class variance method. The algorithm assumes that image pixels[2] can be divided into background and objects according to a threshold value and then calculates the best threshold value to distinguish these two types of pixels, so as to maximize the pixel differentiation. According to the characteristics that crack pixels in pavement images (as an application) are continuous and "darker" than their surroundings, Cheng et al. [5] proposed a new road crack detection algorithm based on fuzzy logic. The algorithm first determines the brightness membership function (a mathematical terminology used in fuzzy sets) for gray levels in the different images, transforms the fuzzy

---

[2] A pixel has a varying intensity and it is the smallest addressable element in a digital image. In a colored image, the color is usually represented by three component intensities such as red, green, and blue. A pixel is also the smallest addressable element in a dot matrix display device to be computationally manipulated. An image can be described as a "signal" because an image contains information that can be extracted or interpreted using signal processing techniques, referred to as image processing techniques. It is worth noting that a signal is any quantity conveying information or broadcasting a message in the physical world between two observers and is measurable through time over space. A one-dimensional (1D) signal is typically measured over time, e.g., an audio signal. Accordingly, a digital image is a two-dimensional (2D) signal defined using a function $I = f(x, y)$ where x and y are the two coordinates and I gives the pixel intensity at point (x, y) of the image.

(a) Folding arm type inspection vehicle

(b) Truss type inspection vehicle

Fig. 1.2 Two commonly used bridge inspection vehicles

image from the gray-level domain to the binary crack domain according to the membership function, then checks the connectivity of the darker pixels to eliminate the isolated points and finally uses the image projection method (e.g., a forward image projection model can project a three-dimensional (3D) scene into the image plane.) to classify the cracks. Through computer experiments, they found that when threshold segmentation is performed on the real-time collected images, the image quality is affected by external factors and a single threshold cannot accurately distinguish the crack from the background and thus proposed an algorithm for real-time threshold adaptation. However, even if the threshold changes in real-time, a large amount of noise inevitably appears during the recognition and it remains necessary to conduct a manual intervention to reduce the noise. To address the issue of uneven illumination of the image, which may occur when the light does not evenly illuminate the object across the field of view, Fujita et al. [7] proposed to adopt the image smoothing method[3] in the image preprocessing, then used a linear filtering method[4] based on the Hessian matrix[5] to enhance the crack information, and finally used the threshold segmentation method to extract the crack information. This algorithm can successfully extract crack information from concrete images, but due to the fixed threshold method, it lacks flexibility and is accordingly narrowly applicable.

In 1996, Heath et al. [9] compared four different edge detection algorithms, including the Canny operator [14], the Sobel operator [15], the Nalwa-Bainford operator [16], and the Sarkar-Boyer operator [17]. The results show that the detection effect of each algorithm changes with varying degrees according to whether the parameters in the algorithm change during the calculations and the Nalwa-Bainford operator is shown to be the least affected. In addition, if the same algorithm is used for images with different backgrounds, the performance varies greatly. Martin et al. [10] proposed an algorithm that uses local images to detect and locate boundaries in natural scenes. The algorithm is directly based on the original brightness, color, and texture of the image. Therefore, it is more accurate than other models based on gray-scaled images. However, there exists a large number of features in the image when training the data-driven model and these features need to be labeled manually, which is not suitable for actual crack detection in practice.

Huang and Xu [11] judged the crack seeds[6] based on the grid cell analysis of adjacent sub-blocks of the image and connected them according to the direction of the information. This method considers that the contrast between the crack and the background is strong and the crack has a certain width and length to eliminate the influence of any noise. In addition, Sun et al. [13] proposed a backtracking method using a priority-based path growing from automatically selected seeds. This method can track the linear features with weak contrast in the image and accordingly can reflect the local features around the selected seeds.

The detection accuracy, efficiency, and robustness of the above mentioned methods do not yet meet the engineering requirements for practical applications and decision-making. Therefore, more intelligent and accurate methods are urgently needed for the purpose of practical applications of vision-based SHM.

---

[3] Image smoothing is a digital image processing technique to reduce or suppress image noises, e.g., using pixels neighborhood averaging of their intensities.

[4] In image enhancement and signal processing, linear filtering is one of the most powerful methods, where part of the signal frequency spectrum is modified by the transfer function of the filter.

[5] The Hessian matrix is a square matrix of second-order partial derivatives of a scalar-valued function, or scalar field. It describes the local curvature of a function of many variables.

[6] In the algorithm, a pavement image is divided into a grid of 8×8 pixel cells and each cell is classified as a non-crack cell or a crack cell based on the statistics of the gray scales of the cell pixels. A crack cell can be regarded as a *seed* for crack formation.

1.3   Artificial Intelligence in Vision-Based SHM

With the development of high-performance computing devices, e.g., graphics processing unit (GPU), AI technologies are growing rapidly, which paved the way to benefit several real-world engineering applications. Simply stated, AI technology is defined as algorithms and methods that can learn by mimicking the learning behavior of human beings and then realize automation for certain repetitive tasks. Nowadays, machine learning (ML) and deep learning (DL) are popular and thought of as two concrete realization methods of AI. The objective of ML and DL implementation is that as computers perform labor-intensive repetitive tasks, they simultaneously "learn" from performing those tasks. ML, founded on pattern recognition, has already undergone substantial development over the past twenty years [18]. Moreover, ML is a general topic with a broad scope and DL is one specific branch of ML, which has significant representation abilities for learning complex patterns by a large number of parameters in the so-called deep neural networks.

Both ML and DL technologies have been developed rapidly in recent decades, especially in the CV domain [19]. Inspired by the above mentioned conventional CV methods, Sect. 1.2, and to address concerns related to the need for tedious manual and non-autonomous efforts, several studies [20–23] attempted to catalog key vision features based on human knowledge with respect to damage patterns, e.g., edges of the crack and texture of the reinforced concrete (RC) cover spalling area, and then developed algorithms and criteria for determining the damage state based on these features. It is noted that features extracted based on human knowledge are known as *handcrafted features* and the procedure of finding these features is referred to as *feature engineering*. Using handcrafted features for decision-making aligns well with the concept and procedure of AI, especially ML. Accordingly, ML algorithms are used to learn hidden relationships between the extracted features from data and the corresponding damage patterns. Handcrafted features are concise, but they may be limited according to the image recognition perspective, where in some cases, useful damage features are abstract and exist in high-dimensional spaces. Humans are limited by having only a sense of low-dimensional features, e.g., locations, colors, or edges. Due to the complexity and diverse representations of the image data, handcrafted features may not align well with the general cases of practical engineering applications and usually only work for limited scenarios.

DL is being considered as a solution to the limitations mentioned in the above paragraph. Unlike conventional pipelines of feature engineering and ML, the DL approach uses the concept of *representation learning* to replace manual feature engineering. Convolutional neural networks (CNNs) have been at the heart of recent advances in DL. The CNN no longer requires handcrafted low-level features or feature engineering, where millions of parameters inside a typical network are constructed by stacking different types of learning layers. In this way, the network is capable of learning countless mid- to high-level image representations with input data obtained from pixel matrix (tensor) data. By stacking more layers, the network becomes deep, which forms the deep CNN. There are many well-designed deep

CNN architectures, e.g., VGGNet [24], InceptionNet [25], and ResNet [26], and they have demonstrated great performance improvements with substantial increases in network depth. Although CNN has been developed since the 1990s to solve handwritten digit recognition tasks [27, 28], its recent applications on a much wider scale can be attributed to advances in computing technology, such as the availability and affordability of high-performance GPUs. In the past few years, great strides have been made by the DL research community, achieving state-of-the-art results in many vision-recognition tasks compared with traditional approaches [19]. Following the trends of AI and the corresponding improvements in DL approaches, it is timely to implement state-of-the-art DL technologies in Civil Engineering applications and evaluate their potential benefits, particularly for vision-based SHM.

The introduction of AI and specifically DL into the field of SHM is not a straightforward task. Until now, vision-based SHM applications have not fully benefited from these technologies, even with the interest in this topic is ever-increasing. There are three key research demands (elaborated upon in the following three sub-sections) that need to be addressed prior to the broader adoption of AI technology in vision-based SHM. These are:

1. Verify the *feasibility* of AI models in SHM recognition tasks.
2. Improve the *adaptability* of AI models in real and complex environments.
3. Strengthen the *resiliency* of the AI paradigm for intelligent disaster prevention[7].

1.3.1   Feasibility

The structural recognition tasks based on vision are novel and challenging. Unlike the natural image data in the traditional CV field, structural images usually have the characteristics of containing a large amount of information, high complexity, and high-level abstraction (i.e., semantics). From another perspective, the application of AI to Structural Engineering is hamstrung by several reasons, which are attributed to the following factors: (1) lack of general automated detection principles or frameworks based on domain knowledge, (2) lack of benchmark[8] datasets with well-labeled large amounts of data, and (3) lack of validation experiments in real engineering scenarios. Therefore, the feasibility of using AI in vision-based SHM, represented by the accuracy and efficiency of the trained AI models, should be verified first.

Many past studies were conducted within a limited scope and were lacking a systematic approach without paying attention to procedures of post-disaster reconnaissance or field

---

[7] The goal here is "preventing a hazard or an extreme event from turning into a disaster".

[8] In CV and DL, a benchmark is a standardized set of tasks or datasets used to evaluate the performance of algorithms or models.

investigation.[9] Instead of superficially borrowing technologies from computer science, more insights into the engineering domain knowledge should be brought into building a general and systematic vision detection framework, which is strongly founded on: (1) defining the objectives, i.e., key detection tasks critical for post-events considering inter-task relationships, (2) defining the outputs, i.e., label categories for annotated dataset creation, and (3) developing technical solutions for a robust framework, i.e., implementation of classification, localization, and segmentation algorithms. Therefore, SHM detection tasks can be systematically performed in a consistent manner for engineering design and evaluation toward the key research demands of SHM, as defined above in Sect. 1.3.

In the CV domain, there exist several large-scale open-source datasets, e.g., ImageNet[10] [29], MNIST[11] [27], and CIFAR-10[12] [30], with about 14 million, 70,000, and 70,000 images, respectively, and the training of DL models can indeed benefit from these large-scale image datasets. However, in Structural Engineering, especially in vision-based SHM, there was no such large-scale high-variety open-source image dataset until recently. Most studies related to the image-based SHM using DL conducted validation experiments based on their own custom-labeled small-scale datasets and self-defined tasks. As a result, the numerical results reported in these previous works are incomparable. Moreover, as mentioned above, the labeled images in vision-based SHM are much more different from natural objects in general CV benchmark datasets. Two major differences are: (1) complex visual patterns and (2) abstract semantics of the labels to describe the structural attributes, e.g., damage occurrence, damage type, structural component type, etc. As for item (1), damages in the form of cracks and RC spalling do not have fixed shapes and vary widely. For item (2), labels in tasks such as identifying concrete damage types between shear vs. flexural damage, as defined in [31] for RC structures, must engage domain knowledge within Structural Engineering. For example, flexural damage is attributed to: (i) cracks developed horizontally in vertical columns or vertically in horizontal beams due to the induced bending moments, or (ii) cracks or spalling concentrating at the ends of the component, e.g., a column, due to plastic hinging also induced by bending moments. Both visual patterns (i) and (ii) are significantly different, but they have the same semantics leading to flexural damage. These ambiguous characteristics bring difficulties in training the AI models.

Although tremendous effort has been made in vision-based SHM, especially crack detection for damaged structures, much of this effort was either based on the ideal conditions with

---

[9] An exception is the **St**ructural **E**xtreme **E**vents **R**econnaissance (StEER) Network, https://www.steer.network/, with its set of elaborate goals and developed detailed procedures of the reconnaissance efforts for structural systems following natural hazards.

[10] An image database, which has been instrumental in advancing CV and DL research, where the data are available to researchers for non-commercial use, https://www.image-net.org/.

[11] Modified National Institute of Standards and Technology (MINST) dataset, http://yann.lecun.com/exdb/mnist/.

[12] A 10-classes dataset developed by the Canadian Institute for Advanced Research (CIFAR-10), https://paperswithcode.com/dataset/cifar-10.

well-processed data or focused on a single topic, e.g., crack classification or crack segmentation. There still exists a gap between the developed vision-based technologies and their applications in real-world projects. Therefore, the feasibility of the AI methods in practical SHM scenarios should be verified.

1.3.2   Adaptability

Applying AI technologies to practical engineering projects faces the challenge of adaptability due to complex environments and uncertainties. Firstly, even though a large number of structural images for structural health assessment is feasible to collect in real life, these massive datasets in actual SHM projects or reconnaissance efforts are often unprocessed and unlabeled. When they are applied to the training of ML and DL models, very large laborious manual efforts and resources are required for data annotation and processing. In addition, labeling such a large dataset of images can be cumbersome, time-consuming, and heavily relies on the availability of extensive domain knowledge. As a result, most relevant past studies limited their scope to be conducted within a small amount of labeled data.

Unlike the above mentioned CV benchmark datasets, image datasets collected in actual SHM projects usually have highly imbalanced classes. For example, the data carrying structural damage labels often only account for a small proportion of the total data[13]. Such imbalance originates from the nature of SHM, where structural damages due to either natural deterioration or extreme events such as earthquakes are fortunately less frequent cases. This is attributed to the structural design code objectives and the corresponding high levels of conservatism for safety purposes. The uneven distribution of labels may lead to training a biased model, i.e., always predicting the results as the majority class. Most existing studies attempted to avoid this issue of data imbalance by constructing relatively small but balanced datasets. However, building an ideal data environment (such as uniformly distributed labels, noise-free data, etc.) is not helpful in enhancing the model's adaptability for practical scenarios.

Computing power limitations create another daunting challenge for efficiently adapting AI models to real projects. DL has benefited from the advancement of high-performance GPUs, the lack of which, however, becomes an insuperable obstacle in real-world applications. For example, the limited payload (i.e., carry-on) capacity of a given small unmanned aerial vehicle (UAV) [32] due to budget limits or external environmental factors forbids the deployment of high-performance GPUs and supporting modules on such UAVs for efficient data collection and real-time processing of a large region following an extreme event. Thus, to be able to conduct real-time recognition and inference with limited hardware, the network architecture needs to be degraded to a shallow one based on a very lightweight design,

---

[13] An exception to this observation usually occurs in reconnaissance work where inspectors typically collect images of damaged structures (e.g., due to a major earthquake) of much larger numbers than those of undamaged structures.

e.g., MobileNet [33], which, however, compromises the recognition performance. Besides, to pursue a better recognition performance, the network architectures (e.g., the number of layers and the number of filters, refer to Chap. 4) in past studies were usually designed and tuned specifically for their respective datasets after many trials and errors. It is well-known that tuning a large model's parameters is often costly and may not even be generic enough for adapting the final network to practical applications.

1.3.3 Resiliency

Upon verifying the feasibility and enhancing the adaptability of the AI models, a third important research demand is to effectively build a resilient AI system for SHM recognition tasks. The ultimate goal of such is to realize a resilient response (prevention or limited effect of the shock to the system or assuring rapid recovery to full functionality) in the face of disasters. In this book, the authors propose to add the concept of "resiliency" into the AI-aided SHM system, manifested in the following three objectives: (1) the recognition tasks should be recognized in an effective, efficient, and robust manner, (2) the system and its outcome can easily recover from uncertain situations, e.g., wrong, noisy, or incomplete data, and (3) the trained AI models are explainable.[14]

Structural images typically contain a large amount of information, defined as multiple structural attributes by domain experts, e.g., the scale of the object, the type of the structural component, and the severity of the damage. Each structural attribute may have a set of labels to describe it. For example, the attribute of component type has labels such as "beam", "column", etc. and the attribute of damage pattern is described as "minor damage", "heavy damage", etc. These attributes and their different labels may have hidden relationships with each other, i.e., a certain hierarchy. For example, knowing that a structural RC shear wall is heavily cracked or has significant spalling of the concrete cover of its steel reinforcement is positively correlated to the fact that this wall is indeed damaged. Therefore, comprehensively considering these attributes and their hierarchical labels can be very informative toward rapid post-disaster reconnaissance efforts and ultimately decision-making. This is expected to increase the robustness of the AI recognition models and thus increase the resiliency of the entire AI system. However, based on an extensive literature search, there are no related studies conducted comprehensively to consider multiple structural attributes simultaneously.

The performance of current AI models is easily subjected to the lack of high-quality data and missing information caused by sensor damage, perception blind spots, and other factors typically existing in field measurements. Especially, missing and wrong labels are unavoidable in the annotation process. For example, some attributes in the structural images may be ambiguous causing the human annotators to make mistakes or even skip labeling. For

---

[14] Explainable AI corresponds to AI models where humans can understand the decisions or predictions made by the model. This is an improvement over the "black box" concept in ML where even its designer may not be able to explain why the ML model arrives at a specific decision.

some large-scale datasets, due to time constraints and limited resources, only part of the data or their attributes are labeled while the remaining data are kept unlabeled. These difficulties lead to missing and possibly wrong labels for certain attributes, where conventional ML and DL methods are incapable of effectively handling such situations. These scenarios of missing or wrong labels greatly limit the recognition accuracy and the generalization of the AI models to practical engineering scenarios.

Another big challenge toward a resilient AI-aided SHM system is interpretability or developing the so-called eXplainable AI (XAI). Currently, the working paradigm of AI (i.e., DL herein) models is mostly numerical iterations and calculations based on the available data, which is difficult to be directly understood by researchers and engineers. As a result, it is often regarded as a "black box". The lack of physical meaning and interpretability of the recognition results greatly limits the application of these AI technologies in real-world projects. Such a shortcoming is very critical to the application of AI in structural and building engineering. Although an uninterpretable DL model may achieve acceptable performance on a custom experimental dataset, without a good understanding of its discrimination principle, it may result in wrong predictions on previously unseen (i.e., not part of the original training dataset) real-world datasets. This can lead to serious consequences and wrong decision-making, e.g., failure to identify shear damage in columns may lead to sudden building story collapse during post-disaster reconnaissance and evaluation. Therefore, since decision-making in SHM has major socioeconomic impacts, a clear understanding and explanation of the DL model will result in a more convincing AI utilization for practical engineering applications [34]. However, in the vision-based SHM, there are no relevant studies specifically focusing on visually understanding and explaining the working principles of the DL models in the required recognition tasks.

1.4 Objectives and Scopes

This book provides a comprehensive coverage of the state-of-the-art AI technologies in vision-based SHM. As researchers begin to apply these concepts to the Civil and Structural Engineering domains, especially in SHM, several critical scientific questions need to be addressed: (1) What SHM problems can be solved using AI? (2) What are the relevant AI technologies to SHM? (3) What is the effectiveness of the AI approaches in vision-based SHM? (4) How to improve the adaptability of the AI approaches for practical engineering projects? and (5) How to build a resilient AI-aided disaster prevention and recovery system making use of the vision-based SHM?

The overall technical overview of the different topics covered in this book is illustrated in Fig. 1.3 as they are related to the above mentioned scientific questions. In Part I, Preliminaries: A concise coverage of the development progress of AI technologies in the vision-based SHM is presented. It gives the readers the motivation and background of the relevant research efforts and developments. Subsequently, basic knowledge of ML & DL is introduced, which

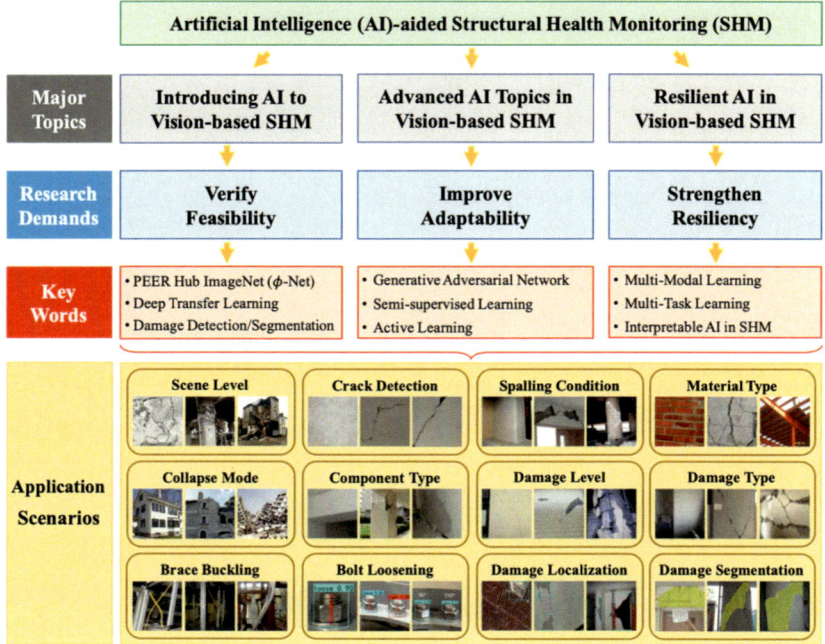

Fig. 1.3 AI-aided computer vision-based SHM

provides the foundation for the readers irrespective of their background. This is followed by three core parts, as discussed below, including verifying the feasibility, improving the adaptability, and strengthening the resiliency of structural systems utilizing the discussed AI methods.

In Part II, Introducing AI to vision-based SHM: To verify the feasibility of the AI methods, the key procedure of the typical AI-aided SHM applications (classification, localization, and segmentation) is explored, including vision data collection, data preprocessing, model training, evaluation, and analysis. Specifically, a general detection framework and dataset, namely, PEER[15] Hub ImageNet (PHI-Net or ϕ-Net), is established and the structural image-based transfer learning (TL) mechanism is proposed, which breaks through the key bottlenecks of the recognition difficulties and improves the effectiveness of the developed AI models.

In Part III, Advanced topics of AI in vision-based SHM: Several new developments, e.g., generative adversarial network, semi-supervised learning (SSL), and active learning (AL), are discussed. These developments aim to address several critical issues in practical engineering projects, e.g., the lack of well-labeled data and the existence of imbalanced labels, to improve the adaptability of the trained AI models.

---

[15] PEER is the **P**acific **E**arthquake **E**ngineering **R**esearch Center, https://peer.berkeley.edu/.

In Part IV, Resilient AI in vision-based SHM: The new concept of "resilient AI" is introduced to establish an intelligently health-monitored structural system. It mainly includes multi-task learning (MTL) and interpretable AI technologies, which are aimed toward increasing the robustness, efficiency, and interpretability of the AI-enabled SHM system and ultimately leading to improved resiliency of the built environment through effective and clearly understood AI-enabled monitoring systems.

1.5   Exercises

1. What is SHM? What is data-driven SHM? How to categorize SHM based on its available data types?
2. Briefly describe the advantages of using AI in SHM.
3. Discuss the role of AI in optimizing maintenance and repair strategies based on available data for the purpose of SHM.
4. Discuss potential future advancements and trends in AI for SHM.

References

1. J. Lynch, K. Loh, A summary review of wireless sensors and sensor networks for structural health monitoring, in *The Shock and Vibration Digest* (2006)
2. Applied Technology Council, *Field Manual: Postearthquake Safety Evaluation of Buildings* (2005)
3. Y. Zheng et al., Multistage semisupervised active learning framework for crack identification, segmentation, and measurement of bridges. Comput. Aided Civil Infrastruct. Eng. **37**(9), 1089–1108 (2022)
4. N. Otsu, A threshold selection method from gray-level histograms. IEEE Trans. Syst. Man Cybern. **9**(1), 62–66 (1979)
5. H.D. Cheng et al., Novel approach to pavement cracking detection based on fuzzy set theory. J. Comput. Civil Eng. **13**(4), 270–280 (1999)
6. H.D. Cheng, X.J. Shi, C. Glazier, Real-time image thresholding based on sample space reduction and interpolation approach. J. Comput. Civil Eng. **17**(4), 264–272 (2003)
7. Y. Fujita, Y. Mitani, Y. Hamamoto, A method for crack detection on a concrete structure, in *18th International Conference on Pattern Recognition (ICPR–9206)*, vol. 3 (IEEE, 2006), pp. 901–904
8. S. Behera, M.N. Mohanty, S. Patnaik, A comparative analysis on edge detection of colloid cyst: a medical imaging approach, in *Soft Computing Techniques in Vision Science* (Springer, 2012), pp. 63–85
9. M. Heath et al., Comparison of edge detectors: a methodology and initial study. Comput. Vis. Image Underst. **69**(1), 38–54 (1998)
10. D.R. Martin, C.C. Fowlkes, J. Malik, Learning to detect natural image boundaries using local brightness, color, and texture cues. IEEE Trans. Pattern Anal. Mach. Intell. **26**(5), 530–549 (2004). https://doi.org/10.1109/TPAMI.2004.1273918

11. Y. Huang, B. Xu, Automatic inspection of pavement cracking distress. J. Electron. Imaging **15**(1), 013017 (2006)

12. Q. Li et al., FoSA: F* seed-growing approach for crack-line detection from pavement images. Image Vis. Comput. **29**(12), 861–872 (2011)

13. C. Sun, P. Vallotton, Priority-based path growing for linear feature detection, in *9th Biennial Conference of the Australian Pattern Recognition Society on Digital Image Computing Techniques and Applications (DICTA 2007)* (IEEE, 2007), pp. 360–365

14. John Canny, A computational approach to edge detection. IEEE Trans. Pattern Anal. Mach. Intell. **6**, 679–698 (1986)

15. N. Kanopoulos, N. Vasanthavada, R.L. Baker, Design of an image edge detection filter using the Sobel operator. IEEE J. Solid-State Circuits **23**(2), 358–367 (1988)

16. V.S. Nalwa, T.O. Binford, On detecting edges. IEEE Trans. Pattern Anal. Mach. Intell. **6**, 699–714 (1986)

17. S. Sarkar, K.L Boyer, Optimal infinite impulse response zero crossing based edge detectors. CVGIP: Image Underst. **54**(2), 224–243 (1991)

18. C.M. Bishop, *Pattern Recognition and Machine Learning* (Springer, New York, 2006)

19. Ian Goodfellow, Yoshua Bengio, Aaron Courville, *Deep Learning* (MIT Press, Cambridge, 2016)

20. D. Feng, M.Q. Feng, Experimental validation of cost-effective vision-based structural health monitoring. Mech. Syst. Signal Process. **88**, 199–211 (2017)

21. M.M. Torok, M. Golparvar-Fard, K.B. Kochersberger, Imagebased automated 3D crack detection for post-disaster building assessment. J. Comput. Civil Eng. **28**(5), A4014004 (2014)

22. C.M. Yeum, S.J. Dyke, Vision-based automated crack detection for bridge inspection. Comput. Aided Civil Infrastruct. Eng. 30(10), 759–770 (2015)

23. H. Yoon et al., Target-free approach for vision-based structural system identification using consumer-grade cameras. Struct. Control Health Monit. **23**(12), 1405–1416 (2016)

24. K. Simonyan, A. Zisserman, Very deep convolutional networks for large-scale image recognition (2014). arXiv:1409.1556

25. C. Szegedy et al., Going deeper with convolutions, in *Proceedings of the IEEE Conference on Computer Vision and Pattern Recognition* (2015), pp. 1–9

26. K. He et al., Deep residual learning for image recognition, in *Proceedings of the IEEE Conference on Computer Vision and Pattern Recognition* (2016), pp. 770–778

27. Y. LeCun et al., Gradient-based learning applied to document recognition. Proc. IEEE **86**(11), 2278–2324 (1998)

28. Y. LeCun et al., Object recognition with gradient-based learning, in *Shape, Contour and Grouping in Computer Vision* (Springer, 1999), pp. 319– 345

29. J. Deng et al., Imagenet: a large-scale hierarchical image database, in *2009 IEEE Conference on Computer Vision and Pattern Recognition* (2009), pp. 248–255

30. A. Krizhevsky, Learning Multiple Layers of Features from Tiny Images, in Technical Report TR-2009 (2009)

31. J. Moehle, *Seismic Design of Reinforced Concrete Buildings* (McGraw Hill Professional, 2014)

32. T. Francesco Villa et al., An overview of small unmanned aerial vehicles for air quality measurements: Present applications and future prospectives. Sensors 16(7), 1072 (2016)

33. A.G. Howard et al., Mobilenets: efficient convolutional neural networks for mobile vision applications (2017). arXiv:1704.04861

34. L. Sun et al., Review of bridge structural health monitoring aided by big data and artificial intelligence: from condition assessment to damage detection. J. Struct. Eng. **146**(5), 04020073 (2020)

Part I
Preliminaries

In Part I, an informative coverage of the rapid advancements of AI technologies within the realm of SHM is undertaken. Firstly, Chap. 2 serves as a foundational introduction, offering a succinct overview of the progress made in the development of AI technologies for vision-based SHM. Four typical directions, namely: (1) structural damage classification, (2) structural damage localization, (3) structural damage segmentation, and (4) structural image generation & augmentation, are introduced in detail. Delving into these topics provides the readers with a clear understanding of the motivation and historical background that have propelled the relevant research in this domain.

To set the stage for the readers, regardless of prior knowledge or background, the fundamentals of ML are introduced in Chap. 3. Starting from the theoretical formulation of the ML problem, the selection of the loss function and the evaluation metrics are discussed. Subsequently, several classical ML algorithms are introduced, namely, logistic regression, support vector machine, decision tree, and random forest.

In Chap. 4, the focus shifts to delve into the complexities and advancements in the field of DL, which is a subset of ML. DL is characterized by the use of variant types of neural networks with multiple layers, such as CNNs, recurrent neural networks (RNNs), and transformers, allowing for the automated extraction of intricate features from the data. The fundamental formulations, classical architecture designs, and practical techniques are introduced. This essential knowledge forms a solid base upon which the readers can build their understanding of the subsequent content related to advanced topics of AI in vision-based SHM.

Vision Tasks in Structural Images

2

2.1 Structural Damage Classification

Structural damage classification is a straightforward application of vision-based SHM, which is defined as follows: Given the structural image input, the trained AI model outputs one damage state among multiple potential choices with their probabilities. The input is usually a colored (i.e., RGB: red, green and blue) image. It is encoded[1] and stored by the computer as a data array or tensor that has a shape of (height, width, 3), where each individual pixel consists of red, green, and blue color components with a value ranging from 0 to 255, with the smaller number closer to zero represents the darker shade (black for a gray-scale image) while the larger number closer to 255 represents the lighter shade (white for a gray-scale image), refer to Fig. 2.1. There are many choices for the selection of the AI models and algorithms, e.g., feedforward neural network, CNN, and transformer, which are discussed in detail in later chapters of this book. With the data array, the AI model generates discrete outputs corresponding to a predefined number of classes. As illustrated in Fig. 2.2, one typical scenario is the *binary* damage detection to identify whether the structure or one of its structural components is damaged or not, corresponding to the damaged state and the undamaged state, respectively. Besides the simple task of damage detection, the damage quantification task can be realized by increasing the number of classes, e.g., a 4-class classification can cover the range from undamaged state to minor, moderate, or heavy damage states.

---

[1] In computer language, "encoding" operation is defined as a process of putting a sequence of characters (letters, numbers, punctuations, or certain symbols) into a specialized format for efficient transmission or storage. In this book, especially for vision-based problems, encoding structural images refers to the process of converting RGB images into a numerical representation with digits.

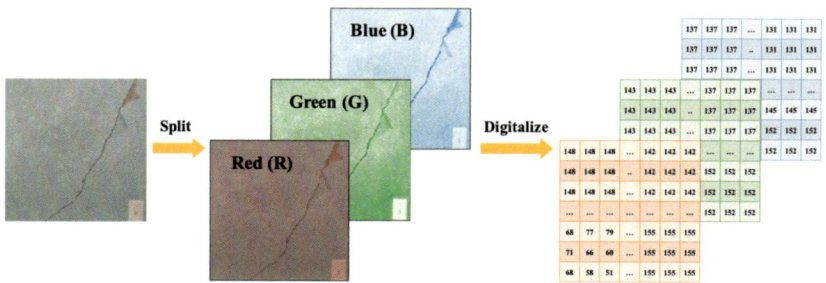

Fig. 2.1   RGB structural image

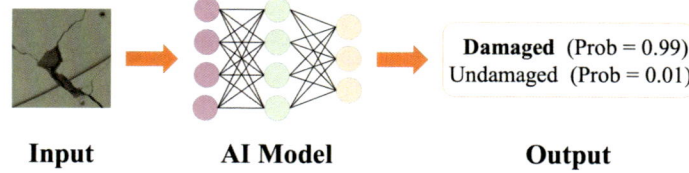

Fig. 2.2   Image-based structural damage classification

There has been an increasing trend of applying CNN models in structural damage classification, especially in concrete crack detection and steel corrosion detection. Soukup and Huber-Mörk [1] applied CNN to detect steel surface defects of the railway, which is a binary classification problem. Cha et al. [2] used a deep CNN to detect concrete cracks as a binary classification without calculating the defect features. Zhang et al. [3] designed a new CNN architecture, namely, CrackNet, for pavement crack detection at the pixel level, also as a binary classification. Vetrivel et al. [4] combined deep CNN and 3D point cloud data to detect façade and roof damage of buildings, using labels for undamaged and damaged states. Besides the binary classification cases, Gao and Mosalam [5, 6] performed a study using TL techniques along with various deep CNNs and defined multiple benchmark classification tasks, with extensions to more complex tasks involving more classes, e.g., 3-class damage level quantification and 4-class damage type identification.

In summary, many early studies have mainly focused on the binary problem of whether the structure and/or its structural components are damaged or not and analyzed the images at the pixel (material) level.[2] In contrast, images collected by engineers in reconnaissance efforts and SHM problems are usually related to the object (structural component) level or even represent the structural (whole or major part of the structural system) level. This makes utilizing current CNN models more difficult to adopt without first conducting an elaborate image preprocessing. Furthermore, in decision-making based on the reconnaissance efforts, conducting more complex tasks is essential and requires further studies for a comprehensive

---

[2] This level and others, namely, object and structural levels, are discussed in Chap. 5.

evaluation, e.g., finding the location of the damage and determining the damage severity level and its type.

2.2   Structural Damage Localization

In the CV domain, besides image classification, object detection is an important application aiming to detect instances of semantic objects of a certain class (e.g., people, trees, or buildings) in images or videos. When extending this concept to SHM, the objective becomes finding damaged areas and further identifying the content within these found areas. This is known as structural damage detection or damage localization. Specifically, the damage location is usually represented by the key coordinates of a bounding box of the target damage, which is the outer contour of the damaged area, as illustrated in Fig. 2.3. In general, there are three common scenarios in the damage localization tasks: (1) single damage area localization, (2) multiple damage areas localization, and (3) multiple damage type identification and localization. Figure 2.3a and b are examples of detecting spalling (SP) damage, i.e., loss of cover material, usually concrete covers of reinforcing cages in RC components, where Fig. 2.3a detects a single spalling concrete cover area in the corner of a RC column and Fig. 2.3b detects multiple (two) areas of plaster spalling of a masonry wall. In Fig. 2.3c, multiple types of damage, i.e., SP and cracking (CR), are identified as well as their respective locations.

There are three types of widely used definitions for bounding box coordinates:

- Definition "1" using (x_1, y_1, x_2, y_2): horizontal (x) and vertical (y) coordinates of top-left and bottom-right corners of the bounding box, refer to Fig. 2.4a.
- Definition "2" using $(x_1, y_1, W_{box}, H_{box})$: x and y coordinates of the top-left corner together with the width and height of the bounding box, refer to Fig. 2.4b.

 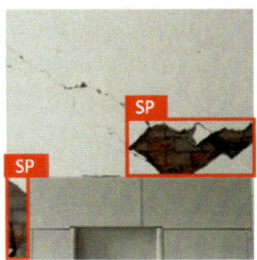 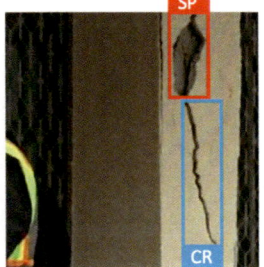

(a) Single spalling area   (b) Multiple spalling areas   (c) Multiple damage types

Fig. 2.3 Examples of structural damage localization

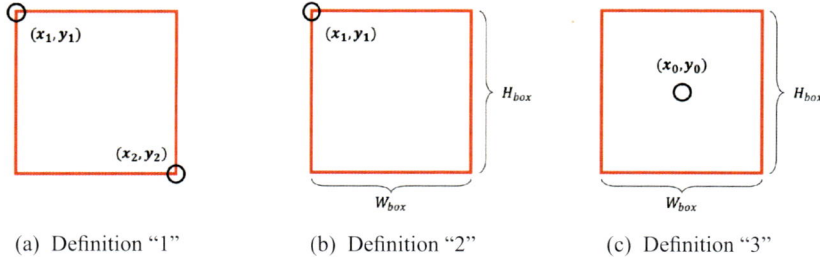

(a) Definition "1"   (b) Definition "2"   (c) Definition "3"

Fig. 2.4 Three common bounding box coordinate definitions

- Definition "3" using $(x_0, y_0, W_{\text{box}}, H_{\text{box}})$: x and y coordinates of the center together with the width and height of the bounding box, refer to Fig. 2.4c.

In practical implementation, these three definitions are easily transformable from one definition to the other.

Since 2014, Girshick et al. [7] proposed a region-based CNN (R-CNN) and pushed object detection research into a new DL-driven era. Until now, there are two major types of detection methods, namely, one-stage and two-stage. The two-stage method usually first extracts the object locations represented by several candidate region proposals (bounding boxes) and then conducts classification through CNN to obtain the class information and its corresponding probability. On the contrary, the one-stage method does not require generating candidate region proposals and instead computes the object locations and their corresponding class probabilities directly from the image pixels. Several representative algorithms of the one-stage method are "you only look once" (YOLO) [8], "single shot detection" (SSD) [9], and RetinaNet [10]. YOLO, in particular, continues to evolve into a series of algorithms. It now has more than 8 variants, where YOLO v8 [11] is the latest version released in January 2023. Similar to YOLO, R-CNN is not only a milestone but also the ancestor of the two-stage detection method, e.g., Fast R-CNN [12], Faster R-CNN [13], and Mask R-CNN [14]. In practice, the two types of detection methods work differently for different scenarios. Usually, region-based methods (two-stage) are superior in the classification and localization accuracy, while end-to-end (one-stage) methods are superior in the computational speed.

Inspired by the above mentioned methods, there have been emerging studies of damage localization in vision-based SHM starting from 2018. For the one-stage method, both the SSD and YOLO series have been applied. Maeda et al. [15] adopted SSD in road damage detection, which obtained high accuracy and good run-time speed on a smartphone. Chen et al. [16] proposed a hybrid detection framework, utilizing both SSD and YOLO in a sequential manner to localize cantilever joints and their fasteners to diagnose defects of the fasteners. Qiu et al. [17] proposed an automated pavement distress analysis system based on the YOLO v2 network to identify road anomalies in need of urgent repairs. Yang et al. [18] introduced a *candidate mechanism* into an automatic bolt loosening detection framework, where YOLO v3 and v4 with their variants were simultaneously implemented and compared as candidates. They showed that this approach improved both the accuracy and robustness of the obtained localization results. With the evolution of YOLO, newer versions of YOLO were lately put into research and applications [19–21].

For the two-stage methods, the majority of the published studies utilize the R-CNN series of available methods. Cha et al. [22] proposed a region-based algorithm, which utilizes the Faster R-CNN, for detecting multiple damage types on the concrete surface. Ju et al. [23] developed a deep network based on the Faster R-CNN for crack detection, which achieves robust performance but requires intense computational resources. Similar to Faster R-CNN, Xue and Li [24] proposed a region-based network adopting fully CNN (FCN[3]) to more efficiently classify and localize tunnel lining defects in geotechnical engineering. Zheng et al. [25] proposed a multi-stage concrete detection and measurement framework, where Mask R-CNN [14] is adopted to locate the concrete cracking and spalling areas.

Conducting a detection/localization task is not as straightforward as a classification task. The former usually requires more sophisticated and well-designed models and training process. In the CV domain, it is sometimes treated as the downstream task[4] from the classification, e.g., using pre-trained parameters from the classification task to fine-tune the model for detection and localization tasks [14, 26].

2.3   Structural Damage Segmentation

The structural damage segmentation task originates from the semantic and instance[5] segmentation tasks in the CV domain, which aims to categorize each pixel in an image into a class or object and is used to recognize a collection of pixels that form distinct classes. Specifically, semantic segmentation only distinguishes and segments objects with different class labels (e.g., all cracks may be marked blue without distinguishing between different instances in the cracking category), while instance segmentation further segments objects of different instances within the same class (e.g., in the cracking category, different crack types may be marked differently such that flexural cracks are marked green while shear cracks are marked red). Herein, the structural damage segmentation task is defined to quantify the damage by finding the whole damaged area, where each pixel has its own label, and regions of pixels with the same label are grouped and segmented as one object (class). An example is illustrated in Fig. 2.5, where there is a large spalling of plaster and bricks along with a wide

---

[3] The main difference between FCN and CNN is that FCN replaces the fully-connected layer in CNN (refer to Chap. 4) with a convolution layer operation.

[4] If there exist multiple tasks, a downstream task is the one the analyst actually wants to solve, which is usually provided with less quantity of labeled data, e.g., object detection and segmentation. The AI models pre-trained on other tasks can be helpful in such a downstream task.

[5] In ML, an instance is a realization of the data (input-output pair) described by a number of attributes (often called features), where one attribute can be a class label.

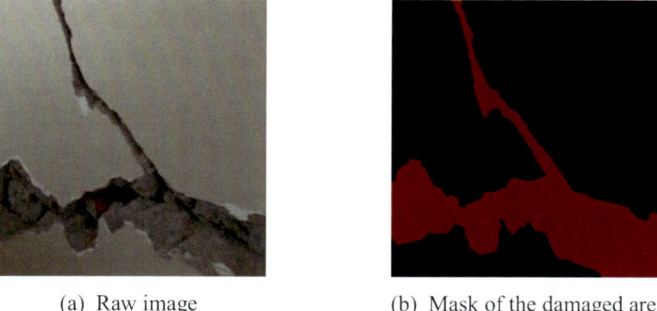

(a) Raw image   (b) Mask of the damaged area

Fig. 2.5 Structural damage segmentation (one mask for two damage types)

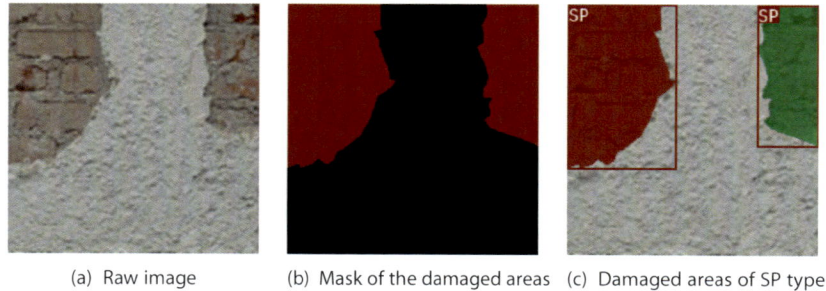

(a) Raw image   (b) Mask of the damaged areas   (c) Damaged areas of SP type

Fig. 2.6 Structural damage segmentation (two masks for a single damage type)

crack on the wall surface. The raw image can be categorized into two parts: (i) damaged region and (ii) background, where in this case the damaged region is marked in red and the background is in black. Another example including multiple damaged areas is presented in Fig. 2.6, where there are two discrete spalling areas of plaster loss (i.e., single damage type, namely, SP) in a brick masonry wall. In practice, especially in data preparation, only the damaged region is labeled pixel-wise and such region is denoted as the *mask*. There could be several damaged regions and accordingly several masks in the same image. The objective of the DL-driven segmentation is to train the AI models to generate masks close to the ground truth. Unlike the localization task, which gives rough rectangular regions for rapid damage positioning, the segmentation task can provide the exact outline and region information of the damaged areas for quantification of the severity of the damage on the structural integrity and performance assessment.

Since 2015, there has been a rapid development of DL-driven algorithms and models in segmentation tasks. Long et al. [27] proposed a FCN that takes input with arbitrary size and produces a corresponding output that categorizes different objects with segmented masks. FCN is the first network supporting end-to-end training and it started the era of DL-driven methods for solving semantic segmentation problems. Subsequently, an encoder-decoder

(EN-DE)-type[6] network structure was proposed and rapidly adopted in model design, where U-Net [28] and SegNet [29] are two commonly used models. Similar to the YOLO series, a large number of variants of U-Net-like models were developed based on the original U-Net [28], e.g., V-Net [30], U-Net++ [31, 32], and Attention U-Net [33]. Because FCN-type models sometimes generate relatively rough segmentation results (ignoring the spatial consistency[7] between pixels), the dilated/atrous convolution operation[8] was proposed and utilized in the model design, which yielded the DeepLab series models, i.e., DeepLab v1/v2/v3/v3+ [34–37]. Furthermore, to combine the rich information from different layers of a deep CNN, the Pyramid Scene Sparse Network (PSP-Net) [38] was developed, which proposed a special pyramid pooling module along with the usage of atrous convolution. In fact, the damage segmentation shares similarities with the localization task, i.e., both tasks are engaged to find the location of damage in the image, compare Figs. 2.3b, c, 2.4, 2.5 and 2.6. Moreover, both tasks can be treated as downstream ones of the classification task [14, 26]. Therefore, some algorithms and AI models developed for damage localization tasks are found to also work well in damage segmentation tasks, e.g., Mask R-CNN [14].

In vision-based SHM, there have been increasing studies and applications in damage segmentation. Zhang et al. [39] applied FCN to detect cracks and leakages in a metro shield tunnel. Dung et al. [40] established FCN with a VGG-based EN to detect concrete cracking. Making use of FCN, Bang et al. [41] proposed a pixel level crack detection method based on an EN-DE network using ResNet-152[9] (deep CNN) [42] as the EN network. Extended from FCN, U-Net[10] was also adopted in concrete crack segmentation [43, 44], showing more robust and accurate performance than FCN in some scenarios. Besides, Ji et al. [45] developed an integrated approach, which adopts Deeplab v3 to achieve automatic crack detection in asphalt pavement at the pixel level, and then conducts crack quantification based on the detected cracks. Sun et al. [46] enhanced DeepLab v3+ by proposing a multi-scale attention module in its DE for road pavement crack detection, which achieved satisfactory results on both benchmark datasets and pavement images collected in practice. As mentioned

---

[6] The autoencoder (AE) consists of two parts: an encoder (EN) and a decoder (DE). The EN compresses the data from a higher-dimensional space to a lower-dimensional space (also called the *latent space*), while the DE does the opposite, i.e., converts the latent space back to the higher-dimensional space.

[7] In image segmentation, there is a high probability that a point in the image and points in its surrounding neighborhood have the same category attribute. FCN ignores such global information leading to the possibility of spatial inconsistency.

[8] Dilated convolution is a variant of convolution (the later is discussed in Chaps. 3 and 4), where the kernel of the former is inflated using a dilation factor. Atrous convolution is similar to the standard convolution except that the kernel is inflated, i.e., the weights of an atrous convolution kernel are spaced apart using a dilation factor making the convolution layer more sparse.

[9] ResNet is the abbreviation of Residual Network. It has multiple variations with a different number of layers and can be easily customized by adding or reducing layers. Herein, ResNet-152 has 152 layers, which is thought of as a deep network.

[10] A U-shaped architecture consists of a specific EN-DE scheme.

in the localization task, Zheng et al. [25] utilized Mask R-CNN for locating the damage, where Mask R-CNN itself can simultaneously generate masks for segmenting damaged areas, i.e., cracking and spalling. The numerical experiments indicated that, in this study, Mask R-CNN outperformed other classical segmentation methods such as FCN and PSP-Net and achieved promising results in quantifying damage for the tested bridge column specimens in a laboratory environment and for actual bridge girders in practical bridge health monitoring projects.

2.4   Structural Image Generation and Augmentation

Besides the above three mainstream tasks (e.g., classification, localization and segmentation) in vision-based SHM, there has been an increasing interest in structural image generation and augmentation in recent years. Even though this topic seems not directly related to structural health assessment and evaluation, it can be treated as an important auxiliary technique to contribute to enhancing the performance in other mainstream tasks. The original objective of this auxiliary task is to alleviate the negative impact of the shortage of labeled data, especially if the different classes in the available dataset are not balanced in terms of the relative amounts of the available data in each class.

Until now, vision-based SHM applications have not fully benefited from the data-driven DL technologies and one of the major reasons is the lack of a large dataset for training, especially labeled data. Conventionally, data augmentation (DA) is one of the common strategies to mitigate data deficiency issues in many DL applications. In DA, the number of training samples is increased via a series of affine transformations[11] or other preprocessing operations, e.g., translation, flip, scale, whitening, and adding noise. Besides, TL is another classical technique to mitigate the dependency on large-volume data and has been widely adopted in various research and industry applications [47]. TL applies the knowledge from source domains to target domains which might be related but different. The major advantage of TL is that it can relax the requirement that training a deep CNN needs a large amount of data, through tuning part of the network parameters from the pre-trained model in the source domain with few labeled data to be used in the target domain for fine-tuning of this pre-trained model, which might lead to a good performance for the target dataset. However, both DA and TL have significant curtailments. The former can only generate highly correlated data, which does not increase the data variety and sometimes might lead to worse results. On the other hand, the latter requires fine-tuning a deep pre-trained network with a large number of parameters, which not only requires a pre-trained open-source model but is also sometimes computationally expensive. An alternative to DA and TL is the generative model-based methods, which are becoming popular in CV applications. These methods generate new and unseen data and can pursue more variety in the data, where generative adversarial

---

[11] Affine transformation is a linear mapping method that preserves points, straight lines, and planes.

network (GAN) [48] is one of the common approaches nowadays and achieves state-of-the-art performance.

GAN, first introduced in [48], works as a generative model that can be utilized to generate new data from an approximated data distribution. Different from other generative models (e.g., linear and quadratic discriminant analysis, Gaussian mixture model, and restricted Boltzmann machine [49, 50]), GAN holds two different networks playing against each other in a zero-sum game[12] while in operation. One network, the *generator*, produces synthetic data from some learned distributions by feeding in some latent random variables[13] and the other network, the *discriminator*, judges the true identity of the data in a batch flow to determine whether such data are real or fake (synthetic). Two types of loss are computed for the generator and the discriminator and optimized iteratively until the two networks converge to optima. In this zero-sum game, the *minmax solution*[14] [51] is the Nash equilibrium [52], which indicates that the discriminator can no longer distinguish whether the data are from a real dataset or are synthetic. Following the above theories and principles, specific GAN models can be designed for generating realistic structural images. The working mechanism of the GAN model is illustrated in Fig. 9.1.

Other than the vanilla (original) GAN proposed by Goodfellow et al. [48], the rapidly growing GAN research resulted in many types of GANs by modifying the original GAN's architecture, loss function, and training strategy. In order to improve the generative performance of the vanilla GAN, conditional GAN [53] was proposed by conditioning the models on additional information (e.g., class labels). Radford et al. [54] proposed a class of CNNs called deep convolutional GAN (DCGAN) which shares the same architectural constraints as the vanilla GAN and has been proven to be powerful for unsupervised learning. The architectural constraints can contribute to the stable training process across a wide range of datasets. One major shortcoming of the DCGAN is *mode collapse*, where the generator produces synthetic images with extremely low variety. To address this problem and keep a balance in training the generator and the discriminator, Wasserstein GAN (WGAN) is presented by introducing Wasserstein distance[15] to the loss function [55]. More details about GANs are introduced in Chap. 9.

For vision-based SHM, the GAN itself is an auxiliary technique, because its outputs (synthetic images) do not directly align with the mainstream task targets, e.g., damage category, damage location, and damage mask. In addition, the GAN is known to be difficult to train and evaluate its performance in generating structural images for the purpose of SHM.

---

[12] Zero-sum game is a mathematical representation in *game theory* of a situation involving two sides, where the result is an advantage for one side and an equivalent loss for the other.

[13] A latent (Latin for lie hidden) variable can only be inferred indirectly through a mathematical model from other directly observable or measured variables.

[14] Minimizing the possible loss for a worst case (maximum loss) scenario.

[15] The Wasserstein distance (or metric) is a distance function defined between probability distributions on a given metric space.

Therefore, it is valuable to tackle the following three research problems related to GANs applications in SHM: (1) whether the GAN can be applied in vision-based SHM to generate realistic synthetic structural images, (2) how to evaluate the goodness of the generated structural images, and (3) how to use the generated synthetic data, aiming to alleviate the labeled data deficiency, for enhancing the AI model performance in SHM detection tasks.

Before the birth of GAN, using generative models as data augmentation or auxiliary techniques for model performance enhancement was rarely reported in vision-based SHM. It should be noted that based on an extensive literature search, there are no GAN-related studies conducted for vision-based SHM before 2019. Since 2019, researchers in this domain started to explore the potential of GAN for SHM research and practical applications [56–59]. From the perspective of utilization, the GAN-relevant studies can be grouped into two directions: (1) treating GAN as a data augmentation method and utilizing its generated synthetic data to enlarge the existing dataset, or (2) merging GAN into the AI model pipeline as an auxiliary module to facilitate the training. For the first direction, Gao and Mosalam [56] in 2019 proposed a new GAN model, named leaf-bootstrapping GAN (LB-GAN), which successfully generated high-quality synthetic structural images and validated the feasibility of using GAN in SHM vision tasks. In addition, to effectively and quantitatively evaluate the quality of the synthetic images to complement human inspection, self-inception score (SIS) and generalization ability (GA) were also proposed in [56], where it was found that directly mixing GAN-synthesized data with real data for training may render worse performance. To overcome this issue, a union training strategy, namely, synthetic data fine-tuning (SDF), was proposed and applied to GAN models, which demonstrated a significant performance enhancement in damage classification tasks with limited training data. Besides classification, Maeda et al. [57] applied GAN to road damage localization. They combined a progressive growing GAN (PGGAN) [60] along with Poisson blending[16] and artificially generated road damage images used as new training data to improve the accuracy of road pothole localization tasks. For the second direction, Gao et al. [59] proposed a SSL GAN pipeline along with the balanced-batch sampling technique (refer to Chap. 9 for details), namely, balanced semi-supervised GAN (BSS-GAN) to address the imbalanced label issue in several vision-based SHM practical scenarios. BSS-GAN merges the GAN with the ordinary classification classifier and the discriminator in the GAN part not only judges the difference between real and synthetic images but also conducts multi-class classification (e.g., spalling vs. non-spalling of the concrete cover in RC columns) simultaneously. In other words, the GAN becomes one part of the whole pipeline. A series of experiments with respect to the detection of the RC cover spalling was designed and conducted in [59] and the

---

[16] Poisson blending aims to copy the gradients of the source image into the destination image by minimizing the sum squared error of the gradients in the first image with the gradient of the combined (source + target) image.

results demonstrated the effectiveness and robustness of the BSS-GAN under the low-data and imbalanced-class regimes.

2.5 Exercises

1. Discuss the definitions and differences of classification, localization, and segmentation tasks in SHM.
2. Explain the process of a CNN for image-based structural damage classification.
3. What are the difficulties in conducting image-based structural damage localization and segmentation?
4. List a few classical AI models for classification, localization, and segmentation.

References

1. D. Soukup, R. Huber-Mörk, Convolutional neural networks for steel surface defect detection from photometric stereo images, in *International Symposium on Visual Computing* (Springer, 2014), pp. 668–677
2. Y.-J. Cha, W. Choi, O. Büyüköztürk, Deep learning-based crack damage detection using convolutional neural networks. Comput.-Aided Civil Infrast. Eng. **32**(5), 361–378 (2017)
3. A. Zhang et al., Automated pixel-level pavement crack detection on 3D asphalt surfaces using a deep-learning network. Comput.-Aided Civil Infrast. Eng. **32**(10), 805–819 (2017)
4. A. Vetrivel et al., Disaster damage detection through synergistic use of deep learning and 3D point cloud features derived from very high resolution oblique aerial images, and multiple-kernel-learning. ISPRS J. Photogr. Remote Sens. **140**, 45–59 (2018)
5. Y. Gao, K.M. Mosalam, Deep transfer learning for image-based structural damage recognition. Comput.-Aided Civil Infrast. Eng. **33**(9), 748–768 (2018)
6. Y. Gao, K.M. Mosalam, PEER Hub ImageNet: a large-scale multiattribute benchmark data set of structural images. J. Struct. Eng. **146**(10), 04020198 (2020)
7. R. Girshick et al., Rich feature hierarchies for accurate object detection and semantic segmentation, in *Proceedings of the IEEE Conference on Computer Vision and Pattern Recognition* (2014), pp. 580–587
8. J. Redmon et al., You only look once: unified, real-time object detection, in *Proceedings of the IEEE Conference on Computer Vision and Pattern Recognition* (2016), pp. 779–788
9. W. Liu et al., Ssd: single shot multibox detector, in *Computer Vision– ECCV 2016: 14th European Conference, Amsterdam, The Netherlands, October 11–14, 2016, Proceedings, Part I 14* (Springer, 2016), pp. 21–37
10. T.-Y. Lin et al., Focal loss for dense object detection, in *Proceedings of the IEEE International Conference on Computer Vision* (2017), pp. 2980–2988
11. G. Jocher, A. Chaurasia, J. Qiu, YOLO by Ultralytics. Version 8.0.0. (2023). https://github.com/ultralytics/ultralytics
12. R. Girshick, Fast r-cnn, in *Proceedings of the IEEE International Conference on Computer Vision* (2015), pp. 1440–1448

13. S. Ren et al., Faster r-cnn: towards real-time object detection with region proposal networks, in *Advances in Neural Information Processing Systems* 28 (2015)
14. K. He et al., Mask r-cnn, in *Proceedings of the IEEE International Conference on Computer Vision* (2017), pp. 2961–2969
15. H. Maeda et al., Road damage detection and classification using deep neural networks with smartphone images. Comput.-Aided Civil Infrast. Eng. **33**(12), 1127–1141 (2018)
16. J. Chen et al., Automatic defect detection of fasteners on the catenary support device using deep convolutional neural network. IEEE Trans. Instrum. Meas. **67**(2), 257–269 (2017)
17. Z. Qiu et al., Automatic visual defects inspection of wind turbine blades via YOLO-based small object detection approach. J. Electr. Imaging **28**(4), 043023–043023 (2019)
18. X. Yang et al., Deep learning-based bolt loosening detection for wind turbine towers. Struct. Control Health Monit. **29**(6), e2943 (2022)
19. R. Vishwakarma, R. Vennelakanti, Cnn model & tuning for global road damage detection, in *2020 IEEE International Conference on Big Data (Big Data)* (IEEE, 2020), pp. 5609–5615
20. G. Guo, Z. Zhang, Road damage detection algorithm for improved YOLOv5. Sci. Rep. **12**(1), 1–12 (2022)
21. Q. Li et al., Insulator and damage detection and location based on YOLOv5, in *2022 International Conference on Power Energy Systems and Applications (ICoPESA)* (IEEE, 2022), pp. 17–24
22. Y.-J. Cha et al., Autonomous structural visual inspection using regionbased deep learning for detecting multiple damage types. Comput.-Aided Civil Infrast. Eng. **33**(9), 731–747 (2018)
23. J. Huyan et al., Detection of sealed and unsealed cracks with complex backgrounds using deep convolutional neural network. Autom. Constr. **107**, 102946 (2019)
24. Y. Xue, Y. Li, A fast detection method via region-based fully convolutional neural networks for shield tunnel lining defects. Comput.-Aided Civil Infrast. Eng. **33**(8), 638–654 (2018)
25. Y. Zheng et al., Multistage semisupervised active learning framework for crack identification, segmentation, and measurement of bridges. Comput.-Aided Civil Infrast. Eng. **37**(9), 1089–1108 (2022)
26. Z. Liu et al., Swin transformer: hierarchical vision transformer using shifted windows, in *Proceedings of the IEEE/CVF International Conference on Computer Vision* (2021), pp. 10012–10022
27. J. Long, E. Shelhamer, T. Darrell, Fully convolutional networks for semantic segmentationin, in *Proceedings of the IEEE Conference on Computer Vision and Pattern Recognition* (2015), pp. 3431–3440
28. O. Ronneberger, P. Fischer, T. Brox, U-net: convolutional networks for biomedical image segmentation, in *Medical Image Computing and Computer-Assisted Intervention–MICCAI 2015: 18th International Conference, Munich, Germany, October 5–9, 2015, Proceedings, Part III 18* (Springer, 2015), pp. 234–241
29. V. Badrinarayanan, A. Kendall, R. Cipolla, Segnet: a deep convolutional encoder-decoder architecture for image segmentation. IEEE Trans. Pattern Anal. Mach. Intell. **39**(12), 2481–2495 (2017)
30. F. Milletari, N. Navab, S.-A. Ahmadi, V-net: fully convolutional neural networks for volumetric medical image segmentation, in *Fourth International Conference on 3D Vision (3DV)* (IEEE, 2016), pp. 565–571
31. Z. Zhou et al., Unet++: a nested u-net architecture for medical image segmentation, in *Deep Learning in Medical Image Analysis and Multimodal Learning for Clinical Decision Support: 4th International Workshop, DLMIA 2018, and 8th International Workshop, ML-CDS 2018, Held in Conjunction with MICCAI 2018, Granada, Spain, September 20, 2018, Proceedings 4* (Springer, 2018), pp. 3–11
32. Z. Zhou et al., Unet++: redesigning skip connections to exploit multiscale features in image segmentation. IEEE Trans. Med. Imaging **39**(6), 1856–1867 (2019)

33. O. Oktay et al., Attention u-net: learning where to look for the pancreas (2018). arXiv:1804.03999
34. L.-C. Chen et al., Semantic image segmentation with deep convolutional nets and fully connected crfs (2014). arXiv:1412.7062
35. L.-C. Chen et al., Deeplab: semantic image segmentation with deep convolutional nets, atrous convolution, and fully connected crfs. IEEE Trans. Pattern Anal. Mach. Intell. **40**(4), 834–848 (2017)
36. L.-C. Chen et al., Rethinking atrous convolution for semantic image segmentation (2017). arXiv:1706.05587
37. L.-C. Chen et al., Encoder-decoder with atrous separable convolution for semantic image segmentation, in *Proceedings of the European Conference on Computer Vision (ECCV)* (2018), pp. 801–818
38. H. Zhao et al., Pyramid scene parsing network, in *Proceedings of the IEEE Conference on Computer Vision and Pattern Recognition* (2017), pp. 2881–2890
39. A. Zhang et al., Deep learning-based fully automated pavement crack detection on 3D asphalt surfaces with an improved CrackNet. J. Comput. Civil Eng. **32**(5), 04018041 (2018)
40. C.V. Dung et al., Autonomous concrete crack detection using deep fully convolutional neural network. Autom. Constr. **99**, 52–58 (2019)
41. S. Bang et al., Encoder–decoder network for pixel-level road crack detection in black-box images. Comput.-Aided Civil Infrast. Eng. **34**(8), 713–727 (2019)
42. K. He et al., Deep residual learning for image recognition, in *Proceedings of the IEEE Conference on Computer Vision and Pattern Recognition* (2016), pp. 770–778
43. Z. Liu et al., Computer vision-based concrete crack detection using U-net fully convolutional networks. Autom. Constr. **104**, 129–139 (2019)
44. J. Liu et al., Automated pavement crack detection and segmentation based on two-step convolutional neural network. Comput.-Aided Civil Infrast. Eng. **35**(11), 1291–1305 (2020)
45. A. Ji et al., An integrated approach to automatic pixel-level crack detection and quantification of asphalt pavement. Autom. Constr. **114**, 103176 (2020)
46. X. Sun et al., Dma-net: deeplab with multi-scale attention for pavement crack segmentation. IEEE Trans. Intell. Transp. Syst. **23**(10), 18392–18403 (2022)
47. S.J. Pan, Q. Yang, A survey on transfer learning. IEEE Trans. Knowl. Data Eng. **22**(10), 1345–1359 (2009)
48. I. Goodfellow et al., Generative adversarial nets, in *Advances in Neural Information Processing Systems* (2014), pp. 2672–2680
49. C.M. Bishop, *Pattern Recognition and Machine Learning* (Springer, New York, 2006)
50. H. Li, *Statistical Learning Methods (in Chinese)* (Tsinghua University Press, 2012)
51. D.-Z. Du, P.M. Pardalos, *Minimax and Applications*, vol. 4 (Springer Science & Business Media, 2013)
52. I. Goodfellow, NIPS 2016 tutorial: generative adversarial networks (2016). arXiv:1701.00160
53. M. Mirza, S. Osindero, Conditional generative adversarial nets (2014). arXiv:1411.1784
54. A. Radford, L. Metz, S. Chintala, Unsupervised representation learning with deep convolutional generative adversarial networks (2015). arXiv:1511.06434
55. M. Arjovsky, S. Chintala, L. Bottou, Wasserstein gan (2017). arXiv:1701.07875
56. Y. Gao, B. Kong, K.M. Mosalam, Deep leaf-bootstrapping generative adversarial network for structural image data augmentation. Comput.-Aided Civil Infrast. Eng. **34**(9), 755– 773 (2019)
57. H. Maeda et al., Generative adversarial network for road damage detection. Comput.-Aided Civil Infrast. Eng. (2020)
58. K. Zhang, Y. Zhang, H.D. Cheng, Self-supervised structure learning for crack detection based on cycle-consistent generative adversarial networks. Comput. Civil Eng. **34**(3), 04020004 (2020)

59. Y. Gao, P. Zhai, K.M. Mosalam, Balanced semisupervised generative adversarial network for damage assessment from low-data imbalanced-class regime. Comput.-Aided Civil Infrast. Eng. **36**(9), 1094–1113 (2021)
60. C. Bowles et al., Gan augmentation: augmenting training data using generative adversarial networks (2018). arXiv:1810.10863

Basics of Machine Learning

3

In the past decades, ML has achieved significant advances in both theory and applications. Generally speaking, ML has three major categories: *supervised* learning, *unsupervised* learning, and *reinforcement* learning based on the data characteristics. Supervised learning trains a model that can learn and infer the mapping function between the input data and their corresponding labels. Compared with the other two categories, supervised learning is the most active branch in ML research and is widely used in many current ML applications.

This book mainly emphasizes supervised learning and some background related to unsupervised learning is also covered as it relates to related topics. This chapter introduces a brief review of the basics of ML focusing on the theoretical and mathematical fundamentals.

3.1 Supervised Learning

3.1.1 Basic Concepts and Notations

In supervised learning, the collections of all possible values for input and output are defined as *input space* and *output space*, respectively, where both spaces can be either finite or the entire Euclid space. Each input-output pair (i.e., a data realization) is called *instance*, as mentioned in Chap. 2, and it is usually represented by the *feature vector*. The space formed by the feature vectors is called the *feature space* and each dimension in the feature space corresponds to one feature. The input space and the feature space can either share the same space or be different. If it is the latter, one instance can be mapped from the input space to the feature space through some mapping or transfer functions.

Statistically, the input and output can be thought of as two random variables in the input (feature) space and output space, where input and output variables are denoted by X and Y, respectively. The values taken by the input and output variables are denoted by x and

© The Author(s), under exclusive license to Springer Nature Switzerland AG 2024
K. M. Mosalam and Y. Gao, *Artificial Intelligence in Vision-Based Structural Health Monitoring*, Synthesis Lectures on Mechanical Engineering,
https://doi.org/10.1007/978-3-031-52407-3_3

y, respectively. Any of these variables can be either a vector or a scalar. For example, the vector form of the ith instance of the d-dimensional x is represented as follows:

$$x_i = \left(x_i^1, x_i^2, \ldots, x_i^j, \ldots, x_i^d \right)^T, \tag{3.1}$$

where x_i^j represents the jth feature of the ith instance of x.

Random variables X and Y can be of different types, i.e., continuous or discrete, to form different prediction tasks and scenarios. For example, if the output variable Y is discrete, the prediction task is defined as *classification*, which is a major focus of this book. Based on different variable types, other tasks, e.g., regression,[1] can also be formulated.

In supervised learning, a collection of the n pairs (x_i, y_i), known as a *sample* or a *sample point*, is first constructed as the *dataset*, D, and expressed as follows:

$$D = \{ (x_1, y_1), (x_2, y_2), \ldots, (x_n, y_n) \}. \tag{3.2}$$

To train and evaluate the intended model that relates the output to the input, the dataset D is further split into the *training set*, *validation set*, and *test set*. The training set is used to train the model "trainable" parameters to find the optimal mapping function between the input and output pairs used in the training. The validation set is used to find the best model configuration among multiple candidates from varying the "pre-defined" model parameters, called "hyper-parameters". The test set is used to evaluate the model on the unseen data (not used in the training), which reflects the model's true generalization ability. In many cases, e.g., research explorations, the split of the datasets can be simplified to only have training and test sets to better exploit the available limited data.

3.1.2 Hypothesis Model

The key idea in the supervised learning is to make the model learn the mapping function from input to output, i.e., the major objective is to find the best model to capture the mapping relationship and such model is called the *hypothesis model* and denoted by $h(x; D)$. Statistically, the hypothesis model is also a random variable, which depends on arbitrary input x and the dataset D.

Based on specific learning algorithms and application scenarios, the hypothesis model can be represented as either a probabilistic model or a deterministic one. For specific instances x and y, the probabilistic model is to learn the *conditional probability* $P(y|x)$ of the output y based (i.e., conditioned) on the input x, which can also be thought of as observations. On the other hand, the deterministic model is to learn the *decision function* $y = f(x)$.

---

[1] In statistical modeling, regression analysis uses statistical processes to estimate the mathematical relationships between a *dependent* variable and one or more *independent* variables.

3.2   Classification

The classification is an important learning task and is based on the data type, where the output Y is a discrete variable. Suppose a dataset D is collected, where each input x_i, a d-dimensional data (feature) vector, is assigned one discrete class label $c_i \in \{1, 2, \ldots, K\}$ among the K possible classes. There exist many well-known classification algorithms and models [1] and they are discussed in detail in subsequent sections of this chapter.

In a classification problem, the model is also known as the *classifier* in the literature. With the input, the classifier outputs a K-dimensional vector of the posterior probability of the class label, $P(y|x)$. The predicted class label, c^*, is the one having the highest probability among all the K classes as follows:

$$c^* = \underset{c_i \in \{1,2,\ldots,K\}}{\mathrm{argmax}} \ P(y = c_i|x). \tag{3.3}$$

This can be solved using a "naïve Bayes classifier". It is a supervised learning algorithm for classification problems and is naïve because the occurrence of a feature is independent of the occurrence of other features. It is based on *Bayes' rule* to determine the probability of a hypothesis with prior knowledge as follows:

$$P(A|B) = \frac{P(B|A)P(A)}{P(B)}, \tag{3.4}$$

where $P(A|B)$ is the *conditional posterior probability* of the hypothesis A on the observed event B, $P(B|A)$ is the *likelihood probability* of the evidence (observed event) given that the probability of the hypothesis is true, $P(A)$ is the *prior probability* of the hypothesis before observing the evidence, and $P(B)$ is the *marginal probability* of the evidence.

3.3   Loss Function

During the training process, to maximize the model performance, the classifier and its parameters are expected to be optimized based on the *objective function*. It is common to conduct minimization on this objective function, which is usually referred to as the *cost function* or the *loss function* instead.

The loss function evaluates the degree of the prediction error made by a trained model between the prediction $\hat{y} = f(x)$ and the *ground truth*,[2] i.e., y, denoted by $L(f(x), y)$. There are many types of loss functions that can be utilized for different tasks. For example, in the classification problem, the commonly used loss functions are the 0–1 loss, the logarithmic loss (usually using the natural logarithm "ln" function), and the cross-entropy loss. On

---

[2] In ML, ground truth is factual data that have been observed, labeled, or measured and can be analyzed objectively. Here, it refers to the true label.

the other hand, in the regression problem, commonly used ones are the mean absolute error
(MAE), the mean square error (MSE), the root mean square error (RMSE), and the somewhat
related coefficient of determination (R^2). The definitions of these common loss functions
are as follows:

- 0–1 loss:

$$L\left(f(x), y\right) = \begin{cases} 0 & y = f(x), \\ 1 & y \neq f(x). \end{cases} \tag{3.5}$$

- Cross-entropy loss:

$$L\left(f(x), y\right) = H\left(\hat{y}, y\right) = - \sum_{\text{all classes}} p(y) \ln\left(q\left(\hat{y}\right)\right), \tag{3.6}$$

  where $\hat{y} = f(x)$. Moreover, $p(y)$ and $q(\hat{y})$ are the probability distributions of the true
  labels (ground truth) and the predictions, respectively.
- Cross-entropy loss (Binary classification):

$$L\left(f(x), y\right) = -y \ln\left(q\left(\hat{y}\right)\right) - (1 - y) \ln\left(1 - q\left(\hat{y}\right)\right), \tag{3.7}$$

  where the ground truth y is either 0 or 1.
- MAE (measures the *average of the absolute residuals* between the outputs from the
  dataset and the model predictions):

$$\text{MAE} = \frac{1}{n} \sum_{i=1}^{n} \left| y_i - \hat{y}_i \right|, \tag{3.8}$$

  where n is the size of the dataset with the discrete output (ground truth) ith values, y_i,
  and the corresponding model predictions $\hat{y}_i$.
- MSE (measures the *variance* of these residuals):

$$\text{MSE} = \frac{1}{n} \sum_{i=1}^{n} \left(y_i - \hat{y}_i\right)^2. \tag{3.9}$$

- RMSE (measures the *standard deviation* (Std) of these residuals):

$$\text{RMSE} = \sqrt{\text{MSE}} = \sqrt{\frac{1}{n} \sum_{i=1}^{n} \left(y_i - \hat{y}_i\right)^2}. \tag{3.10}$$

- R^2 (represents the proportion of the variance in the output that is explained by the *linear
  regression model* and $R^2 \in [0.0, 1.0]$ where a value closer to 0.0 or 1.0 indicates a
  respective poor or good *fit* of the model to the ground truth data):

$$R^2 = 1.0 - \frac{\sum_{i=1}^{n} \left(y_i - \hat{y}_i\right)^2}{\sum_{i=1}^{n} \left(y_i - \bar{y}\right)^2},$$

(3.11)

where $\bar{y}$ is the mean value of the discrete output variable y_i and $i \in \{1, 2, \ldots, n\}$.

3.4 Evaluation Metrics

Some metrics related to the discussion in the subsequent chapters are introduced in the following sections. These metrics are used for evaluating the goodness of the trained models using ML techniques.

3.4.1 Overall Accuracy

In the classification tasks, there are many metrics to evaluate the goodness of the results. The most common one is the *overall accuracy* (referred to as accuracy for brevity). Starting from the binary classification, several definitions are made first. In statistics, condition *positive* corresponds to what is defined as the positive state (i.e., class) in the data, e.g., in the context of SHM, this may correspond to the state (i.e., label) of the structure is *damaged*. Similarly, condition *negative* corresponds to what is defined as the negative class in the data, e.g., this may correspond to the label of the structure is *undamaged*. On the other hand, *true* and *false* are two judgments of the model prediction, and together with the positive and negative classes, they yield four possible outcomes: true positive (TP), true negative (TN), false positive (FP), and false negative (FN).

A TP means that the model correctly predicts the positive class, e.g., if the column is damaged, and the model also predicts it is indeed damaged. On the other hand, FP represents the outcome that the model incorrectly predicts the negative class, e.g., if the column is undamaged (negative), but the model predicts it as damaged (positive). Similarly, a TN means that the model correctly predicts the negative class and a FN means that the model incorrectly predicts the positive class.

Based on these defined four outcomes, in binary classification, the *overall accuracy* is computed as Eq. 3.13. Furthermore, in many practical scenarios, the number of classes is larger than two, i.e., multi-class classification, and the *overall accuracy* is defined as the ratio between the number of correct predictions to the total number of data used for prediction, Eq. 3.12. As its name suggests, the overall accuracy represents a general view of the accuracy of the model, but it may not be informative to reflect the true situation under a biased dataset, where the influence of the majority class dominates the minority class, i.e., merely guessing all samples from the majority class yields a misleadingly high accuracy. For example, if 90% images are labelled as in the undamaged state of a building and only 10% images are damage-related, a simple classifier that predicts all images are in the undamaged state yields a 90% overall accuracy. The value is indeed promising, but the result is nevertheless dangerous. The reason is that, in practice, the main objective of the SHM is to accurately identify the structural damage state and herein the overall accuracy can not correctly satisfy such a goal based on a biased dataset. Therefore, some other metrics are needed to complement the results of the overall accuracy in some cases like in the case of a biased dataset.

$$\text{Accuracy} = \frac{\text{Number of correct predictions}}{\text{Number of total predictions}}, \tag{3.12}$$

$$= \frac{\text{TP} + \text{TN}}{\text{TP} + \text{TN} + \text{FP} + \text{FN}}. \tag{3.13}$$

3.4.2 Confusion Matrix

To investigate the biased dataset issue, discussed above, a confusion matrix (CM) [2] is usually constructed. Let us start with the binary classification with the above mentioned four outcomes. As illustrated in Fig. 3.1, the rows and columns in the CM represent the ground truth, e.g., the first row represents the number of data whose true labels are positive (e.g., damaged). On the other hand, the columns in the CM represent the model predictions, e.g., the first column represents the number of data predicted to be positive (e.g., damaged) by the ML model or algorithm. By normalizing the CM entries with the number of ground truth of each class, the diagonal entries become the *true positive rate* (TPR), Eq. 3.14, and the *true negative rate* (TNR), Eq. 3.15, which are used for evaluating the accuracy of detecting TP and TN outcomes, respectively. In addition, the normalized CM can be applied to multi-class problems, i.e., $K > 2$, for which the recall (defined in the next paragraph) for each class is evaluated and placed on the diagonal cells.

Fig. 3.1 Confusion matrix

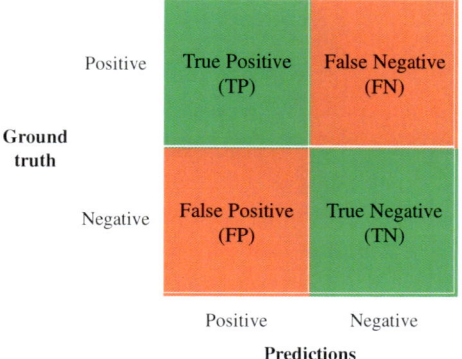

$$TPR = Recall = Sensitivity = \frac{TP}{TP + FN}. \tag{3.14}$$

$$TNR = \frac{TN}{TN + FP}. \tag{3.15}$$

Utilizing the relationships between the above mentioned four outcomes leads to other useful metrics in a binary classification, namely, *recall* and *precision*. Recall or *sensitivity* is a different way of calling the TPR, Eq. 3.14, which measures the model recognition ability toward the positive cases, and is contrasted with the precision. Suppose the computed recall is 0.75, this implies that the model can identify 75% of the buildings that are indeed damaged, refer to the numerical example in Sect. 3.4.3. Precision is calculated from Eq. 3.16, which presents the model's ability to identify the true positive data among all data predicted as positive. Take the same example of binary classification in building damage identification. Suppose the computed precision of the model is 0.6, this implies that if the building is predicted as damaged by the model, it is expected to be correct 60% of the time, refer to the numerical example in Sect. 3.4.3.

$$Precision = \frac{TP}{TP + FP}. \tag{3.16}$$

Both recall (in the context of the CM, recall is the TP normalized by the first row sum of the CM) and precision (in the context of the CM, precision is the TP normalized by the first column sum of the CM) are useful to evaluate the model performance in a biased dataset, where the former and latter are more appropriate for minimizing the FN and FP, respectively [3]. Due to the nature of the SHM, damage occurring in a structure is less frequent but can be very dangerous and may lead to significant loss (causalities (i.e., life loss), monetary (i.e., economic loss), or downtime (i.e., business interruption)). Failing to detect damages (i.e., more FN) bears severer consequences than wrongly recognizing undamaged samples as damaged (i.e., more FP). Accordingly, the first focus of the trained model should be to minimize the FN, measured by the recall. In practice, engineers may tolerate the possibilities of the FP and try to eliminate the FN as much as possible. In other words, a model with a high TPR is thought to be more conservative and it is more desirable if accompanied by high accuracy and precision. On the other hand, solely using precision sometimes is inappropriate because due to a large number of negative (undamaged) data, a small drop in TNR will cause a large increase in the FP, which overwhelms the TP and leads to a misleadingly low precision value. Furthermore, TNR can be an alternative to the recall metric, which takes the FP into account and measures the accuracy of the TN detection.

Extending from binary to a multi-class classification (i.e., the number of classes is larger than 2), the CM is still a good and reliable metric. The CM, in particular, by normalizing each of its rows, the entries along its diagonal represent the recall for each class. The normalized CM provides an easy evaluation of the model recognition ability toward the specific classes. More concrete examples are shown in the computer experiments presented in the later chapters of this book, while a simple numerical example is included in the next section for completeness.

3.4.3 Numerical Example

Assume a field reconnaissance effort following an earthquake generated CV data for 20 buildings with 8 of them labeled as damaged (the positive class of a classifier) and the remaining labeled as undamaged (the negative class of the classifier). A ML algorithm that uses CV was able to predict 10 (instead of the correct number of 8) of these buildings as damaged, where 6 were truly damage and the remaining 4 were truly undamaged (i.e., falsely damaged). This implies that the classifies produced TP = 6 and FP = 4. Moreover, the truly undamaged 12 buildings, out of the total dataset of 20 buildings, obviously had 2 buildings falsely classified as damaged, leading to FN= 2. Accordingly, this binary classifier identified the remaining fourth outcome of this dataset as TN= $20 - (6 + 4 + 2) = 8$, which are basically the buildings that were not classified as damaged and they were indeed truly undamaged in this dataset.

The metrics in the previous section are computed for this example. From the 10 positive class predictions (i.e., output of the algorithm indicating 10 damaged buildings), we have 6 true positive predictions, i.e., the precision (ratio of the model findings of the true positive class within the total model findings of this positive class) is computed as follows:

$$\text{Precision} = \frac{\text{TP}}{\text{TP} + \text{FP}} = \frac{6}{6 + 4} = 60\%.$$

From the actual 8 ground truth of the positive class in the dataset (i.e., the 8 damaged buildings), we have 6 true positive predictions, i.e., the recall (ratio of model findings of the true positive class within the total ground truth dataset of this positive class) is computed as follows:

$$\text{Recall} = \frac{\text{TP}}{\text{TP} + \text{FN}} = \frac{6}{6 + 2} = 75\%.$$

Finally, the ML algorithm overall accuracy (ratio of all correct model predictions of both positive and negative classes within the total ground truth dataset) is computed as follows:

$$\text{Accuracy} = \frac{\text{TP} + \text{TN}}{\text{TP} + \text{TN} + \text{FP} + \text{FN}} = \frac{6 + 8}{6 + 8 + 4 + 2} = 70\%.$$

3.4.4 F_β Score and Others

Besides CM and the metrics discussed above, the F_β score is another suitable metric for imbalanced binary classification problems [3]. F_β weights and combines both precision and recall scores into a single measurement, Eq. 3.17. Based on different β values, F_β measures the varying importance of recall over precision. When $\beta = 1$, both recall and precision are weighted equally (F_1 score). The β factor has real-world interpretations and using the $F_\beta - \beta$ graph is easy to interpret and communicate to engineers or stakeholders [3], e.g., a larger β value can be treated as a parameter that emphasizes the cost of mislabeling a damaged (case of FP) sample compared to producing a false alarm. According to [3], $\beta = 2$ is a common value (F_2 score). However, in SHM applications, the low tolerance of missing damage calls requires a higher β value, e.g., $\beta = 3$ for F_3 score.

$$F_\beta = \left(1 + \beta^2\right) \frac{\text{Precision} \times \text{Recall}}{\left(\beta^2 \times \text{Precision}\right) + \text{Recall}}. \tag{3.17}$$

For the numerical example in Sect. 3.4.3, we will compute the F_β scores for three values of $\beta = 1,\ 2$ and 3. Making use of the above results for Precision and Recall, we have the following,

$$F_1 = \left(1 + 1^2\right) \frac{0.60 \times 0.75}{\left(1^2 \times 0.60\right) + 0.75} = 0.667,$$

$$F_2 = \left(1 + 2^2\right) \frac{0.60 \times 0.75}{\left(2^2 \times 0.60\right) + 0.75} = 0.714,$$

$$F_3 = \left(1 + 3^2\right) \frac{0.60 \times 0.75}{\left(3^2 \times 0.60\right) + 0.75} = 0.732.$$

Beyond the above mentioned metrics, there are other ones used for the classification problems, e.g., the receiver operating characteristic (ROC) curve and the "area under the curve" (AUC) of the ROC. These metrics can be found in [1, 4] and other statistics-related references.

3.5 Over-Fitting and Under-Fitting

The main objective of training a ML model is to adapt it toward new and previously unseen data. The ability of the model to recognize these unobserved data is called *generalization*. In several ML research activities and practical applications, as discussed in Sect. 3.1.1, the collected dataset is usually divided into: (a) training set, (b) validation set, and (c) test set or simply training and test sets. Thus, a good generalization is reflected by the fact that the ML model can achieve: (1) good performance on the training set (small training error) and (2) equivalently good performance on the unseen test set (small test error and small gap between the training and test errors). Equivalently, having a small error is the same as having a high accuracy.

There are two common challenges in ML applications, namely, *over-fitting* and *under-fitting*. The over-fitting issue leads to a large gap between training and test errors, i.e., the test error can be much larger than the training error. Moreover, over-fitting indicates that the model lacks generalization and it usually over-emphasizes the training data with a small

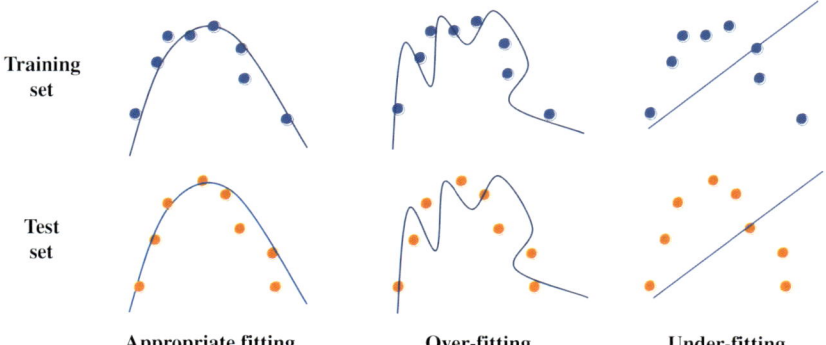

Training
set

Test
set

Appropriate fitting   **Over-fitting**   **Under-fitting**

Fig. 3.2 Illustration of appropriate fitting, over-fitting, and under-fitting of a hypothetical dataset and its fitted models

training error and may even memorize them instead of learning the true distribution of the data. On the other hand, under-fitting is attributed to the model being unable to achieve sufficiently small errors on either the training set or the test set. Examples of appropriate fitting, over-fitting, and under-fitting using the same dataset are presented in Fig. 3.2 for illustration.

3.6   Cross-Validation

In ML models or algorithms, besides the model parameters to be found during training, there exists a set of parameters that need to be determined before training based on prior knowledge, which are referred to as *hyper-parameters*. The use of these hyper-parameters usually directly contributes to the model performance, e.g., some poor choices of the hyper-parameters may cause over- and under-fitting issues.

In order to obtain a set of optimal hyper-parameters to avoid loss of generality, usually *cross-validation*, also known as *k- fold cross-validation*, is applied. In the cross-validation, all data is randomly and equally split into k folds. For each iteration of the cross-validation process, $k - 1$ folds with a new combination of folds are used as a training set and the remaining fold is assigned as a test set until each fold is tested once. Conventionally, there are two adopted patterns in cross-validation under different conditions: (1) separate all data to be training (T) and test (T) sets when the collected dataset has not specified the test dataset (TT pattern) and (2) if the collected dataset has already been separated into training and test sets, then one further separates the training set only into training and validation (V) sets (TVT pattern). Two examples of 5-fold ($k = 5$) cross-validations for the two patterns TT and TVT are illustrated in Fig. 3.3a and b, respectively.

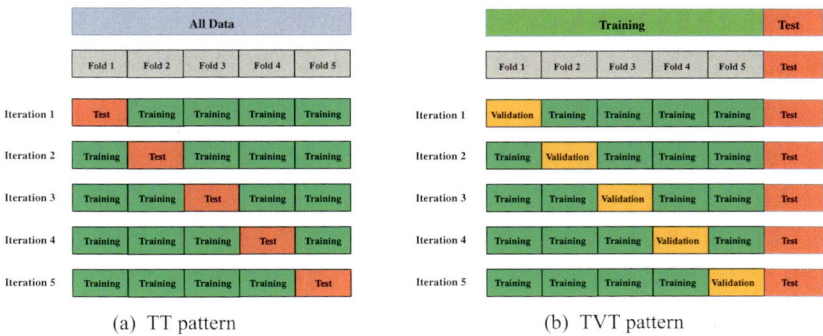

Fig. 3.3 Examples of 5-fold cross-validations

3.7 Logistic Regression

Logistic regression (LR) is one of the basic ML algorithms for classification, which has a direct probabilistic interpretation. Starting from the setting of a binary classification, the two classes are represented by 0 and 1. The model conducts a linear mapping, $f(x) = w^T x$, on the $(d + 1)$-dimensional input feature space of x considering an additional *bias* term, where $w \in \mathbb{R}^{d+1}$, $\mathbb{R}$ indicates the space of real numbers, and superscript T is for the transpose. Subsequently, through a *sigmoid transformation*, Eq. 3.18, the output is converted into a number within the range of 0.0–1.0, i.e., $\sigma(w^T x)$. Assume having a dataset of size n, i.e., number of data instances as defined in Sect. 3.1.1, for each single *data (feature) vector* x_i with corresponding *ground truth class label vector* y_i (with values of 0 or 1 for the binary classification problem), the output represents the probability of the given data to be classified as 1, Eq. 3.19. Based on a defined threshed, ϵ, usually 0.5, the predicted class label $\hat{y}_i$ is computed by a decision rule, as given by Eq. 3.20.

$$\sigma(z) = \frac{1}{1 + e^{-z}} = \frac{e^z}{e^z + 1}. \tag{3.18}$$

$$p(y_i = 1 | x_i) = \sigma\left(w^T x_i\right), \quad i \in \{1, 2, \ldots, n\}. \tag{3.19}$$

$$\hat{y}_i = \begin{cases} 1 & \sigma\left(w^T x_i\right) \geq \epsilon, \\ 0 & \text{otherwise.} \end{cases} \tag{3.20}$$

From Eq. 3.18, it is clear that the model relies on the mapping function $z = w^T x$. Thus, the objective of the LR is to find the *optimal parameters vector* w. As mentioned above, the cross-entropy loss for binary classification (Eq. 3.7) can be one suitable loss function for optimizing the LR model parameters. This equation is reproduced in Eq. 3.21 by plugging Eq. 3.19 into Eq. 3.7 to represent the probabilities, making use of the property, $-\ln x = \ln \frac{1}{x}$, and summing up the contributions from all n data instances.

$$L(w) = \sum_{i=1}^{n} \left\{ y_i \ln \left[\frac{1}{\sigma\left(w^T x_i\right)} \right] + (1 - y_i) \ln \left[\frac{1}{1 - \sigma\left(w^T x_i\right)} \right] \right\}. \qquad (3.21)$$

Moreover, for each input data x_i and ground truth label y_i of the binary classifier, Eq. 3.21 can be simplified to represent the loss for each single data point as follows:

$$L_i = \begin{cases} -\ln\left[\sigma\left(w^T x_i\right)\right] & y_i = 1, \\ -\ln\left[1 - \sigma\left(w^T x_i\right)\right] & y_i = 0. \end{cases} \qquad (3.22)$$

From Eq. 3.22, if $\sigma\left(w^T x_i\right) = 1.0$, there is zero loss to penalize the model implying that the model perfectly recognizes the class of $y_i = 1$. However, this is a rare case. In practice, the probabilities obtained by the model are usually less than 1.0 leading to non-zero values of the loss. This practical situation does not compromise the optimization procedure because the trend is still the same, i.e., a higher probability of the identified class always leads to a lower value of the loss.

3.7.1   One-Hot Vector Encoding

Extending the LR from a binary classification to a multi-class setting, some modifications are needed. Firstly, *one-hot vector encoding* is used to represent the class label. Suppose the total number of classes is K, the ith class is encoded as $[0, 0, \ldots, 1, \ldots, 0]^T$, which consists of zeros in all elements with the exception of a single 1 for the ith entry. For example, if the total number of classes is 3, e.g., identifying structural component objects as beam, column, and other, the labels for the three classes are encoded as $[1, 0, 0]^T$, $[0, 1, 0]^T$, and $[0, 0, 1]^T$.

Secondly, the weight representation, w, is modified to a matrix form associated with every class, namely, $W \in \mathbb{R}^{K \times (d+1)}$, where the kth row of the W matrix is w_k^T, $k \in \{1, 2, \ldots, K\}$, which is similar to that in the binary classification case. Therefore, for $x_i \in \mathbb{R}^{d+1}$, the new mapping output is represented in Eq. 3.23, which is also known as the *logits*.

$$W x_i = \left[w_1^T x_i, w_2^T x_i, \ldots, w_K^T x_i \right]^T = [z_1, z_2, \ldots, z_K]^T. \qquad (3.23)$$

3.7.2   *Softmax* Function

The *softmax* function, Eq. 3.24, can be used to replace the *sigmoid* function. For x_i, the probability vector for all K classes is computed in Eq. 3.24. After the *softmax* operation, all entries are between 0.0 and 1.0 and also sum up to 1.0, representing the *probability distribution* of the class classification.

$$softmax\left(\left[z_1, z_2, \ldots, z_i, \ldots, z_K\right]\right) = \left[\frac{e^{z_1}}{Se}, \frac{e^{z_2}}{Se}, \ldots, \frac{e^{z_i}}{Se}, \ldots, \frac{e^{z_K}}{Se}\right], \quad (3.24)$$

where $Se = \sum_{j=1}^{K} e^{z_j}$.

Based on the one-hot encoding, both predictions and ground truth are encoded in vector forms, which act as the distributions from the statistical perspective. Denote $\hat{P}_i$ as the target distribution from the ground truth y_i of the ith data instance and its label is class l from all the K classes. Thus, $P(\hat{P}_i = l) = y_i[l]$ represents the probability of the lth class label ($y_i[l]$ is the lth entry of the label vector y_i), which is 1 and the only non-zero entry in the $\hat{P}_i$. Moreover, the entire distribution $\hat{P}_i$ is concentrated around the label l. Similarly, denote Q_i as the estimated distribution from the above *softmax* function, the predicted probability of the lth class is thus $P(Q_{i=l}) = e^{z_l}/Se$.

3.7.3 Kullback–Leibler Divergence Loss Function

Unlike binary LR, the loss function in the multi-class case is evaluated by the difference between the target distribution $\hat{P}_i$ and the estimated distribution Q_i, which is typically computed by Kullback–Leibler (KL) divergence, $D_{\mathrm{KL}}\left(\hat{P}_i\|Q_i\right)$. Therefore, the optimal weight matrix W^* can be obtained by minimizing the KL divergence using all n data samples or instances, as shown in Eq. 3.25.

$$
\begin{aligned}
W^* &= \operatorname*{argmin}_{W} L(W) = \operatorname*{argmin}_{W} \sum_{i=1}^{n} D_{\mathrm{KL}}\left(\hat{P}_i\|Q_i\right), \\
&= \operatorname*{argmin}_{W} \sum_{i=1}^{n} \sum_{j=1}^{K} P\left(\hat{P}_i = j\right) \ln\left(\frac{P\left(\hat{P}_i = j\right)}{P\left(Q_i = j\right)}\right), \\
&= \operatorname*{argmin}_{W} \sum_{i=1}^{n} \sum_{j=1}^{K} y_i[j] \ln\left(\frac{y_i[j]}{softmax\left(\left[w_1^T x_i, \ldots, w_K^T x_i\right]\right)[j]}\right), \\
&= \operatorname*{argmin}_{W} \left[-\sum_{i=1}^{n} \sum_{j=1}^{K} y_i[j] \ln\left(\frac{e^{w_j^T x_i}}{\sum_{k=1}^{K} e^{w_k^T x_i}}\right)\right].
\end{aligned}
\quad (3.25)
$$

Note that in Eq. 3.25, if $\hat{P}_i = j$, we have $y_i[j] = 1$ and conversely if $\hat{P}_i \neq j$, we have $y_i[j] = 0$. Accordingly, we can generalize to write $y_i[j] = \delta_{\hat{P}_i, j}$, where $\delta_{\hat{P}_i, j}$ is the

Kronecker delta.[3] However, the optimization of the cross-entropy loss and KL divergence loss in LR has no analytic closed-form solution. Thus, to minimize these loss functions, some algorithms, e.g., gradient descent, are used as discussed in Sect. 4.1.3.

3.8   Support Vector Machine

The support vector machine (SVM) is one of the most popular ML algorithms, which is applicable to classification, regression, or other tasks. It was first developed by Vladimir Vapnik and co-workers at AT&T Bell Laboratories in 1995 [5] and has received considerable attention in both industry applications and research endeavors.

SVM falls into the scope of supervised learning. Through constructing a *hyper-plane* or a set of hyper-planes in a high dimensional space, the data can become separable by these planes based on their class labels. The hyper-plane is also known as *decision boundary*, H, and the data points closest to H are denoted as the *support vector (SV) points* and the minimum distance from the decision boundary H to any of the SV points is known as the *margin*, M. Herein, H is expressed as $H = \{w^T x - b = 0\}$, where the vector w is perpendicular to the hyper-plane and b is the offset of the hyper-plane from the origin along the normal vector w. It is worth mentioning that in Sect. 3.7 for LR and in comparison with the SVM formulation herein, the linear mapping $f(x) = w^T x$ for LR is defined on the $(d + 1)$-dimensional input feature space of x considering the additional *bias* term, b, as part of the weight vector, w, i.e., $w \in \mathbb{R}^{d+1}$ and x is augmented with a unit component corresponding to b, i.e., $w \leftarrow \{w, b\}$ and $x \leftarrow \{x, 1\}$.

Similar to the LR discussion, starting from the binary classification, the setup of the SVM problem is as follows: A training set $D = \{(x_1, y_1), \ldots, (x_i, y_i), \ldots, (x_n, y_n)\}$, where $x_i \in \mathbb{R}^d$ and y_i is labeled -1 or $+1$ (two classes), find a $(d - 1)$-dimensional hyper-plane (decision boundary) H to separate the two classes. In general, a good separation is achieved by a large margin, which indicates a lower generalization error of the classifier.

3.8.1   Hard-Margin SVM

In the hard margin SVM, the classifier is trained to find the hyper-plane H that maximizes the margin M and is subjected to several constraints as follows:

---

[3] The Kronecker delta is a function of two non-negative integer variables, i and j, expressed as follows:

$$\delta_{i,j} = \begin{cases} 0 & \text{if } i \neq j, \\ 1 & \text{if } i = j. \end{cases}$$

Fig. 3.4 Illustration of the distance D by projecting $(z - x_0)$ on w

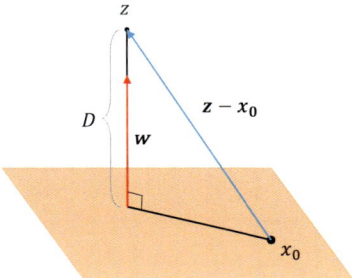

1. All points classified as $+1$ are on the positive side of H and the distance to H is larger than M.
2. All points classified as -1 are on the negative side of H and the distance to H is also larger than M.
3. The value of the margin M is non-negative.

From Fig. 3.4, taking one arbitrary point z not located on the hyper-plane, H, and another arbitrary point x_0 on H, i.e., $w^T x_0 = b$, the distance D from z to H is the length of the projection of the vector $(z - x_0)$ to the vector w (perpendicular to H) obtained by the dot product $w^T (z - x_0)$, which is normalized by the length of the vector w, i.e., its Euclidean ($\mathcal{L}_2$) norm $\|w\|_2$, as shown in Eq. 3.26. Thus, the distance from any points x_i to H in the training set is $(w^T x_i - b)/\|w\|_2$.

$$D = \frac{w^T (z - x_0)}{\|w\|_2} = \frac{w^T z - w^T x_0}{\|w\|_2} = \frac{w^T z - b}{\|w\|_2}. \tag{3.26}$$

With the representation of D and considering the above mentioned three constraints, the optimization problem can be formulated as in Eq. 3.27.

$$\max_{w,b,M} \quad M,$$

$$\text{s.t.} \quad y_i \frac{w^T x_i - b}{\|w\|_2} \geq M \quad \forall\, i, \tag{3.27}$$

$$M \geq 0.$$

To eliminate the variable M and for further simplification, new variables $w' = \dfrac{w}{\|w\|_2 M}$ and $b' = \dfrac{b}{\|w\|_2 M}$ are introduced and now Eq. 3.27 is equivalent to:

$$\max_{w',b'} \quad \frac{1}{\|w'\|_2},$$

$$\text{s.t.} \quad y_i \left(w'^T x_i - b' \right) \geq 1 \quad \forall\, i. \tag{3.28}$$

In the domain of ML and optimization, minimizing the loss is preferable over maximizing the gain. To make the objective function more precise and its solution more numerically stable and robust, Eq. 3.28 is rewritten as a minimization along with renaming the variables, squaring the objective function, and scaling with the constant multiplier $\frac{1}{2}$, we obtain,

$$
\begin{aligned}
\min_{w,b} \quad & \frac{1}{2}\|w\|^2, \\
\text{s.t.} \quad & y_i\left(w^T x_i - b\right) \geq 1 \quad \forall\, i.
\end{aligned}
\tag{3.29}
$$

Finally, the objective of the hard-margin SVM is to solve the optimization problem in Eq. 3.29. It is noted that the norm in Eq. 3.29 can be taken other than the Euclidean one. Therefore, subscript 2 of the used norm is dropped in Eq. 3.29 and also in the sequel for generality.

3.8.2   Soft-Margin SVM

In real applications, the available data are usually not linearly separable. Accordingly, the hard-margin SVM may have no solution. Moreover, hard-margin SVM is sensitive to the outlier data points,[4] which makes the classifier less generalized. To address these two issues, soft-margin SVM is introduced. Through modifying the constraints of the hard-margin SVM using a *slack variable* ξ (denote ξ_i as one realization of ξ for the ith data point, where $i \in \{1, 2, \ldots, n\}$, which is a scalar and assigned a specific optimal value of the slack variable, as derived in the next section), some data points can now violate the constraints, by actually constructing a less-strict version of the hard-margin SVM. Therefore, the new objective is established as follows:

$$
\begin{aligned}
\min_{w,b,\xi} \quad & \frac{1}{2}\|w\|^2 + C\sum_{i=1}^{n} \xi_i, \\
\text{s.t.} \quad & y_i\left(w^T x_i - b\right) \geq 1 - \xi_i \quad \forall\, i, \\
& \xi_i \geq 0 \qquad\qquad\qquad\quad \forall\, i,
\end{aligned}
\tag{3.30}
$$

where C is a hyper-parameter, which needs to be pre-determined by the analyst. It is usually firstly determined by prior knowledge and then tuned by cross-validation.

---

[4] In statistics, an outlier is a data point that differs significantly from other observations. An outlier may be due to variability in the measurements, an indication of novel data, or it may be the result of experimental errors.

3.8.3   Dual SVM and Kernel Trick

Instead of directly solving Eq. 3.30, it is more desirable to reformulate it and solve its dual optimization problem. This is because of the following:

1. The dual problem transfers the inequality constraints into equality constraints, which are easier to deal with.
2. It is easier in the dual problem to introduce the so-called "kernel trick" to deal with situations involving nonlinearly separable data.
3. The dual problem changes the complexity of the SVM problem, where the primal problem is related to the "unknown" dimension of the feature weight vector w, i.e., d, but the more manageable dual problem is related to the "known" number of the data points, i.e., n.

3.8.3.1 Dual SVM

First, the primal and dual variables are defined for solving the SVM problem. The primal variables are w, b, and ξ for the constraints in Eq. 3.30. On the other hand, the introduced dual variables are α and β, whose realizations for the ith data point are α_i and β_i, respectively. The *Lagrangian* for the soft-margin SVM is as follows:

$$
\begin{aligned}
L(w, b, \xi, \alpha, \beta) &= \frac{1}{2}\|w\|^2 + C\sum_{i=1}^{n}\xi_i + \\
&\quad \sum_{i=1}^{n}\alpha_i\left((1-\xi_i) - y_i\left(w^T x_i - b\right)\right) + \sum_{i=1}^{n}\beta_i\left(0 - \xi_i\right), \\
&= \frac{1}{2}\|w\|^2 - \sum_{i=1}^{n}\alpha_i y_i\left(w^T x_i - b\right) + \sum_{i=1}^{n}\alpha_i + \sum_{i=1}^{n}(C - \alpha_i - \beta_i)\,\xi_i.
\end{aligned}
\tag{3.31}
$$

Thus, the dual problem is stated as follows:

$$
\begin{aligned}
\max_{\alpha, \beta \geq 0} g(\alpha, \beta) = \min_{w, b, \xi} \frac{1}{2}\|w\|^2 &- \sum_{i=1}^{n}\alpha_i y_i\left(w^T x_i - b\right) + \\
&\sum_{i=1}^{n}\alpha_i + \sum_{i=1}^{n}(C - \alpha_i - \beta_i)\,\xi_i.
\end{aligned}
\tag{3.32}
$$

The optimal dual variables can be obtained by using *Karush–Kuhn–Tucker (KKT) conditions*. For the stationarity conditions of KKT, the optimal primal values should satisfy the following:

$$
\frac{\partial L}{\partial w} = \frac{\partial L}{\partial b} = \frac{\partial L}{\partial \xi} = 0.
\tag{3.33}
$$

According to the three conditions in Eq. 3.33, the optimal values (quantities with $*$ as a superscript) are related as follows:

1. $w^* = \sum_{i=1}^{n} \alpha_i^* y_i x_i$.
2. $\sum_{i=1}^{n} \alpha_i^* y_i = 0$.
3. $C - \alpha_i^* - \beta_i^* = 0$, i.e., $0 \le \alpha_i^* \le C$.

Replacing the variables with optimal ones in Eq. 3.31, a more compact form of the objective function is obtained as follows:

$$L(w, b, \xi, \alpha^*, \beta^*) = \frac{1}{2}||w||^2 - \sum_{i=1}^{n} \alpha_i^* y_i w^T x_i + \sum_{i=1}^{n} \alpha_i^*. \tag{3.34}$$

Moreover,

$$\begin{aligned} g(\alpha^*, \beta^*) &= \min_{w,b,\xi} L\left(w, b, \xi, \alpha^*, \beta^*\right), \\ &= L\left(w^*, b^*, \xi^*, \alpha^*, \beta^*\right), \\ &= \frac{1}{2}\left\|\sum_{i=1}^{n} \alpha_i^* y_i x_i\right\|^2 - \sum_{i=1}^{n}\left(\alpha_i^* y_i x_i^T \left(\sum_{j=1}^{n} \alpha_j^* y_j x_j\right)\right) + \sum_{i=1}^{n} \alpha_i^*, \tag{3.35} \\ &= \sum_{i=1}^{n} \alpha_i^* - \frac{1}{2}\sum_{i=1}^{n}\sum_{j=1}^{n} \alpha_i^* y_i x_i^T x_j y_j \alpha_j^*. \end{aligned}$$

Finally, the dual optimization problem can be defined as follows:

$$\begin{aligned} \max_{\alpha} \quad & \sum_{i=1}^{n} \alpha_i - \frac{1}{2}\sum_{i=1}^{n}\sum_{j=1}^{n} \alpha_i y_i x_i^T x_j y_j \alpha_j, \\ \text{s.t.} \quad & \sum_{i=1}^{n} \alpha_i y_i = 0, \\ & 0 \le \alpha_i \le C, \quad i = 1, 2, \ldots, n. \end{aligned} \tag{3.36}$$

Once the optimal solution $\alpha^* = \left(\alpha_1^*, \alpha_2^*, \ldots, \alpha_i^*, \ldots, \alpha_n^*\right)^T$ is obtained, plug it back into the above three conditions leading to the optimal values to determine the optimal primal variables w^* (Eq. 3.37), b^* (Eq. 3.38), and the specific choices of ξ_i (Eq. 3.39).

$$w^* = \sum_{i=1}^{n} \alpha_i^* y_i x_i, \tag{3.37}$$

$$b^* = y_j - w^{*T}x_j = y_j - \sum_{i=1}^{n} \alpha_i^* y_i x_i^T x_j, \tag{3.38}$$

$$\xi_i^* = \begin{cases} 1 - y_i \left(w^{*T}x_i - b^* \right) & \text{if } \alpha^* = C, \\ 0 & \text{otherwise,} \end{cases} \tag{3.39}$$

where b^* is obtained using one single data point $\left(x_j, y_j\right)$ with α_j^* that satisfies the condition $0 \leq \alpha_j^* \leq C$. In addition, b^* can be taken as the mean of all points whose α_j^*'s satisfy this condition.

With the known optimal values of w^* and b^*, the hyper-plane is also known from the following:

$$w^*x + b^* = 0. \tag{3.40}$$

Therefore, the prediction $\hat{y}$ for any input data x is as follows:

$$\hat{y} = \text{sgn}\left(w^*x + b^*\right), \tag{3.41}$$

where the sgn($\bullet$) function is defined as follows:

$$\text{sgn}(z) = \begin{cases} +1 & z > 0, \\ 0 & z = 0, \\ -1 & z < 0. \end{cases} \tag{3.42}$$

3.8.3.2 Kernel Trick

The above formulation of the SVM is based on the assumption that the data is linearly separable. However, in reality, this may be difficult to satisfy. Moreover, the vector/matrix computations in the above formulation are not efficient. The so-called "kernel trick" using kernel function is introduced and found to be effective in addressing these issues.

Using a linear SVM classifier for the raw data (feature) space is usually very limiting. However, if some mapping (transformation) is performed on the raw data, the SVM may still work effectively in the augmented feature space. Suppose an arbitrary mapping function, $\phi(\bullet)$, can be found such that it can elevate the raw data (feature) space χ to a higher dimensional augmented feature space, $\mathbb{H}$. Therefore, the transformed data $\phi(x)$ in the augmented space may become linearly separable. However, it is not easy to find such a mapping function.

In the "kernel trick", a kernel function $k(x, z)$ is defined between two inputs, x and z. It contains the *inner product* of the implicit mapping $\varphi(\bullet)$ between these two inputs. Assume there exists a mapping from χ to $\mathbb{H}$, the "kernel trick" leads to an approximation of the inner product of the nonlinear mapping $\phi(\bullet)$ as follows: $k(x, z) = \varphi(x) \cdot \varphi(z) \approx \phi(x) \cdot \phi(z)$. With the defined kernel function, the analyst can directly conduct computations instead of explicitly applying the mapping function $\phi(\bullet)$. In addition, using the "kernel trick" reduces the computational complexity where using a kernelized method is proven to be more efficient than using a non-kernelized method for some cases [1].

Now, we go back to solving the SVM optimization problem. First, changing Eq. 3.36 to a vector form, the objective function is rewritten as follows:

$$\sum_{i=1}^{n} \alpha_i - \frac{1}{2} \sum_{i=1}^{n} \sum_{j=1}^{n} \alpha_i y_i x_i^T x_j y_j \alpha_j = \alpha^{*T} \mathbf{1} - \frac{1}{2} \alpha^{*T} Q \alpha^*, \tag{3.43}$$

where $\mathbf{1}$ is the unit vector whose all entries are 1 with the same dimension of α^* and $Q_{ij} = y_i x_i^T x_j y_j$.

From Eqs. 3.36 to 3.42, the only operation related to the data x is the inner product, which is convenient to directly apply the "kernel trick". Through constructing the kernel function $k(\bullet, \bullet)$, $k(x_i, x_j)$ is computed instead of the original inner product. The full mapping matrix from the raw input data $X = [x_1, x_2, \ldots, x_n]^T$ to the augmented feature space $\mathbb{H}$ is called the *Gram matrix* and is expressed as follows:

$$K(X, X) = \begin{pmatrix} k(x_1, x_1) & k(x_1, x_2) & \cdots & k(x_1, x_n) \\ k(x_2, x_1) & k(x_2, x_2) & \cdots & k(x_2, x_n) \\ \vdots & \vdots & \ddots & \vdots \\ k(x_n, x_1) & k(x_n, x_2) & \cdots & k(x_n, x_n) \end{pmatrix}. \tag{3.44}$$

A valid kernel function should be positive semi-definite (PSD), which is difficult to validate using an arbitrarily limited training dataset. Thus, in practice, several empirical kernel functions are usually used, e.g., linear kernel (Eq. 3.45), polynomial kernel (Eq. 3.46), and Gaussian kernel (Eq. 3.47), also known as the radial basis function (RBF) kernel.

$$k(x, z) = x \cdot z + c, \tag{3.45}$$

where c is an optional constant.

$$k(x, z) = (\alpha(x \cdot z) + c)^p, \tag{3.46}$$

where the adjustable parameters are slope α, constant c, and polynomial degree p.

$$k(x, z) = \exp\left(-\frac{\|x - z\|^2}{2\sigma^2}\right), \tag{3.47}$$

where the adjustable parameter σ plays a major role in the performance of the kernel and should be carefully tuned to the problem at hand.

3.9 Decision Tree and Random Forest

Decision tree (DT) and random forest (RF) are two of the most popular ML algorithms and have been widely applied in both classification and regression problems. DT and briefly RF are introduced in this section due to the increased demand in SHM to use ML for solving several classification problems.

3.9.1 Decision Tree Model

For classification, the DT forms a tree-like classifier (i.e., model), which consists of nodes and directed edges. The type of the node is either an *internal node* or a *leaf node*, where the former represents the feature and the latter represents the class label. Starting from the root node of the tree, the classifier recursively keeps partitioning the input space into nodes based on certain criteria until arriving at the leaf node. Finally, the input data space is separated into several regions.

Suppose the input data x has 3 features, namely, x^1, x^2, and x^3 and the target output describes two states of a building structure, namely, *damaged state* and *undamaged state*. Figure 3.5 illustrates the procedure of how the DT works. Starting from the root node, if the first feature of the data satisfies the assumed condition of $x^1 \geq 0.01$, the data is further examined in the next node by the third feature x^3 on the left branch; otherwise, it is examined by the condition of the second feature x^2 on the right branch. The rectangle boxes represent the leaf nodes and the intersection of all conditions from the root node to one leaf node forms one region for a particular class label. For example, the edges in the left-most branch form one decision region for the data labeled as damaged state, where features are satisfying the conditions of $\{x^1 \geq 0.01\} \cap \{x^3 \geq 0.35\} \cap \{x^2 \geq 2.55\}$. Exploring the tree branches by all the data, the full DT model is established and the regions for both class labels are also determined. Once the new data come into the model, by matching the features with the conditions and the determined regions, the predicted class label is obtained.

3.9.2 Feature Selection

In each node, the DT actually works with a feature-value pair, (x^i, v), and the data are separated into two classes by examining whether the feature x^i exceeds the threshold value v or not. According to the example shown in Fig. 3.5, the split results may be influenced by how the features x^i are selected and split. Therefore, multiple tree models can lead to similar separation results where a good feature selection criterion is required. The *information gain* is one of the most commonly used criteria in the DT algorithm.

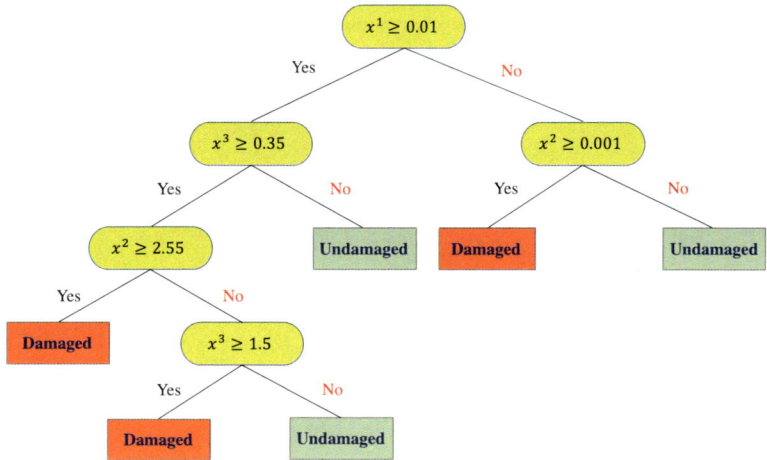

Fig. 3.5 Illustration of a hypothetical DT model for three-dimensional data

3.9.2.1 Information Gain

For convenience and readability, some basics related to *entropy* are introduced herein. In information theory and statistics, the entropy is a measure that represents the uncertainty of the random variable. Suppose a random variable X is discrete, which has N possible outcomes, i.e., $x_1, x_2, \ldots, x_N$ (not to be confused with the data points expressed in Eq. 3.2) and its distribution is as follows:

$$P(X = x_i) = p_i, \quad i = 1, 2, \ldots, N, \tag{3.48}$$

where p_i equals the ratio between the number of the data belonging to the outcome x_i and the total number of the data. The entropy of X, denoted by $H(X)$, is expressed by Eq. 3.49 where here the base of the log function is 2, which is the convention in the *information theory*.

$$H(X) = -\sum_{i=1}^{N} p_i \log_2 (p_i). \tag{3.49}$$

Suppose there exist two random variables, X and Y, and their joint probability distribution is known as follows:

$$P\left(X = x_i, Y = y_j\right) = p_{ij}, \quad i = 1, 2, \ldots, N; \ j = 1, 2, \ldots, M. \tag{3.50}$$

The *conditional entropy* $H(X|Y)$, which represents the uncertainty of the variable X under the condition of knowing the variable Y, is expressed as follows:

$$H(X|Y) = -\sum_{j=1}^{M} P\left(Y = y_j\right) H\left(X|Y = y_j\right). \tag{3.51}$$

If the probabilities used in the entropy and the conditional entropy are obtained by statistical estimation based on the data, they are called the *empirical entropy* and the *empirical condition entropy*, respectively.

The information gain, $IG(X; Y)$, which is also known as the *mutual information*, is defined as follows:

$$IG(X; Y) = H(X) - H(X|Y). \tag{3.52}$$

Under the working principle of the DT, the objective of each node can be thought of as maximizing the information gain after the feature-splitting operation, where Y in Eq. 3.52 represents such a split event of the operation. Considering the example of the feature split of the root node, shown in Fig. 3.5, Y has two possible outcomes, i.e., $M = 2$ in Eqs. 3.50 and 3.51, namely, $\{x^1 \geq \xi\}$ and $\{x^1 < \xi\}$, $\xi = 0.01$ herein. These outcomes can further be represented by $Y = 1$ and $Y = 0$, respectively. Thus, the information gain after the split in the root node is presented as follows:

$$\begin{aligned} IG(X; Y) &= H(X) - \left[P\left(x^1 \geq \xi\right) H\left(X|x^1 \geq \xi\right) + P\left(x^1 < \xi\right) H\left(X|x^1 < \xi\right) \right], \\ &= H(X) - \left[P(Y = 1)H(X|Y = 1) + P(Y = 0)H(X|Y = 0) \right]. \end{aligned} \tag{3.53}$$

Similarly, based on Eq. 3.53, the calculation of the information gain continues in a *greedy* way (i.e., making a locally optimal choice at each stage with the intent of finding a global optimum, refer to Sect. 4.1.2 for more explanation about the "Greedy algorithm"), where each step results in the maximum information gain, until the computation reaches the leaf nodes or the designated depth in the DT, defined by the analyst. In addition, the DT using the above approach for the information gain computation is called ID3 (Iterative Dichotomize 3) [6], but it only works for datasets with discrete features.

3.9.2.2 Gini Index

Based on different ways of computing the information gain and conducting feature selection, there are different types of DT algorithms other than ID3, e.g., C4.5 [7], and classification and regression tree (CART) [8], where C4.5 overcomes the shortcomings of ID3 and CART is a more general algorithm. It is worth noting that the information gain in CART is alternatively computed by the *Gini index* (also known as *Gini impurity*), which is a measure of the information uncertainty.

Suppose we have a K-class classification problem and the probability of the data which belongs to the kth class is denoted by $p_k = |l_k|/|CT|$, computed by the ratio between the number of data labeled for the kth class, $|l_k|$, and the number of data in this collection, $|CT|$. Accordingly, $\sum_{k=1}^{K} p_k = 1.0$. The Gini index (developed by Corrado Gini in 1912 [9]) is calculated as follows:

$$Gini(CT) = \sum_{k=1}^{K} p_k(1 - p_k) = 1 - \sum_{k=1}^{K} p_k^2. \tag{3.54}$$

Let the selected feature be A and the collection CT be separated into I parts by a certain threshold of the feature, namely, CT_1, CT_2,..., CT_I. Thus, the conditional Gini index is computed as follows:

$$Gini(CT|A) = \sum_{i=1}^{I} p_i \, Gini\,(CT_i) = \sum_{i=1}^{I} \frac{|l_i|}{|CT_i|} \, Gini\,(CT_i). \tag{3.55}$$

The information gain using the Gini index under the condition of feature A is presented in Eq. 3.56, which is also known as the *Gini information gain*.

$$Gini\; IG = Gini(CT) - Gini(CT|A). \tag{3.56}$$

For each node, the Gini information gain is computed for all features and the one that has the largest gain is selected as the split feature in the current node. Compared with the above mentioned entropy-based information gain, using the Gini index may achieve better computational efficiency than using entropy due to avoiding the logarithm operations.

3.9.3   Random Forest Model

The DT only represents one single tree, which may be limited or suffering from over-fitting issues. Intuitively, if multiple trees are utilized at the same time, they may yield more robust results. As a result, RF is proposed by Leo Breiman [10], which is an *ensemble learning* algorithm that performs predictions by using multiple DTs and the output can be the *mode* of the classes in a classification problem or the *mean* of the prediction in a regression problem, Fig. 3.6. These statistical quantities are computed from the used multiple individual DTs in the RF.

Fig. 3.6 Prediction procedure of the RF

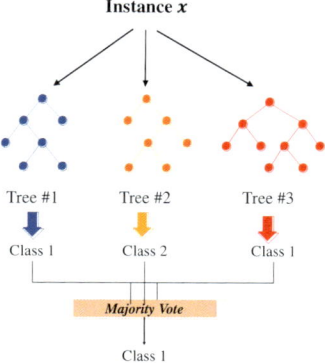

There are many ways to conduct the RF computations according to many past studies. This brief section only introduces one conventional method, namely, *bagging*, which also stands for *bootstrap aggregation*.[5] While conducting the bagging, the major steps are as follows:

- Construct subsets with size m from the raw dataset whose size is n.
- Train T weak models (classifier or regressor) on the subsets.
- Assemble the single models for prediction using a majority vote.

For example, suppose $T = 50$ DTs are used in the RF and the number of the total data in the dataset is $n = 500$. Firstly, 50 subsets are constructed through random sampling with replacement from the raw dataset and the size of the subsets is, e.g., $m = 400$. Then, these 50 DTs are trained on the different subsets of size 400. While in prediction, the test data are passed to all DTs and there will be 50 outputs from the 50 trees. For a classification problem, the final decision on the class label is the majority vote among all outputs, e.g., if for $T = 50$, 47 trees predict the building is undamaged, but the remaining 3 trees predict it is damaged, the final prediction based on the majority vote is undamaged. The bagging method treats each tree equally and there are many other ways to improve this method for better performance. These other ways yield different RF methods, e.g., Boosting, AdaBoost, etc. More details can be found in references [1, 4].

3.10 Exercises

1. Define over-fitting and under-fitting in the context of ML and provide examples of each.
2. Describe strategies to mitigate over-fitting and under-fitting, such as regularization and cross-validation.
3. Define a CM and explain its components (true positives, true negatives, false positives, and false negatives).

---

[5] From a single sample of size N, one can estimate the mean. To reason about the population, we need a sense of the variability of the mean that we have computed. The simplest bootstrap method involves taking the original data set and sampling from it to form a new sample (called a "resample" or a bootstrap sample) that is also of size N. The bootstrap sample is taken from the original data by using sampling with replacement (e.g., we might resample 5 times from $[1, 2, 3, 4, 5]$ and obtain $[2, 5, 4, 4, 1]$). Assuming N is sufficiently large, for all practical purposes, there is virtually zero probability that it will be identical to the original "real" sample. This process is repeated a large number of times (typically 1,000 or 10,000 times) and for each of these bootstrap samples, we compute its mean (each of these is called a "bootstrap estimate"). We now can create a histogram of the bootstrap means. This histogram provides an estimate of the shape of the distribution of the sample mean from which we can answer questions about how much the mean varies across the samples. This method, described herein for the mean, can also be applied to estimate almost any other statistic or estimator).

4. Describe commonly used performance metrics derived from a CM, such as precision, recall, F_1 score, and accuracy. Discuss their applications and interpretations.
5. In LR, explain how the $logodds$[6] are used to model the $probability$ of a binary event.
6. Describe the $sigmoid$ ($logistic$) function and how it maps log odds to probabilities.
7. In SVM, explain the kernel trick and how it allows SVM to handle nonlinearly separable data.
8. Describe the concept of a soft margin in SVM and its role in handling data that is not linearly separable.
9. Define entropy and explain how it measures the impurity or disorder in a dataset.
10. Describe the information gain and its role in DT construction. Provide a formula for calculating the information gain.

References

1. C.M. Bishop, *Pattern Recognition and Machine Learning* (Springer, New York, 2006)
2. R. Kohavi, F. Provost, Confusion matrix. Mach. Learn. **30**(2–3), 271–274 (1998)
3. N.V. Chawla, N. Japkowicz, A. Kotcz, Special issue on learning from imbalanced data sets. ACM SIGKDD Explor. Newslett. **6**(1), 1–6 (2004)
4. H. Li, *Statistical Learning Methods (in Chinese)* (Tsinghua University Press, 2012)
5. C. Cortes, V. Vapnik, Support-vector networks. Mach. Learn. **20**(3), 273–297 (1995)
6. J.R. Quinlan, Induction of decision trees. Mach. Learn. **1**(1), 81–106 (1986)
7. J.R. Quinlan, *C4. 5: Programs for Machine Learning* (Elsevier, 2014)
8. L. Breiman et al., *Classification and Regression Trees* (CRC Press, 1984)
9. C. Gini, Variabilitá e mutabilitá, in *Reprinted in Memorie di metodologica statistica*, ed. by E. Pizetti, T. Salvemini (Libreria Eredi Virgilio Veschi, Rome, 1912)
10. L. Breiman, Random forests. Mach. Learn. **45**(1), 5–32 (2001)

---

[6] The relationships between the $odds$, the $logodds$, and the $probability$ of an outcome are explained with a numerical example. If the $probability$ of an event is 0.2, then $odds = 0.2/0.8 = 0.25$; $logodds = \ln(odds) = \ln(0.25) = -1.39$; $probability = odds/(1.0 + odds) = 0.25/(1.0 + 0.25) = 0.2 = \exp(logodds)/(1.0 + \exp(logodds)) = \exp(-1.39)/(1.0 + \exp(-1.39)) = 0.2$.

Basics of Deep Learning

4

As a subset of ML, DL is attracting rapidly increasing interests nowadays. However, the study of DL dates back to the 1940s and it has undergone multiple periods of research and development [1]. The concept of DL originated from the study in neuroscience and its objective was to simulate the mechanisms of the human brain to understand and interpret data, e.g., images, texts, and sounds.

A DL model is thought of as a composition of multiple levels of functions and is usually represented by stacking multiple layers which consist of special units, i.e., *neurons*. Currently, the *neural network (NN)* has three main types, namely, the *convolutional neural network (CNN)*, the *recurrent neural network (RNN)* (a special case of which is the *long short-term memory (LSTM) network)*, and the *transformer*. Through stacking layers to a certain depth of the network, the number of model parameters significantly increases. As a result, the model gains a deep representation of the learning ability to learn complex patterns of the data. The word "deep" in the name of DL partially refers to this characteristic. In this chapter, fundamental knowledge of the DL is introduced.

The original version of this chapter has been revised: Figures 4.8, 4.9, 4.10 has been corrected. A correction to this chapter can be found at https://doi.org/10.1007/978-3-031-52407-3_14

© The Author(s), under exclusive license to Springer Nature Switzerland AG 2024,
corrected publication 2024
K. M. Mosalam and Y. Gao, *Artificial Intelligence in Vision-Based Structural Health Monitoring*, Synthesis Lectures on Mechanical Engineering,
https://doi.org/10.1007/978-3-031-52407-3_4

4.1   Neural Network

From the long history of development, there are many ways to name the NN, e.g., artificial NN (ANN), feed-forward NN, fully-connected NN (FC-NN), and multi-layer perceptron[1] (MLP). These different names indeed have slight differences in modeling, but nowadays researchers are usually unifying them into NN with the following general characteristics:

1. A NN has at least one layer.
2. All neurons between the layers of the NN are linked together but are unconnected within the same layer.
3. All connections of the NN are feed-forward and have no recurrence.

Through a large number of connected neurons activated by certain nonlinear functions, NN gains a powerful representation ability, which can somewhat model the working mechanisms of a biological brain to process the input information and to realize AI. To provide a theoretical background for the approximation ability of the NN, the *universal approximation theorem* was developed by Kurt Hornik [2], which states that: "A feed-forward network with a single hidden layer containing a finite number of neurons can approximate continuous functions". It also states that: "Simple neural networks can represent a wide variety of interesting functions when given appropriate parameters". Thus, instead of using one layer, NN is extended from the theory to the current practice and research where multi-layer NN is adopted, i.e., deep NN, which is thought to have the capability to well-approximate more complex functions or patterns of data.

To obtain the best approximation, the large number of parameters in the neurons are trained with the input data through several iterations with the objective of minimizing the defined *loss function*. Usually, the NN is trained iteratively and the workflow in NNs consists of two passes: (i) forward pass and (ii) backward pass. A standard architecture of the NN usually consists of three parts: (a) an input layer, (b) multiple hidden layers, and (c) an output layer. An example of a three-layer NN (or two-hidden layer NN) is illustrated in Fig. 4.1. Note that the input layer is typically not counted in the designation of the NN.

---

[1] The word perceptron is rooted in "perception" because of its sensing ability. It was invented by Frank Rosenblatt in January 1957 at Cornell Aeronautical Laboratory. In ML, the perceptron is the basic building block of the NN as an algorithm for supervised learning of the binary linear classifiers. It consists of four parts: input values, weights and a bias, a weighted sum, and nonlinear activation, e.g., a Heaviside step function as defined below:

$$H(x) = \begin{cases} 1 & \text{if } x \geq 0, \\ 0 & \text{if } x < 0. \end{cases}$$

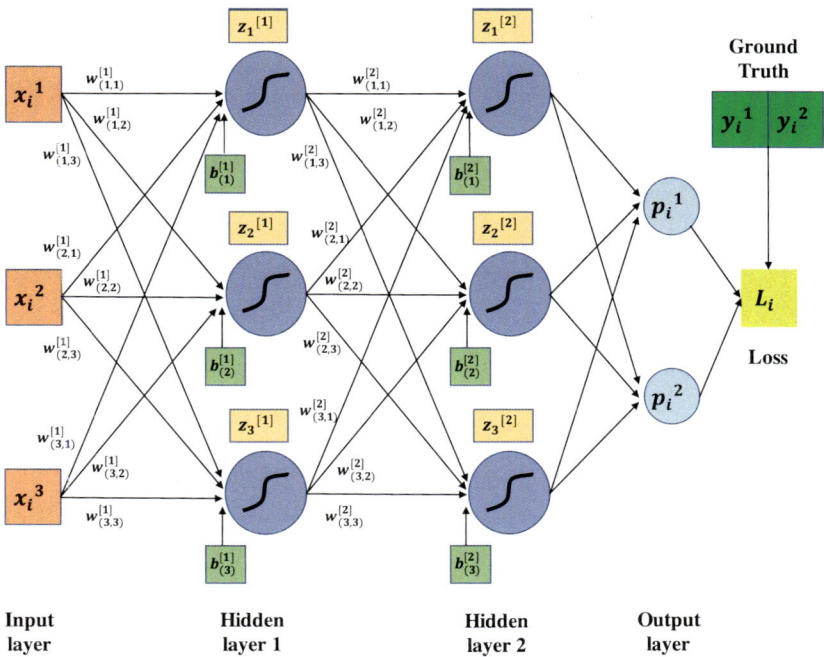

Fig. 4.1 Example of a three-layer NN with the ith input vector

4.1.1 Forward Propagation

For the supervised learning problems, the forward pass between layers is as follows:

1. Flatten the input (features) to a vector as the labeled input for the input layer.
2. Multiply entries of this vector by weights on the edges between nodes in two different (and adjacent) layers.
3. Sum over weighted entries pointing to the node and also add a bias term for each node in the next layer.
4. Activate the results obtained in step (3) via the activation functions to produce outputs, which are treated as the inputs for the next layer.
5. Repeat steps (1)–(4) until the forward pass arrives at the final output layer, where the outputs usually represent probability values.
6. Compute the loss by the outputs in step (5) and the ground truth (input labels).

For a multi-class classification problem, suppose the input data ($\mathbf{x}$) and labels ($\mathbf{y}$) are arranged as a tensor pair, i.e., ($\mathbf{x}, \mathbf{y}$), and assume the NN has a total of M layers. For example, the data tensor $\mathbf{x} \in \mathbb{R}^{N \times W \times H \times C}$ can be from stacking N images, $\{x_1, x_2, \ldots, x_N\}$, where each image has width W pixels, height H pixels, and color channels C, e.g., $C = 1$ for the gray scale in black and white images, $C = 3$ for RGB channels in colored images, or $C = 4$ for RGB channels + a depth channel for colored images taken with a depth camera. On the other hand, the label tensor $\mathbf{y} \in \mathbb{R}^{N \times K}$ is stacked by $\{y_1, y_2, \ldots, y_N\}$ after one-hot encoding according to a total of K classes. It is noted that the $(\mathbf{x}, \mathbf{y})$ tensor pair can be given a subscript $i \in \{1, 2, \ldots, N\}$ to indicate different sets (a total of N data points) of this pair as shown in Fig. 4.1. Denote the input of the lth layer as $\mathbf{I}^{[l]}$, the linear operation (multiplication by weights and addition of a bias term) and subsequent nonlinear activation can be represented as functions $f^{[l]}\left(\mathbf{I}^{[l]}\right) = \left(\mathbf{W}^{[l]}\right)^T \mathbf{I}^{[l]}$ and $\sigma^{[l]}(\bullet)$, respectively. The output of $f^{[l]}(\bullet)$ is denoted by $\mathbf{z}^{[l]}$. In its basic form, the linear operation, according to Fig. 4.1, performs the computation for a single neuron unit as: $z = \mathbf{w}^T \mathbf{x} + b$ (e.g., $z_1^{[1]} = \sum_{j=1}^{3} w_{(j,1)}^{[1]} x_i^j + b_{(1)}^{[1]}$ for the first neuron in the first hidden layer), where $\mathbf{w}$ and $\mathbf{x}$ can be augmented to include b (the bias term) and the constant 1 in an additional feature channel (dimension), respectively, to give $\mathbf{W}$ and $\mathbf{I}$, refer to Sect. 3.8 for a similar augmentation discussion. Here, z is the input to the activation function to produce the output of a particular hidden layer, which is the input to the subsequent layer. Some common activation functions $\sigma^{[l]}(\bullet)$ are *sigmoid* (Eq. 3.18), *rectified linear unit (ReLU)* [3] (Eq. 4.1), *leaky ReLU (LReLU)* [4] (Eq. 4.2), *exponential linear unit (ELU)* [5] (Eq. 4.3), *softmax* (Eq. 3.24), and *hyperbolic tangent (tanh)* (Eq. 4.4), where *ReLU* is commonly adopted in most layers and *sigmoid* and *softmax* are usually applied in the final layer to transfer the output into probability values.

$$ReLU(z) = \max(0, z). \tag{4.1}$$

$$LReLU(z) = \begin{cases} z & z \geq 0, \\ \alpha z & z < 0, \end{cases} \tag{4.2}$$

where $\alpha > 0$ is the negative slope hyper-parameter and usually taken as $\alpha = 0.01$.

$$ELU(z) = \begin{cases} z & z \geq 0, \\ \alpha \left(\exp(z) - 1\right) & z < 0, \end{cases} \tag{4.3}$$

where $\alpha > 0$ is a hyper-parameter controlling the value to which an ELU saturates for negative inputs, e.g., $\alpha = 1.0$ was used in [5].

$$\tanh(z) = \frac{e^z - e^{-z}}{e^z + e^{-z}}. \tag{4.4}$$

The output after activation (assuming it is a *sigmoid* function, $\sigma(\bullet)$) is the so-called *feature*, $\mathbf{A}^{[l]} = \sigma^{[l]}\left(f^{[l]}\left(\mathbf{I}^{[l]}\right)\right)$ and is used as the input for the $(l + 1)$th layer (assuming the NN has a total of M layers), i.e., $\mathbf{I}^{[l+1]} = \mathbf{A}^{[l]}$. The forward activation pass propagates through the entire network until the signals (all N data points) reach the final layer. The whole procedure can be heavily parametrized by a combination of function compositions,

Eq. 4.5, denoted by $S(\bullet)$, where $S(\mathbf{x}) \in \mathbb{R}^{N \times K}$ if feeding data $\mathbf{x}$ corresponds to the K classes defined in $\mathbf{y}$.

$$S(\mathbf{x}) = \sigma^{[M]}\left(f^{[M]}\left(\sigma^{[M-1]}\left(f^{[M-1]}\left(\dots f^{[2]}\left(\sigma^{[1]}\left(f^{[1]}(\mathbf{x}) \right) \right) \dots \right) \right) \right) \right). \tag{4.5}$$

Finally, the total loss of the NN can be computed based on the defined loss function $L(\bullet)$, refer to Sect. 3.3, with $S(\mathbf{x})$ and label $\mathbf{y}$, i.e.,

$$L = L\left(S(\mathbf{x}), \mathbf{y} \right). \tag{4.6}$$

4.1.2   Backward Propagation

The training objective of the NN is to minimize the loss and to find the optimal parameters in the layers, which are denoted by $\mathbf{W}$. However, the optimization in the NN is usually *non-convex*[2] and the optimization problem is difficult to solve using conventional algorithms. As a result, a gradient-based updating algorithm, namely, *gradient descent* [1], was developed by updating the model parameters with the gradient of the loss with respect to these parameters in a *greedy*[3] way and finally one arrives at relatively "good" local optima. It is worth noting that it is impractical to obtain the global optima in most cases of the NNs. However, usually some local optima are equivalently as good as the global optima. Thus, it is reasonable and acceptable to use these "good" local optima as the way to practically terminate the NN training cycles.

The gradient descent algorithm is summarized in the following steps:

1. Initialize parameters randomly or from certain distributions, e.g., Gaussian.
2. Update the parameters iteratively by subtracting the incremental change of the parameters in terms of the product of the gradient of the loss with respect to the parameters and an analyst-defined scalar *learning rate* hyper-parameter, η.
3. Repeat step (2) until the loss converges to a minimum or the number of iterations reaches a designated maximum value.

Take an example of an updating procedure of the parameters in the lth layer (refer to Sect. 4.1.1 for definitions of all the parameters and mathematical operations), denoted by

---

[2] A non-convex function has some "valleys" (local minima) that are not as deep as the overall deepest "valley" (global minimum). Optimization algorithms can get stuck in the local minimum and it can be hard to tell when this takes place.

[3] Greedy is an algorithmic paradigm that builds up a solution piece by piece, always choosing the next piece that offers the most obvious and immediate benefit. Therefore, the problems, where a chosen locally optimal solution also leads to the global solution, are the best fit for Greedy.

$\mathbf{W}^{[l]}$. The gradient of loss with respect to $\mathbf{W}^{[l]}$, i.e., $\nabla_{\mathbf{W}^{[l]}} L$ or $\left(\dfrac{\partial L}{\partial \mathbf{W}^{[l]}}\right)^T$, is computed using the *chain rule*[4] as follows:

$$
\begin{aligned}
\nabla_{\mathbf{W}^{[l]}} L &= \left(\frac{\partial L}{\partial \mathbf{W}^{[l]}}\right)^T, \\
&= \left(\frac{\partial L}{\partial \mathbf{A}^{[M]}} \cdot \frac{\partial \mathbf{A}^{[M]}}{\partial \mathbf{z}^{[M]}} \cdot \frac{\partial \mathbf{z}^{[M]}}{\partial \mathbf{A}^{[M-1]}} \cdots \frac{\partial \mathbf{z}^{[l+1]}}{\partial \mathbf{A}^{[l]}} \cdot \frac{\partial \mathbf{A}^{[l]}}{\partial \mathbf{z}^{[l]}} \cdot \frac{\partial \mathbf{z}^{[l]}}{\partial \mathbf{W}^{[l]}}\right)^T, \\
&= \left(\frac{\partial L}{\partial \mathbf{A}^{[M]}} \cdot \left(\sigma^{[M]}\right)' \cdot \left(f^{[M]}\right)' \cdots \left(f^{[l+1]}\right)' \cdot \left(\sigma^{[l]}\right)' \cdot \mathbf{I}^{[l]}\right)^T, \\
&= \left(\mathbf{I}^{[l]}\right)^T \cdot \left(\sigma^{[l]}\right)' \cdot \left(f^{[l+1]}\right)' \cdots \left(f^{[M]}\right)' \cdot \left(\sigma^{[M]}\right)' \cdot \nabla_{\mathbf{A}^{[M]}} L.
\end{aligned}
\tag{4.7}
$$

In each iteration, the parameters $\mathbf{W}^{[l]}$ are updated by the gradients after scaling using the predefined learning rate, η, as follows:

$$
\mathbf{W}^{[l]} \leftarrow \mathbf{W}^{[l]} - \eta \cdot \nabla_{\mathbf{W}^{[l]}} L.
\tag{4.8}
$$

It is noted that the presented approach in Eq. 4.7 to calculate the gradients with respect to the parameters is known as the *backpropagation* algorithm [1].

4.1.3 Gradient Descent Method

Due to possible inefficiencies and limitations of the available computational resources, instead of obtaining the *gradient* based on Eq. 4.7 to use with the *batch gradient descent* algorithm, which is usually effective for *convex* problems, the *stochastic gradient descent (SGD)* algorithm was proposed and is now widely adopted [1]. In SGD (and more generally the *mini-batch gradient descent (MBGD)* algorithm), instead of using *all* the data at one time, in each iteration of the backpropagation, the NN is only updated according to *one small batch* containing $m < N$ training points sampled from the full dataset of size N. More specifically, if $m = 1$, it is called SGD; otherwise, it is called MBGD. Using one sample per iteration in the SGD often leads to a noisy gradient. On the contrary, the MBGD is usually more stable. Moreover, the MBGD takes advantage of parallel computing on the

---

[4] The chain rule differentiates a "composite function" $f(g(x))$ as $f'(g(x)) \cdot g'(x)$ where the prime indicates a derivative with respect to x, e.g., let $f(g(x)) = \sin\left(x^2\right)$ as a composite function constructed from $f(z) = \sin(z)$ and $z = g(x) = x^2$. Thus, $f'\left(\sin\left(x^2\right)\right) = f'(\sin(z)) = \cos(z) \cdot z' = \cos\left(x^2\right).(2x) = 2x \cos\left(x^2\right)$. The chair rule is also extended to multivariate functions, e.g., let $h = g(s, t) = s^2 + t^3$ with functions $s = xy$ and $t = 2x - y$. Thus, $\dfrac{\partial h}{\partial x} = \dfrac{\partial h}{\partial s} \cdot \dfrac{\partial s}{\partial x} + \dfrac{\partial h}{\partial t} \cdot \dfrac{\partial t}{\partial x} = 2s \cdot y + 3t^2 \cdot 2 = 2(xy)y + 6(2x - y)^2 = 2xy^2 + 24x^2 - 24xy + 6y^2$.

vector/matrix form of the batch of $m > 1$ training points, which is more efficient than the similar computation in the SGD but needs m iterations. For brevity, in some conventional literature, both SGD and MBGD are simply unified as SGD with a specified batch size $m \geq 1$.

In the SGD, the gradient in the kth iteration is denoted by $\mathbf{G}_k$, which shares the same dimension as the parameters, $\left(\mathbf{W}^{[l]}\right)^{(k)}$, and is computed as follows:

$$\mathbf{G}_k = \frac{1}{m} \sum_{i=1}^{m} \nabla_{\mathbf{W}^{[l]}} L\left(s\left(x_i\right), y_i\right), \tag{4.9}$$

where $s\left(x_i\right)$ is computed from Eq. 4.5. Accordingly, the *updating rule* in Eq. 4.8 is revised herein for the kth iteration as shown in Eq. 4.10. In the SGD, the learning rate $\eta^{(k)}$ is usually changeable during the course of training based on the *learning history* and the strategy or plan used to adjust it is defined as the *learning schedule*.

$$\left(\mathbf{W}^{[l]}\right)^{(k+1)} \leftarrow \left(\mathbf{W}^{[l]}\right)^{(k)} - \eta^{(k)}\mathbf{G}_k. \tag{4.10}$$

It is noted that $\mathbf{G}_k$ is a random variable and its expectation satisfies $E\left[\mathbf{G}_k\right] = \left(\nabla_{\mathbf{W}^{[l]}} L\right)^{(k)}$. Thus, the gradient obtained in each iteration is an unbiased estimate of the true gradient and the final result is thought to be close to the one using the true gradient with more *repetition* of sufficient number of iterations. Moreover, in the SGD, instead of randomly sampling m training points, it is suggested to use the term *epoch* to conduct training in a more well-organized order, where one training epoch means to go over all the training data once [1]. At the beginning of each epoch, the sequence of the data fed into the NN is *shuffled randomly* and then the SGD is performed batch by batch until looping over all data points, where the batch size maintains the m value except for the last batch size, where its size, m_{last}, satisfies the condition $m_{\text{last}} \leq m$. Shuffling and repeating continue until the NN loss converges or the designated number of training epochs is reached. In addition, the number of total iterations, N_{iter}, in this case for all epochs, can be easily computed based on the number of all the data points, N, the number of the training epochs, N_{epoch}, and the "constant" (with the possible exception of m_{last}) batch size, m, as follows:

$$N_{\text{iter}} = N_{\text{epoch}} \cdot \left\lceil \frac{N}{m} \right\rceil, \tag{4.11}$$

where $\lceil x \rceil$ is the ceiling operation,[5] which maps $x \in \mathbb{R}$ to the least integer $\geq x$.

---

[5] For a real number x, two integers, i and j, and the set of integers, $\mathbb{Z}$, the *floor* and *ceiling* functions are respectively defined as follows:

$$\lfloor \times \rfloor = \max\{i \in \mathbb{Z} | i \leq x\}, \tag{4.12}$$

$$\lceil x \rceil = \min\{j \in \mathbb{Z} | j \geq x\}. \tag{4.13}$$

4.1.4   Dropout

When using more neurons and stacking more layers, the NN becomes a deep structure, namely, deep NN, and it obtains a more powerful representation ability through a significantly increased number of parameters in the NN model. However, when dealing with small datasets, using a deep NN is likely to trigger over-fitting issues. To address this, *dropout* is one of the possible choices, which was shown to be very efficient for relieving over-fitting in deep NNs [6]. The key idea of dropout is to randomly drop neurons along with their connections between layers during training and the dropped ones are neither activated in the forward nor the backward propagation. However, during the test process, the original complete network is used, refer to Fig. 4.2. It is noted that the dropout is only performed in every training epoch and the selection for the dropped connection is random and independent of the previous epoch.

According to [6], the dropout operation leads to a change of the expected output values between the training and the test processes. Thus, the weights need to be adjusted and scaled accordingly during the test process to preserve the expected output values. Denote the dropout rate as p (probability to drop a neuron) and an arbitrary output entry considering dropout is $z = \mathbf{w}^T \mathbf{x} + b$. The expected value of the entry applying dropout during training is $E\left(\text{dropout}(z)\right) = (1 - p) \cdot z + p \cdot 0 = (1 - p) \cdot z$, which is scaled by $(1 - p)$. Note that the z and 0 are for the respective retained and dropped-out entries. Therefore, during the test process, to preserve the expected value, the weights for computing the output entry z need to be multiplied by $(1 - p)$. Equivalently, without adjusting the weights during the test process, the same results can be achieved by normalizing (i.e., dividing) the weights by $(1 - p)$ during training. Empirically, the dropout rate (i.e., its probability), as a hyper-parameter, p, is typically selected from the range of 0.5–0.8 for the hidden layers [6]. A small value of p may not produce enough dropout to prevent over-fitting ($p = 0.0$ corresponds to no dropout). The larger the value of p, the more neurons are possibly dropped during

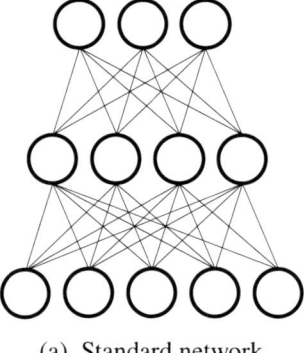
(a) Standard network

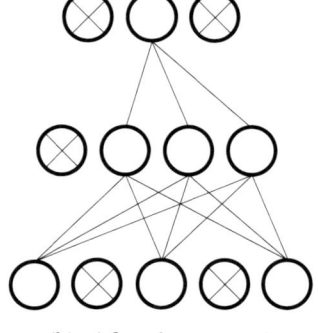
(b) After dropout ($\otimes$)

Fig. 4.2 Dropout operation in the NN model

training. Especially for the case of using DL in a small-scale dataset, the deep NN containing a large number of neurons may suffer from over-fitting due to data deficiency. In this case, a heavy dropout rate (e.g., $p = 0.7$) reduces the trainable parameters in the neurons during training, which may be helpful in alleviating the over-fitting issue.

4.2 Convolutional Neural Network

Similar to the NN, a typical CNN is also composed by stacking multiple layers, but different types of layers are considered besides the fully-connected (FC-)layer used in the NN, e.g., convolutional (Conv) and pooling layers as illustrated in Fig. 4.3. As mentioned in Chap. 1, nowadays DL adopting deep CNN is the state-of-the-art approach for vision-based SHM and several well-known CNN models (i.e., architectures), e.g., VGGNet [7], InceptionNet [8], and ResNet [9], with strong representation ability are widely used in many applications. Before diving into these advanced models, a brief mathematical reasoning and basics about how CNN works are presented in this section.

4.2.1 Convolution

The convolution operation in CNN is performed in the Conv layer by convolving a *filter* (also known as *kernel*) with the input in a manner of a sliding window from left to right and from top to bottom. The output of the Conv layer, sometimes referred to as the *feature map*, shares similar characteristics as the features in the NN but the feature map in CNN is usually a tensor and exists in a more abstract and high-dimensional space. To increase the recognition ability, in each Conv layer, the input is manipulated with multiple filters and then the outputs are activated by some nonlinear activation functions (introduced in Sect. 4.1.1), where *ReLU* is the most commonly used one and accordingly multiple feature maps are generated. Each feature map describes certain information by a matrix in the 2D space and multiple feature maps produced by the filters are stacked together along the new third axis, which enhances the space from 2D to 3D and makes the stacked feature maps represented as a 3D tensor.

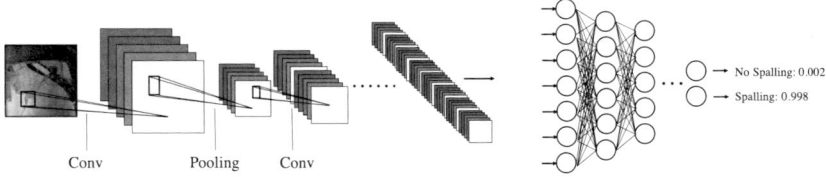

Fig. 4.3 Conv and pooling operations in a CNN for vision-based SHM

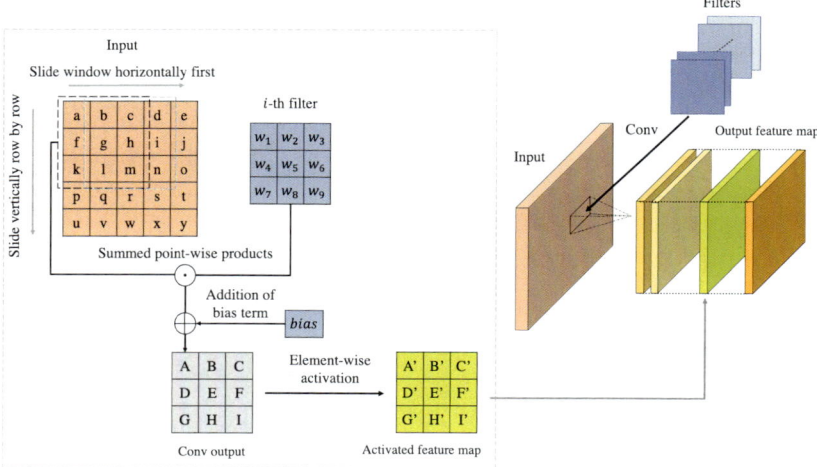

Fig. 4.4 Illustration of the Conv operations in a simple example

The mathematical foundation of CNN is introduced through a simple example as illustrated in Fig. 4.4. Suppose the input is a 5×5 matrix, e.g., representing the pixel map of an image, and it is convolving with the ith filter, whose size is 3×3. Starting from the top left, the first entry of the Conv output "A" is computed by the summed element-wise products of the weight matrix (filter) with the same size of a sub-matrix in the top-left corner of the input and then adding one bias term "$bias$" as follows:

$$A = a \cdot w_1 + b \cdot w_2 + c \cdot w_3 + f \cdot w_4 + g \cdot w_5 + h \cdot w_6 + k \cdot w_7 + l \cdot w_8 + m \cdot w_9 + bias. \tag{4.14}$$

To increase the nonlinearity[6] in the CNN, the Conv output is activated by *ReLU*. Therefore, the first element of the activated feature map "A'" is computed as follows:

$$A' = ReLU(A) = \max(0, A). \tag{4.15}$$

Define the *stride* as the step size of the sliding window, e.g., stride 1 means each time move 1 unit away from the current position. Suppose in the example shown in Fig. 4.4, a stride of 1 is used, then the filter moves by 1 unit and the multiplication and summation operations are repeated for elements (b, c, d, g, h, i, l, m and n) with the same weight matrix

---

[6] Nonlinearity in a NN (and also CNN) layer allows the subsequent layers to build off each other. Two consecutive linear layers have the same power (they can represent the exact same set of functions) as a single linear layer using the applicable *superposition* in this hypothetical case. The introduced nonlinearity makes it possible for the NN model to adapt with a variety of data and to differentiate between the outcomes.

(filter), i.e., w_1 to w_9. Similarly, the element B′ in the active feature map is obtained. Loop and repeat the above mentioned operations for the entire input matrix with the stride of 1 sliding window following the sequence of left to right and top to bottom. Accordingly, a 3×3 activated feature map is generated for the ith filter with the results from "A′" to "I′" shown in Fig. 4.4.

The stride size S is adjustable, which is a hyper-parameter, and using different stride sizes yields different shape of the feature map. Conventionally, the same stride is used for both horizontal and vertical sliding windows. Suppose the shape of the current input is (P, Q), where P and Q represent the size in the horizontal and the vertical directions, respectively, and the size of the filter is F (the conventional filter is a square). The output shape (P', Q') is computed as follows:

$$P' = \frac{(P - F)}{S} + 1,$$
$$Q' = \frac{(Q - F)}{S} + 1. \tag{4.16}$$

However, from Eq. 4.16, using $S > 1$ requires the differences between the shape of the input (P, Q) and the filter size F, i.e., $(P - F)$ and $(Q - F)$, to be divisible by S, which limits the selection of S. To circumvent this limitation, the *padding* technique was developed in [10] and introduced next. Note that when Eq. 4.16 is applied to the example in Fig. 4.4, i.e., $P = Q = 5$, $F = 3$, and $S = 1$, one obtains $P' = Q' = 3$, as shown by the size of the activation feature map in Fig. 4.4.

One Conv layer usually consists of multiple filters, yielding multiple feature maps, where one filter contributes to its corresponding feature map. Suppose there are M filters in Fig. 4.4, the Conv layer produces M 3×3 feature maps. Therefore, by stacking all the feature maps together, the shape of the final layer output is $(N, 3, 3, M)$, where N is the number of input data. As for a deep CNN, N input data are processed via multiple Conv layers and finally end up with the high-dimensional feature maps with the shape of (N, W, H, M), where W and H represent the 2D shape of each feature map in the final Conv layer. To obtain the class probability in a classification problem, the feature maps of the final Conv layer are further flattened into the shape of $(N, W \times H \times M)$. Subsequently, the flattened feature is further connected with neurons in the FC-layers to output the probabilities of the different classes of the feature, as shown in Fig. 4.3 for the example of two-class classification of concrete cover "Spalling" (99.8% probability) and "No Spalling" (0.2% probability) for the shown RC column input image.

In many cases, e.g., RGB images, the input is usually a tensor, whose fourth dimension represents channel information C, where the first three dimensions correspond to the number of data, input image width in pixels, and input image height in pixels. Usually, the fourth dimension of the output increases after the Conv operations, which is equal to the number of applied filters (kernels). Accordingly, in the training, the filters are also expanded into the third dimension to be consistent with the fourth dimension of the input. Suppose the current input size is (N, P, Q, C) and the actual shape of the filters to be initialized and trained is (F, F, C). These filters convolve with all data points of size (P, Q, C) along the first dimension N. Moreover, since all parameters of one Conv layer are weights in the filters and one bias term, it is straightforward to compute the number of parameters N_{para} to be trained in one Conv layer with M $F \times F$ filters as follows:

$$N_{\mathrm{para}} = M \cdot (F \cdot F \cdot C) + 1. \tag{4.17}$$

4.2.2   Padding

From the example shown in Fig. 4.4, it is observed that after one Conv layer, the shape of the feature maps shrinks in the (W, H) plan but expands in the M dimension. Moreover, as mentioned above, using a stride $S > 1$ may lead to indivisible numbers in Eq. 4.16 and using a larger size stride also leads to the shape of output diminishing quickly, which causes difficulties in constructing a deep network. Thus, the technique called *zero padding* is introduced, which works by padding a certain number of "0" elements around the raw input. One example of one unit padding on a raw 3×3 input pixel map is shown in Fig. 4.5.

More generally, if padding U units on each side of the surroundings of the input with shape (P, Q) and using the same $F \times F$ filter, the output shape (P', Q') can be computed as follows:

$$\begin{aligned} P' &= \frac{P + 2U - F}{S} + 1, \\ Q' &= \frac{Q + 2U - F}{S} + 1. \end{aligned} \tag{4.18}$$

Mathematically, padding the "0" elements does not influence the results of the Conv operations and can also maintain the same shape between the input and the output (i.e., $P = P'$ and $Q = Q'$) when using stride $S = 1$ provided that the condition in Eq. 4.19 is satisfied. In addition, the input becomes divisible if padding an appropriate number of the "0" elements. If $S \neq 1$, a similar condition on U to that in Eq. 4.19 for maintaining the same shape between the input and the output can be developed but it also depends on the input size (P, Q), leading to possibly different zero padding in the P and Q directions, i.e., U_P and U_Q, respectively, refer to Eq. 4.20.

Fig. 4.5 Example of one unit padding on a 3×3 input

0	0	0	0	0
0	a	b	c	0
0	d	e	f	0
0	g	h	i	0
0	0	0	0	0

$$2U - F + 1 = 0 \rightarrow U = \frac{F - 1}{2}. \tag{4.19}$$

$$U_P = \frac{(S - 1)P + F - S}{2},$$
$$U_Q = \frac{(S - 1)Q + F - S}{2}. \tag{4.20}$$

4.2.3   Pooling

The pooling layer is another special layer component in deep CNNs. It can reduce the dimensionality of the representation and create invariance to small shifts and distortions [1]. By using pooling layers, which perform down-sampling operations, the CNNs can produce an output that is invariant to small translations in the input image, which is known as *translational equivariance*. In a typical deep CNN design, the pooling layers are usually placed after multiple stacked Conv layers.

There are two types of widely used pooing layers, namely, *max pooling* and *average pooling*. As reflected by their names, max pooling outputs the maximum element within the sliding window and average pooling outputs the average value of all elements in the sliding window. Usually, the same value is used for both the sliding window and the stride size, which avoids overlapping in the pooling layer. The most common combination is to use 2 for both the window size and stride size, where using a larger size leads to a more rapid dimensionality reduction. One example with respect to both max pooling and average pooling is shown in Fig. 4.6 using 2×2 window size and stride $S = 2$. For example, considering the shown position of the pooling filter, the element of value "3" in the max pooling case is obtained from $\max\{1, 3, -1, 1\} = 3$ and that of value "1" in the average pooling case is obtained from $(1 + 3 + (-1) + 1)/4 = 1$.

Fig. 4.6 Example of max pooling and average pooling

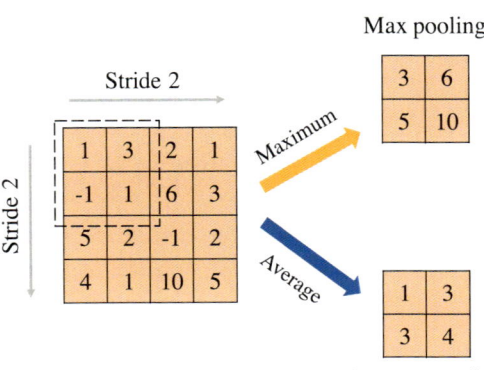

It is noted that max pooling has some positive effects in extracting superior invariant[7] features [11], which might help with identifying specific localized features, e.g., cracks in an image for vision-based SHM. Average pooling, on the other hand, focuses more on maintaining the integrity of the information during the dimensionality reduction.

4.2.4   Batch Normalization

One common issue in DL is that with the increase of the network depth, it becomes more difficult to train and it converges slowly. This issue can be partially attributed to the *internal covariate shift* phenomenon. In a deep network, it is known that the output from the previous layer is used as the input for the current layer. However, with the increase of depth, the distribution of the input for each layer changes and shifts frequently during the iterations of the training process and the distributions of the input between layers may differ significantly. It is noted that *covariate shift* usually relates to the input layer, but here internal covariate shift refers to the distribution shift within the network with respect to the input of the internal layers. To reduce the internal covariate shift, *batch normalization* (BatchNorm) was proposed in [12].

Simply conducting normalization toward the input of each internal layer usually does not work and makes the network learn nothing from the data. Thus, BatchNorm is performed in a slightly different way. Suppose the input batch data from the previous layer is $\{x_1, 2, \ldots, x_m\}$. During training, after normalizing the input, Eqs. 4.21–4.23, BatchNorm (BN) introduces two trainable parameters, namely, the *shift parameter*, β, and the *scale parameter*, γ, to map the normalized input, $\hat{x}_i$, using input BN mean, μ_{BN}, and variance, σ_{BN}^2, back to another distribution, x_i', which is intuitively thought to be suitable for the network training, e.g., changeable model parameters, decreasing loss, and moderate training speed, Eq. 4.24. To be specific, β and γ are automatically learned and adjusted by the SGD algorithm in the training to fit the corresponding distribution, which means that the model can adapt to changes in the input distribution over time, making it more robust to changes in the data. By normalizing the input to each layer, BatchNorm reduces the internal covariate shift and the loss values during training, which is helpful in stabilizing the network. In addition, BatchNorm can help with the vanishing and exploding gradients problem[8] to keep the gradients in a reasonable range, i.e., to avoid too large or too small values, and thus speed up the convergence of the network training. To avoid numerical difficulty through division by zero, a small positive value ϵ (e.g., 10^{-6}) is typically added to the denominator as shown in Eq. 4.23. Finally, during the test for prediction, the BatchNorm layer functions differently from the training and the input mean, μ_{BN}, and variance, σ_{BN}^2, are not computed from the test batch but instead taken as fixed values as obtained from the training.

---

[7] Invariant is a property of a mathematical object, which remains unchanged even after repetitive transformations of the object.

[8] This occurs when the value of the gradient diminishes to zero or explodes to a very large number.

$$\mu_{BN} = \frac{1}{m} \sum_{i=1}^{m} x_i. \tag{4.21}$$

$$\sigma_{BN}^2 = \frac{1}{m} \sum_{i=1}^{m} (x_i - \mu_{BN})^2. \tag{4.22}$$

$$\hat{x}_i = \frac{x_i - \mu_{BN}}{\sqrt{\sigma_{BN}^2 + \epsilon}}. \tag{4.23}$$

$$x_i' = \beta + \gamma \hat{x}_i. \tag{4.24}$$

In summary, the BatchNorm layer realizes the input normalization and it also maps the normalized input back to maintain the original data characteristics, where both mapping parameters (β, γ) are automatically learned by the network itself during training. According to [12], BatchNorm helps improve the gradient flows in the network, allows the use of higher learning rates, η, for faster convergence, and reduces the strong dependence on parameter initialization. Besides accelerating the training, BatchNorm somewhat plays the role of *regularization*, which can be moderately used in conjunction with other regularization methods, e.g., dropout. Empirically, while adding the BatchNorm layer into the network design, the dropout rate can be decreased (i.e., relax the regularization effects of the dropout).

4.3 Deep Convolutional Neural Networks

Similar to the deep NN, by stacking multiple Conv, pooling, and FC-layers in a certain design, a deep CNN is easily constructed and its configuration is known as the CNN architecture. As mentioned above, several deep CNN architectures have been proposed and validated effectively. This section introduces several popular and widely used deep CNN architectures, namely, VGGNet, InceptionNet, and ResNet.

4.3.1 VGGNet

Inherited from the two well-known architectures of LeNet [13] and AlexNet [10], VGGNet[9] uses similar configurations such as multiple *Conv blocks*,[10] two FC-layers representing image features, and one FC-layer for classification usage. Compared with previous network architectures, VGGNet goes much deeper, exemplifying the concept of "very deep networks" [7], with most Conv layers providing a direction for the design and efficient usage of CNNs. The most prominent contribution of the VGGNet architecture is to prove that by applying

---

[9] Visual geometry group (VGG), Department of Engineering Science, University of Oxford, UK.

[10] In this book, a Conv block is defined as several Conv layers ending with a max pooling layer.

a small filter (kernel) size, e.g., 3×3 or even 1×1, and increasing the depth of the network, one can effectively improve the model performance. The pre-trained VGGNet model on ImageNet [14] has very good generalization on other datasets and provides an environment for a TL study, where the DL classifier can learn from a *source domain* to transfer the learning to another *target domain* with fewer data.

The input size used in [7] is fixed at 224×224, which is consistent with the image size, where the pixel intensity in an image channel is reduced by subtracting the mean value of the corresponding RGB channel[11] as the only preprocessing procedure. For the Conv layer, only a small filter size (typically 3×3, i.e., $F = 3$ or even 1×1, i.e., $F = 1$) is used for the Conv operations. In order to maintain the size of the feature maps generated by convolution, the stride is taken as $S = 1$ with proper zero padding (refer to Eq. 4.19), i.e., for $F = 3$ as used for VGGNet, use zero padding of $U = 1$, while performing the convolution. In order to save computational time and pursue better performance of feature translational invariance,[12] after several rounds of convolution, max pooling layers are added to the Conv layers. The stride for the max pooling is set to $S = 2$ with a 2×2 window size. In the original design, after a series of Conv blocks, three FC-layers (also known as *Dense layers*) follow, where each of the first two layers has 4,096 neurons and the third one has 1,000 neurons, matching the 1,000 classes of the detection targets. The nonlinear activation function *ReLU* is applied in all Conv and FC-layers, except the last FC-layer (also known as *softmax layer*) whose activation function is the *softmax* function. According to the different number of filters in the Conv layer and the different number of the Conv layers and the FC-layers, i.e., different depths, VGGNet has several variations, e.g., VGG-16 and VGG-19, where their original network architectures are shown in Fig. 4.7 and "16" and "19" stand for the number of weight layers (Conv and Dense layers) in the network. More details about VGGNet can be found in [7].

4.3.2 InceptionNet

Besides the VGGNet, InceptionNet [8] is another important deep CNN architecture in current DL research and applications. InceptionNet appeared for the first time during the 2014 ImageNet large scale visual recognition challenge (ILSVRC) [14] and it achieved first place in the classification task with a top-5 error rate[13] of 6.7%, which out-performed the VGG-16's error rate of 7.3%. Afterwards, InceptionNet went through several improvements with different versions, starting from the original InceptionNet v1 [8], then v2 [12], v3 [15], and

---

[11] The typical approach to determine the mean RGB colors is to separately add up all R, G and B values and divide each sum by the number of pixels to obtain the mean value of each color channel.

[12] Translational invariance means that if inputs are translated, the CNN is still able to detect the class to which the input belongs.

[13] It is the percentage of times the classifier failed to include the proper class among its top 5 guesses.

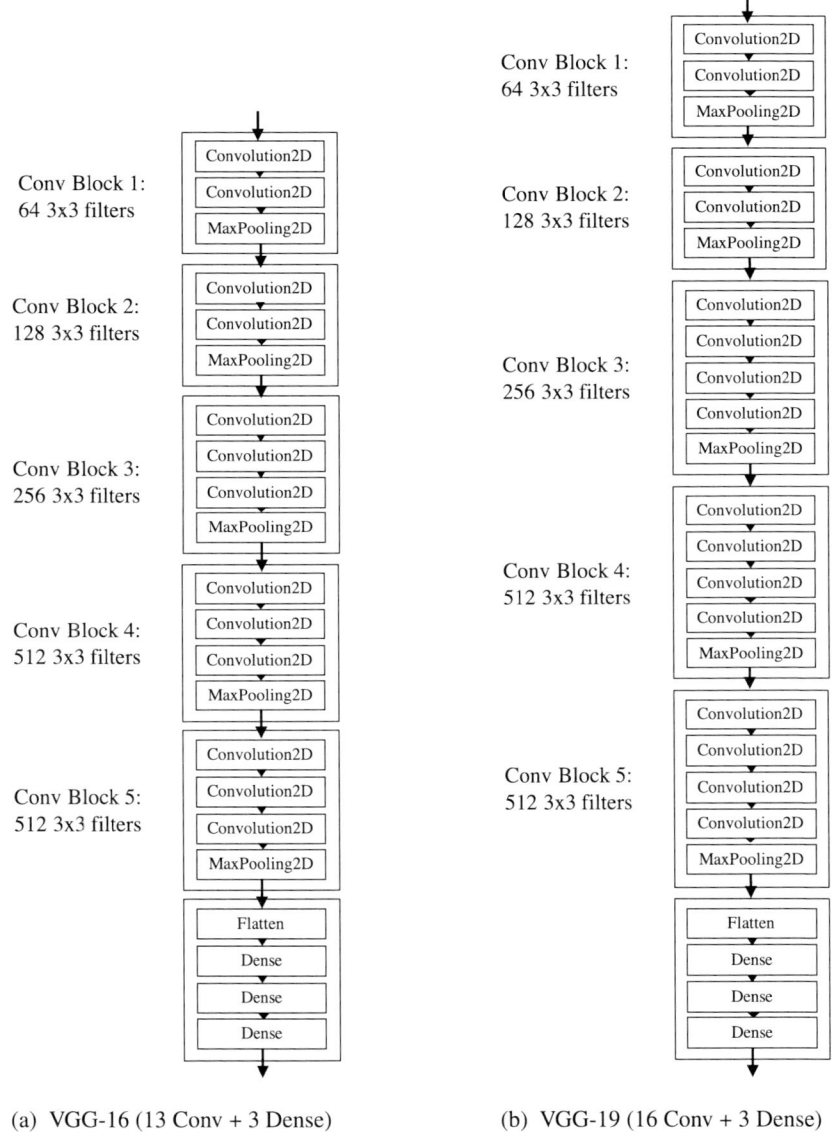

(a) VGG-16 (13 Conv + 3 Dense)   (b) VGG-19 (16 Conv + 3 Dense)

Fig. 4.7 Illustration of the two original VGGNet architectures [7]

v4 [16]. Because of its registration name in the ImageNet [14] challenge 2014, InceptionNet v1 is also referred to as GoogleNet.

Compared with the early DL model, e.g., AlexNet [10], InceptionNet v1 significantly decreased the number of network parameters from 6 billion to only 0.5 billion, which alleviated the high demand on the computational resources. However, the decrease of the number of parameters did not reduce its recognition performance. This is mainly due to the following novel changes from the early DL model designs:

1. Introduce the *inception module* to improve the utilization of parameters.
2. Apply different sizes of filters (kernels) to increase diversity, especially the use of multiple small filters to replace larger ones.
3. Add auxiliary classifiers in some middle layers and add their outputs with small weights into the final output of the classification.
4. Remove the last several FC-layers, containing a large number of parameters, and use a global average pooling (GAP) layer[14] to reduce over-fitting and accelerate training.

Using the *inception module* is the core of the InceptionNet, which is the origin of the name InceptionNet. Essentially, the inception module is a sub-structure of the network, which is similar to the Conv block used in the VGGNet, and the final deep network is stacked by multiple inception modules. Unlike in the CNNs where the input signal (i.e., data) directly passes through the entire network structure, when the signal from a previous layer passes through the inception module, it goes through different branches and the outputs from these branches are concatenated and used for the next layer. One example of an inception module is shown in Fig. 4.8 with the following 4 branches: (1) the first conducts 1×1 Conv operations, (2) the second contains two Conv layers to conduct 1×1 and 3×3 Conv operations, (3) the third contains two layers to conduct 3×3 max pooling and 1×1 Conv operations, and (4) the fourth contains two Conv layers to conduct 1×1 and 5×5 Conv operations. It is noted that a 1×1 filter is widely used in these modules, which can transfer information between channels and also help increase or decrease the dimensions.[15] There are usually multiple different sizes of filters working on passing the input signal and the final module output is actually the concatenation of the outputs from multiple different Conv blocks. These characteristics of

---

[14] Global average pooling computes the average value of each channel across the entire spatial dimensions of the feature map, which is typically used as an alternative to the FC-layer in a CNN.

[15] 1×1 convolution is a pixel-wise weighting of feature maps since each pixel is multiplied by a single number without regard to the neighboring pixels. If one would like to build a deeper network, but there are excessive (e.g., 900) feature maps in an intermediate layer, one could apply 1×1 convolution to reduce the number of feature maps (e.g., 900 to 30). Thus, if there are (900,64,64) feature maps, i.e., 900 representations, where each of them has the size 64 pixels $\times$ 64 pixels, in a convolution layer, we can apply (30,900,1,1) filters, i.e., 30 1×1 convolution filters, each with 900 channels, to reduce the number of feature maps to 30. Thus, (30,64,64) feature maps are produced. The 1×1 convolution can also include summation across channels. Therefore, the 1×1 convolution is a channel-wise pooling and the produced 30 feature maps are "summaries" of all 900 feature maps present in the layer.

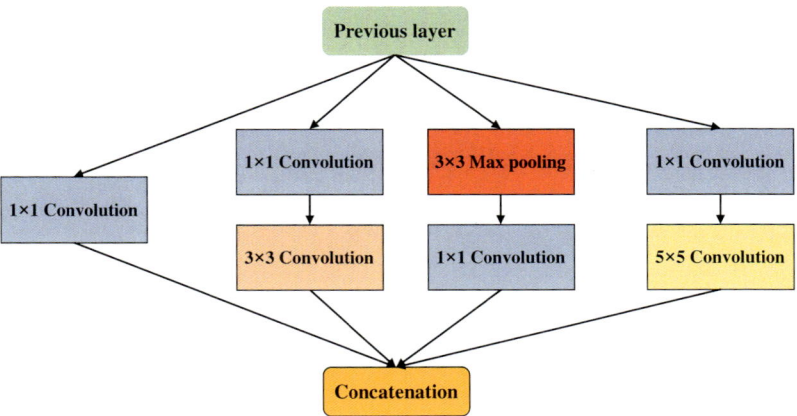

Fig. 4.8 Example of one inception module with four branches

the inception module increase both network complexity for recognition and its adaptability to different scales [8].

Several modifications were made to further improve the performance of InceptionNet v1. In InceptionNet v2 [12], the 5×5 filter is replaced by two-layer convolution with two 3×3 filters. In order to reduce the internal covariate shift, BatchNorm is introduced in v2. Moreover, InceptionNet v3 [15] further optimized the inception module and introduced the concept of "factorization into small convolutions", by changing a relatively large $n \times n$ filter to two-layer convolution with $1 \times n$ and $n \times 1$ filters, which reduced the number of parameters and increased the nonlinearity. InceptionNet v4 [16] combined ResNet [9] (Sect. 4.3.3) within the InceptionNet. This produced the state-of-the-art version achieving the lowest top-5 error on ImageNet [14]. More details about InceptionNet can be found in [12, 15, 16].

4.3.3 ResNet

A deep network has more trainable parameters compared with a shallow one and the increasing complexity usually helps to solve more complex problems and improve performance. Unfortunately, vanishing and exploding gradients and degradation[16] issues occur with increasing network depth, making it difficult to effectively train.

To address the above concern, He et al. [9] developed a new CNN architecture with a *shortcut connection* named ResNet. The layers are reformatted with reference to the inputs through the shortcut connection and the network learns *residual functions*, which differ from those learned by the traditional CNN. These shortcuts act like highways and the

---

[16] By increasing the network depth, the accuracy becomes saturated and then degrades rapidly.

gradients can easily flow back, resulting in faster training and support for stacking more layers. Mathematically, assuming $H(x)$ is any desired mapping function, the traditional way assumes that Conv operations try to fit $H(x)$. On the contrary, in ResNet, the Conv operations are used to fit the residual function, $F(x)$, and then mapping can be represented as the sum of functions of input and residual terms as follows:

$$H(x) = F(x) + h(x). \tag{4.25}$$

Therefore, it is said that ResNet learns *referenced functions* (i.e., with reference to the layer inputs) unlike the unreferenced functions learned by CNN. Two common types of the shortcut connections, their corresponding additive terms, and residual functions are discussed below.

If the additive term $h(x)$ is the original input, expressed as,

$$h(x) = x, \tag{4.26}$$

then the shortcut connection is considered as an *identity mapping*, Fig. 4.9a. From, Eqs. 4.25 and 4.26, we have,

$$\frac{\partial H}{\partial x} = \frac{\partial F}{\partial x} + 1. \tag{4.27}$$

This improves the gradient vanishing issues and the network learns the residual term, $F(x) = H(x) - h(x) = H(x) - x$, in this case, instead of directly mapping the relationship between the input x and the output $H(x)$, as in CNN. A more sophisticated mapping was also designed for the shortcut connection using $h(x)$ as the Conv operations, namely, *Conv shortcut* or *projection shortcut*, where again the network learns the residual term, $F(x)$, refer to Fig. 4.9b.

In the ResNet architecture, it is common to apply 1×1 convolution to match the dimensions between the Conv blocks, where the dimension difference between the Conv blocks is mainly due to the applied max pooling layers, which usually down-samples the dimension by a factor of 2. In the context of ResNet, the CNN sub-structures shown in Fig. 4.9a and b are also denoted as an identity residual unit and a Conv residual unit, respectively. Moreover, He et al. [17] studied more variants of the shortcut configurations and validated the results with detailed derivations.

The example of one variant of ResNet, namely, ResNet-50, is shown in Fig. 4.10, which consists of stacking multiple identity and Conv residual units, where the first unit in each Conv block is set as a Conv unit to match the dimensions between layers and the remaining shortcut units are all identity units. In the shown ResNet-50, the depth 50 (not counting the pooling layers) consists of $1 + 3 \times 3 + 4 \times 3 + 6 \times 3 + 3 \times 3 = 49$ convolution layers and one Dense layer. Similar to VGGNet, there are multiple variations of ResNet based on the depth, e.g., ResNet-50, ResNet-101, and ResNet-152, or the more advanced design, e.g., ResNeXt [18]. Interested readers can find more information about ResNet in [9, 18].

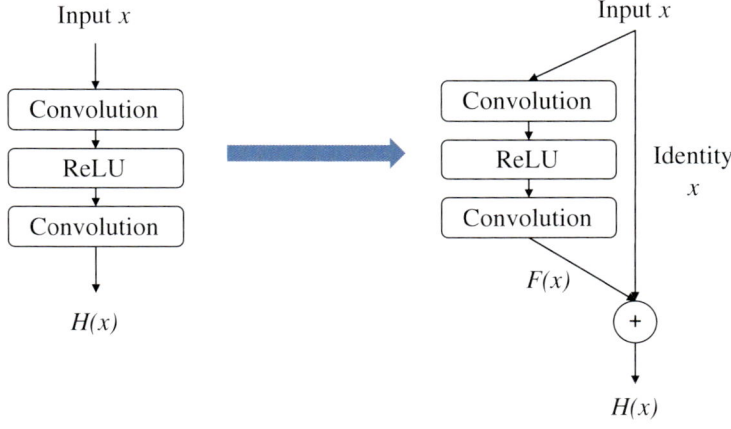

(a) Identity mapping

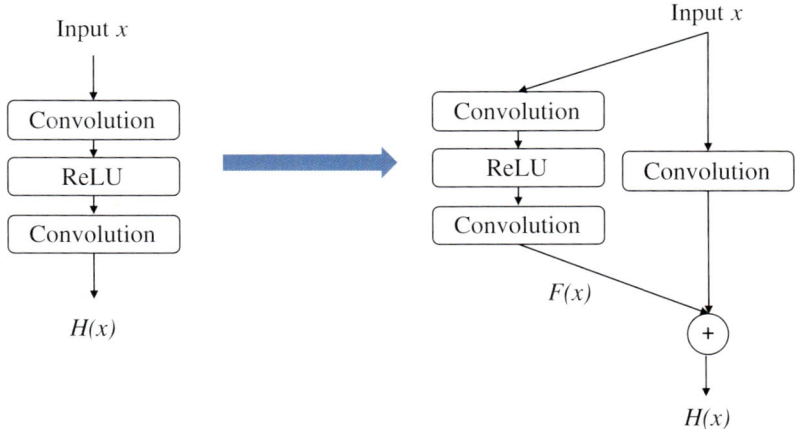

(b) Convolutional (projection) shortcut

Fig. 4.9 Two common types of shortcut connections in ResNet

Fig. 4.10 Illustration of
ResNet-50 (49 Conv +1
Dense) network architecture

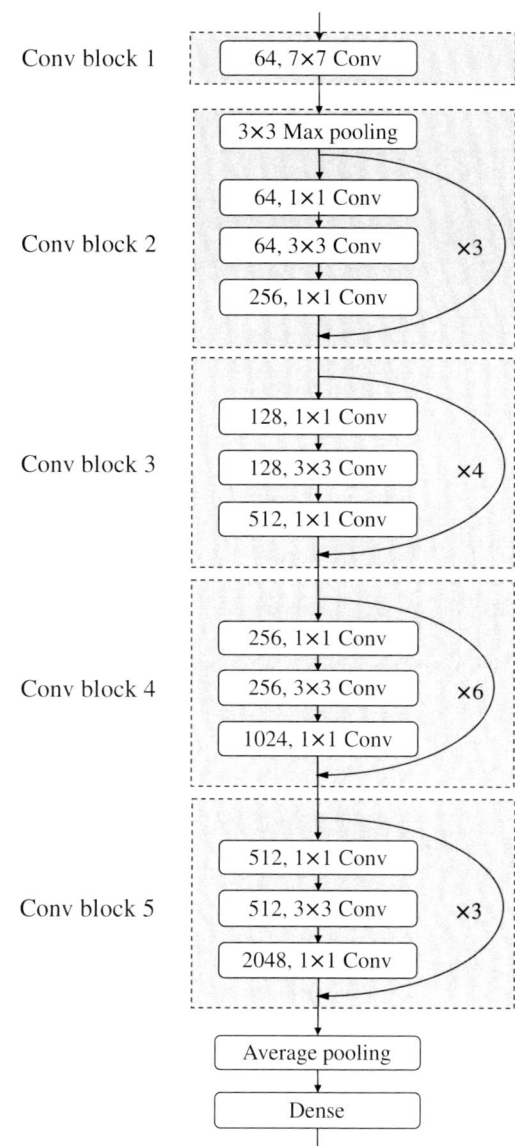

4.4   Recurrent Neural Networks

4.4.1   RNN

RNN is widely used as a DL sequence model. The difference between RNN and ordinary NN (refer to Sect. 4) is that in RNN both the input from the current step and the state information from the previous step are considered in the output of the current step. In this way, RNN captures the *intrinsic correlation* between the current step and the previous steps. Figure 4.11 shows the architecture of a simple RNN, where x_t, h_t, and y_t are, respectively, the input vector, the hidden state vector, and the output vector, all at time step t. The hidden state vector implicitly contains information about the history of all the past elements of the data sequence. Moreover, h_t (Eq. 4.28) and y_t (Eq. 4.29) depend on the current step input vector, x_t, and the previous step hidden state vector, h_{t-1}. In addition, the previous step's hidden state vector, h_{t-1}, depends on the previous step input vector, x_{t-1}, and the hidden state vector in the step before, h_{t-2}. Therefore, the hidden state vector at each step contains sequential information from all previous steps.

$$h_t = f\left(W_{hh}\,h_{t-1} + W_{hx}\,x_t + b_h\right), \tag{4.28}$$

$$y_t = f\left(W_{yh}\,h_t + b_y\right), \tag{4.29}$$

where f is the activation function, W_{hh}, W_{hx}, and W_{yh} are weight matrices as illustrated in Fig. 4.11, and b_h and b_y are bias vectors. Note that in RNN, the weights are shared among steps. In comparison to the ordinary NN, RNN can be seen as a very deep feed-forward network in which all the layers share the same weights [19]. Therefore, one of the common problems in deep NN, namely, *vanishing gradient*, where the gradient becomes vanishingly small and accordingly effectively prevents the weights from changing their values (i.e., the networks fail to learn from the training data), also takes place in RNN. Moreover, because RNN is "very" deep, the problem is exacerbated, thereby modifications to the vanilla RNN architecture are put forward to resolve this problem, as discussed in this section.

There are two basic variants of RNN. The first is the bi-directional RNN, where the output vector at time t, y_t, depends on the input vector at time t, x_t, and the hidden state vectors at

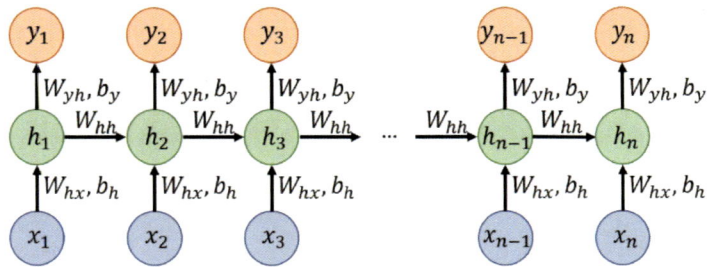

Fig. 4.11 Vanilla RNN

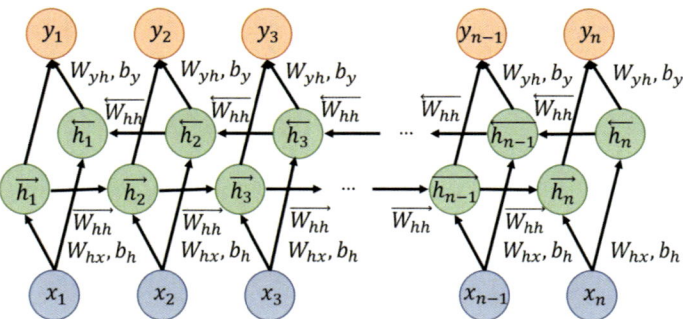

Fig. 4.12 Bi-directional RNN

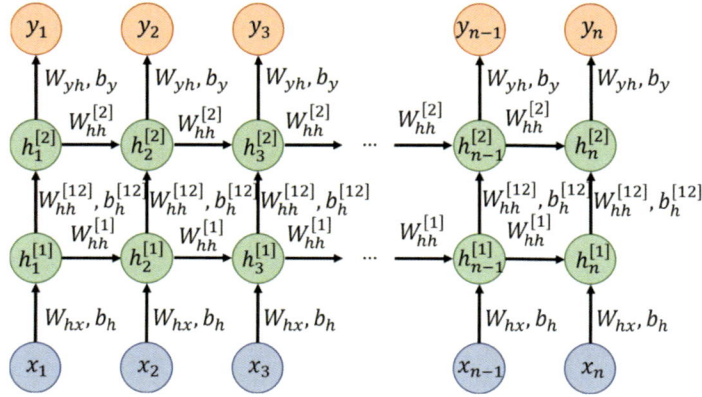

Fig. 4.13 Two-layer (stacked) RNN

the previous step, i.e., time $(t - 1)$, and at the next step, i.e., time $(t + 1)$, namely, h_{t-1} and h_{t+1}, respectively. In this way, the current step not only considers previous steps but also future ones, refer to Fig. 4.12. The second is the multi-layer RNN, or stacked RNN, which stacks layers of RNN cells for each step. The lower layer of each step passes the hidden state vector to the upper layer as the input. Therefore, the hidden state vector not only goes to the next step at the same layer, but also to the same step at the next layer. In this way, the *complexity* and *expressiveness* of the RNN are improved.[17] Generally, the number of layers for a multi-layer RNN is two to four and the three-layer RNN is already considered "deep", refer to Fig. 4.13.

The training process of the RNN is similar to that of the ordinary NN, e.g., using *gradient descent* to minimize the training *loss function*. However, as mentioned above, RNN has some intrinsic problems, e.g., vanishing gradient. NN uses backpropagation to minimize the training loss function and the gradient of the weights comes from the *chain rule* of derivatives

---

[17] Complexity of a system is the varying amount of effort required to analyze and utilize such system. Expressiveness of a system is the ability for the system to define complex sets of objects.

of composite functions. In RNN, the gradient from the output is difficult to backpropagate to affect the weights of earlier layers. Therefore, the final output value is more influenced by the last steps. The vanishing gradient problem makes the RNN hard to capture very long dependencies inside the sequential data. As an intuitive example, the model needs to choose between "was" and "were" to fill in the *BLANK* for this sentence: "The cat, (*a long sentence*), (*BLANK*) full." The correct answer, from human judgment, would be "was", as "cat" is a singular subject. However, the RNN is not good at handling long-range dependencies and it forgets the number of cats mentioned at the beginning of the sentence after processing the *long* sentence in the middle.

Mathematically, in the forward propagation of the RNN model, the output at the final step n (for illustration, the *sigmoid* function (Eq. 3.18) is used as the activation function herein) and making use of Eqs. 4.29 and 4.28, is expressed as follows:

$$
\begin{aligned}
y_n &= \sigma\left(W_{yh}\, h_n + b_y\right), \\
&= \sigma\left(W_{yh}(\sigma(W_{hh}\, h_{n-1} + W_{hx}\, x_n + b_h)) + b_y\right), \\
&= \sigma\left(W_{yh}(\sigma(W_{hh}(\sigma(W_{hh}\, h_{n-2} + W_{hx}\, x_{n-1} + b_h)) + W_{hx}\, x_n + b_h)) + b_y\right), \\
&= \cdots, \\
&= \sigma\left(W_{yh}(\sigma(W_{hh}(\sigma(W_{hh}(\cdots h_2 \cdots) + W_{hx}\, x_{n-1} + b_h)) \right. \\
&\quad \left. + W_{hx}\, x_n + b_h)) + b_y\right), \\
&= \sigma\left(W_{yh}(\sigma(W_{hh}(\sigma(W_{hh}(\cdots (\sigma(W_{hh}\, h_1 + W_{hx}\, x_2 + b_h)) \cdots) \right. \\
&\quad \left. + W_{hx}\, x_{n-1} + b_h)) + W_{hx}\, x_n + b_h)) + b_y\right).
\end{aligned}
\tag{4.30}
$$

Using the chain rule of derivatives of the composite functions, the gradient of the loss is given by:

$$
\begin{aligned}
\frac{\partial L}{\partial W} &= \frac{\partial L}{\partial y_n} \frac{\partial y_n}{\partial h_n} \frac{\partial h_n}{\partial h_{n-1}} \cdots \frac{\partial h_3}{\partial h_2} \frac{\partial h_2}{\partial h_1} \frac{\partial h_1}{\partial W}, \\
&= \frac{\partial L}{\partial y_n} \frac{\partial y_n}{\partial h_n} \left(\prod_{t=2}^{n} \frac{\partial h_t}{\partial h_{t-1}}\right) \frac{\partial h_1}{\partial W},
\end{aligned}
\tag{4.31}
$$

where L is the loss function that measures the difference between the outputs and the *ground truth*. One could update the model parameters by:

$$
W \leftarrow W - \alpha \frac{\partial L}{\partial W},
\tag{4.32}
$$

where α is the *learning rate*, a hyper-parameter that needs to be tuned, similar to η for NN in Eq. 4.8. At step t, we have the following expression:

$$
h_t = \sigma(W_{hh}\, h_{t-1} + W_{hx}\, x_t + b_h).
\tag{4.33}
$$

Therefore, one could compute the derivative of h_t as follows:

$$\frac{\partial h_t}{\partial h_{t-1}} = \sigma'(W_{hh}\, h_{t-1} + W_{hx}\, x_t + b_h) \cdot \frac{\partial}{\partial h_{t-1}}(W_{hh}\, h_{t-1} + W_{hx}\, x_t + b_h),$$

$$= \sigma'(W_{hh}h_{t-1} + W_{hx}x_t + b_h) \cdot W_{hh}, \tag{4.34}$$

where σ' is the derivative of the *sigmoid* function σ with respect to h_{t-1}. Plugging Eq. 4.34 into Eq. 4.31, the back-propagated gradient is determined as follows:

$$\frac{\partial L}{\partial W} = \frac{\partial L}{\partial y_n}\frac{\partial y_n}{\partial h_n}\left(\prod_{t=2}^{n}\sigma'(W_{hh}h_{t-1} + W_{hx}x_t + b_h) \cdot W_{hh}\right)\frac{\partial h_1}{\partial W}. \tag{4.35}$$

Without calculating the gradient explicitly, one could observe that Eq. 4.35 tends to vanish (approaches zero) when n is large. The reason is that the derivative of the activation function, $(\sigma' < 1.0)$.[18] Therefore, the upper bound of the absolute value of the gradient approaches 0.0 when this derivative is multiplied for $(n-1)$ time steps, i.e., from $t = 2$ to n.

4.4.2   GRU

In order to overcome the above mentioned vanishing gradient problem and learn the long-term dependencies, variants of the RNN have been proposed. Among these variants, gated recurrent unit (GRU) is widely used in the literature. A memory cell of the GRU is shown in Fig. 4.14. The GRU tries to overcome the vanishing gradient problem by using a *gating mechanism*, which could regulate the flow of information. Each GRU cell has two gates, which are called the *update gate* and the *reset gate*. The update gate decides what information to throw away and what new information to add from prior steps. The reset gate decides how much past information to forget. Both gates form the standard GRU cell, Fig. 4.14, involving h_{t-1} and x_t to produce r_t for the reset gate and z_t for the update gate. The GRU equations are as follows:

$$z_t = \sigma\left(W_{zx}\, x_t + W_{zh}\, h_{t-1} + b_z\right), \tag{4.36}$$

$$r_t = \sigma\left(W_{rx}\, x_t + W_{rh}\, h_{t-1} + b_r\right), \tag{4.37}$$

$$\tilde{h}_t = \tanh\left(W_{hx}\, x_t + W_{hr}\left(r_t \odot h_{t-1}\right) + b_h\right), \tag{4.38}$$

$$h_t = z_t \odot h_{t-1} + (1 - z_t) \odot \tilde{h}_t, \tag{4.39}$$

where W_{zx}, W_{zh}, W_{rx}, W_{rh}, W_{hx}, and W_{hr} are weight matrices, b_z, b_r, and b_h are bias vectors, z_t and r_t are the update and reset gate activation vectors, respectively, $\odot$ is the

---

[18] One can show that $\sigma'(x) = \sigma(x) \cdot (1 - \sigma(x))$, indicating that $\max_x \sigma'(x) = \sigma'(0.0) = 0.25$ and $\min_x \sigma'(x) = \sigma'(\pm\infty) = 0.0$, i.e., $\sigma'(x) \in [0.0, 0.25]$.

Fig. 4.14 GRU cell

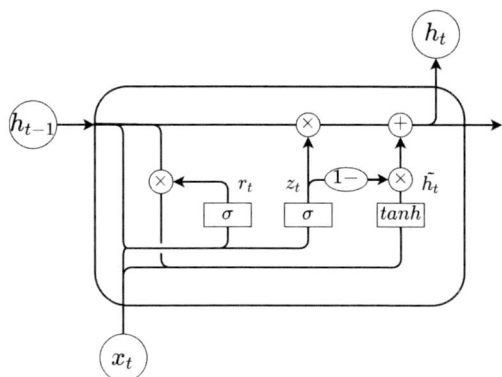

Fig. 4.15 LSTM cell

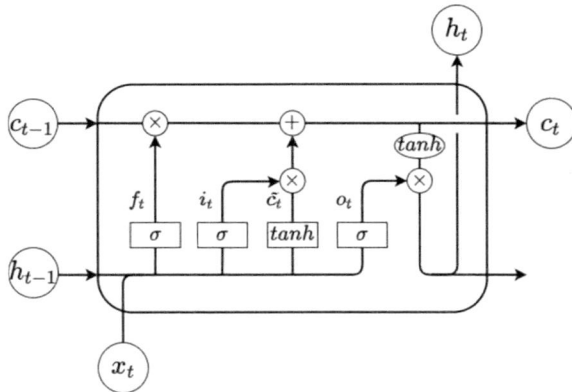

Hadamard product (element-wise multiplication, i.e., for two matrices of the same size A and B, $(A \odot B)_{ij} = A_{ij} \times B_{ij}$), and σ and tanh are the *sigmoid* and *hyperbolic tangent* (Eq. 4.4) activation functions, respectively.

4.4.3   LSTM

The LSTM network was invented by Hocheriter and Schmidhuber [20] in 1997, which is another variant of RNN. Similar to the GRU, it uses the gating mechanism to control the flow of information. The architecture of the LSTM network and a typical cell configuration are shown in Figs. 4.15 and 4.16, respectively.

In comparison to the GRU, the LSTM model has one extra vector, called the *cell state vector*, which is denoted by c_t. Each LSTM cell has 3 gates, which are called the *forget gate*, the *input gate*, and the *output gate*. The forget gate decides what is relevant to keep from prior steps, which looks at h_{t-1} and x_t and outputs a number between 0.0 and 1.0. This number is multiplied by c_{t-1}, as a discount for the cell state in the previous step. The input

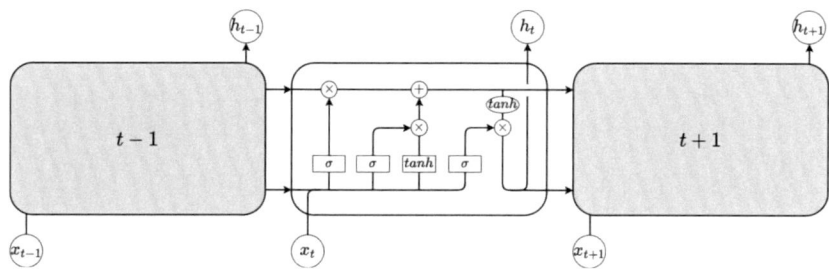

Fig. 4.16 A typical LSTM cell configuration

gate decides what information is relevant to add from the current step, which decides how much new information from h_{t-1} and x_t are added to the current cell state. Finally, the output gate determines what the next hidden state should be, which decides what part of the cell state the model is going to output as the hidden state for the current step. Accordingly, the output at each step depends on x_t, h_{t-1}, and c_{t-1}. An intuitive understanding of the LSTM is as follows: The cell state works like a conveyor belt and it passes through the LSTM cell with limited interactions within the cell. Unless it is intentionally forgotten, the information from the previous steps could be well-preserved. Each memory cell could selectively forget and remember sections of the previous states. The LSTM equations are as follows:

$$f_t = \sigma \left(W_{fx} x_t + W_{fh} h_{t-1} + b_f \right), \tag{4.40}$$

$$i_t = \sigma \left(W_{ix} x_t + W_{ih} h_{t-1} + b_i \right), \tag{4.41}$$

$$o_t = \sigma \left(W_{ox} x_t + W_{oh} h_{t-1} + b_o \right), \tag{4.42}$$

$$\tilde{c}_t = \tanh \left(W_{cx} x_t + W_{ch} h_{t-1} + b_c \right), \tag{4.43}$$

$$c_t = f_t \odot c_{t-1} + i_t \odot \tilde{c}_t, \tag{4.44}$$

$$h_t = o_t \odot \tanh \left(c_t \right), \tag{4.45}$$

where W_{fx}, W_{fh}, W_{ix}, W_{ih}, W_{ox}, W_{oh}, W_{cx}, and W_{ch} are weight matrices, b_f, b_i, b_o, and b_c are the bias vectors, f_t, i_t, and o_t are the respective forget, input, and output gate activation vectors, and $\tilde{c}_t$ is the cell input activation vector.

Despite the complexity of the LSTM model, it is much more frequently used in recent years than the traditional RNN models, because of its ability to capture long-term dependencies. Cho et al. [21] have shown that the hidden state could capture the semantically[19] and syntactically[20] meaningful representation of the data. They proposed an *encoder-decoder*

---

[19] Semantically refers to the study of the relationships between symbols or signs, e.g., words, phrases, sentences, and discourses, and what these elements of data in general mean or stand for.
[20] Syntactically refers to the investigation of the rules, principles, and processes which determine the structure of sentences in human languages or in data in general.

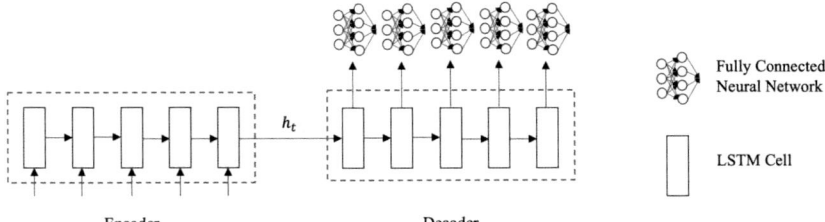

Fig. 4.17 LSTM EN-DE architecture

structure, which is able to learn the mapping from a sequence of an arbitrary length to another sequence. It is seen that the EN in this context is expected to extract the features from the input and learn a good representation of the input data, which is proven by the fact that the original data can be reconstructed using the DE [22]. This structure has been used in natural language processing (NLP), e.g., using an (English) sentence as the EN input and its French-translated sentence as the DE output. Several variants of this original EN-DE architecture were proposed in the literature. One LSTM EN-DE architecture is shown in Fig. 4.17.

The architecture in Fig. 4.17 consists of two LSTM networks that act as an EN and a DE pair. The EN maps a variable-length source (input) sequence to a fixed-length vector, which is the hidden state vector of the last step, h_t, also known as the *latent space vector* (LSV), as the internal representation of the input. On the other hand, the DE maps the vector representation back to a variable-length target (output) sequence. In practice, a shallow FC-NN is often added to the output of the DE. An intuitive understanding of the EN-DE architecture is that the LSV is acting as a "bottleneck". In order for the DE to decode the information of the EN input, the hidden state needs to compress the input without significant information loss. Therefore, the hidden state preserves the information of the input and actions toward the original input could also be performed on the compressed LSV, which is typically easier to manipulate.

4.4.4   RNN Family in Vision Tasks

The RNN family of models has achieved success in many NLP applications, including but not limited to speech recognition [23], text generation [24], and machine translation [21]. In addition, the RNN models can be applied to vision tasks with the combination of CV models, e.g., CNN, which could capture the temporal dependencies of images. Ng et al. [25] used the recurrent CNN for video classification. Another research direction is the introduction of an *attention mechanism* [26] (refer to Sect. 4.5.1), which initially comes with an EN-DE structure and could improve the performance for long inputs for the EN. This mechanism allows the model to automatically search for parts of an input that are relevant to the prediction in the DE, by assigning weights to the EN input and DE output pairs. Higher relevant pairs receive higher scores and the DE output pays more attention to these input parts in the form of the weighted sum of contributions from all the EN inputs.

4.5 Transformers

The transformer does not include any recurrence or convolution operations. Instead, it is constructed by an EN-DE architecture and heavily focuses on the attention mechanisms, e.g., self-attention (SA) and multi-head self-attention (MSA). Compared with the classical CNNs, the SA can access a larger receptive field,[21] which can better capture the hidden relationship between the inputs and between cross-tasks [27]. This is useful for MTL, as discussed in Chap. 12. The transformer was first developed in the NLP field but later was shown to achieve equivalent or even better performance than the CNN in the CV field with multiple variations, e.g., **Vi**sion **T**ransformer (ViT) [28] and Swin transformer[22] [29].

4.5.1 Self-attention

The NNs can receive inputs (e.g., vector or tensor) with varying sizes and there may exist certain relationships between different input vectors or tensors. However, in training conventional networks, the relationship between these inputs cannot be fully utilized, resulting in less optimal results. The SA mechanism can circumvent this issue to make the networks pay attention to the correlation between different parts of the entire input. Consider the example illustrated in Fig. 4.18 where x^i, $i \in \{1, 2, 3, 4\}$ represent the inputs (outputs from the last hidden layer) and through the SA mechanism, the outputs $\hat{x}^j$, $j \in \{1, 2, 3, 4\}$ are obtained, which consider all the inputs.

Mathematically, the procedure of SA is a dot product of the input, x, and the weight, w. Three intermediate variables can be introduced, namely, query (Q), key (K), and value (V) along with three corresponding weights, i.e., w^Q, w^K, and w^V, respectively, which borrow the concept from the domain of *information retrieval* (IR).[23] Q, K, and V have concrete meanings in NLP scenarios, e.g., send Q for searching certain words in the collections of (K, V) and return matching results. For more details, refer to [27].

---

[21] The receptive field is the region of the input image that is mapped back from the current feature map. In CNN, it can be increased in size by many ways, including making the network deeper, i.e., adding more convolutional layers.

[22] Swin stands for **S**hifted **win**dow.

[23] IR is a process of retrieving relevant resources from the collection of available resources, e.g., by using popular search engines, such as Google (https://www.google.com/) or Bing (https://www.bing.com/).

Fig. 4.18 SA mechanism

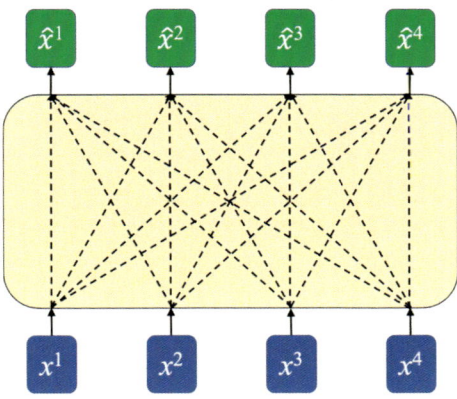

Fig. 4.19 Dot product for Q, K, and V

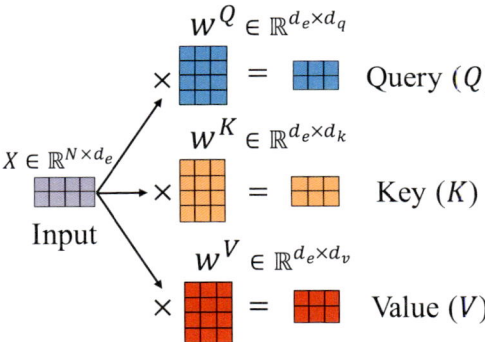

Given one input vector $x^i \in \mathbb{R}^{d_e}$ having d_e-dimensional features, firstly, its corresponding $q^i \in Q$, $k^i \in K$, and $v^i \in V$ are obtained by dot product with the weight matrices $W^Q \in \mathbb{R}^{d_e \times d_q}$, $W^K \in \mathbb{R}^{d_e \times d_k}$, and $W^V \in \mathbb{R}^{d_e \times d_v}$, as shown in Eq. 4.46. In most cases of SA, $d_q = d_k = d_v$. In addition, for the entire input with N vectors, $X \in \mathbb{R}^{N \times d_e}$, the vectorized representation of Eq. 4.46 is expressed in Eq. 4.47 and Fig. 4.19. Subsequently, utilizing the relevance between Q and K, the attention score, a, is usually computed by dot product, where $a_{i,j}$ is determined by the ith row of $Q \in \mathbb{R}^{N \times d_q}$, i.e., q^i, and the jth row of $K \in \mathbb{R}^{N \times d_q}$, i.e., k^j, refer to Eq. 4.48 or Eq. 4.49. The previous operations are repeated for all pairs of Q and K to obtain an attention score matrix, $A \in \mathbb{R}^{N \times N}$. Furthermore, the attention scores can be activated by the *softmax* function to obtain $a' = softmax(a)$ or $A' = softmax(A)$. It is noted that to enhance the nonlinearity, *ReLU* activation can be applied. Finally, the output, $\hat{x} \in \mathbb{R}^{D \times d_v}$, is computed by dot product between the attention score A' and the value V, Eq. 4.50 or Eq. 4.51. Note that, the example shown in Fig. 4.19 has $N = 2$, $d_e = 4$ and $d_q = d_k = d_v = 3$.

$$q^i = \left(W^Q\right)^T \cdot x^i,$$

$$k^i = \left(W^K\right)^T \cdot x^i, \tag{4.46}$$

$$v^i = \left(W^V\right)^T \cdot x^i.$$

$$Q = X \cdot W^Q,$$

$$K = X \cdot W^K, \tag{4.47}$$

$$V = X \cdot W^V.$$

$$a_{i,j} = q^i \cdot \left(k^j\right)^T. \tag{4.48}$$

$$A = Q \cdot K^T. \tag{4.49}$$

$$\hat{x}^i = \sum_j a'_{i,j} \cdot v_i. \tag{4.50}$$

$$\hat{X} = A' \cdot V. \tag{4.51}$$

To avoid the vanishing gradients problem due to the *softmax* function, yielding extremely small gradients, a scaling factor, $\sqrt{d_k}$, is included, which is related to the dimension of K, i.e., d_k [27]. Therefore, the complete vector-form of attention is presented as follows:

$$\text{Attention } (Q, K, V) = softmax\left(\frac{Q \cdot K^T}{\sqrt{d_k}}\right)V. \tag{4.52}$$

The dot product and activation operations in the attention computation in Eqs. 4.46–4.52 show the similarity with the working principle of CNNs. Both the attention and CNN aim to consider not only a pixel but a certain area of the input (images in the CV tasks). However, the CNN considers a fixed square or rectangular receptive field determined by its filter size. On the contrary, the SA determines the shape and type of the receptive field. Therefore, the CNN can be viewed as a special type of the SA under certain restrictive conditions.

4.5.2   Multi-head Self-attention

To allow the attention function to extract more robust information from different representation subspaces, the multi-head attention[24] (MHA) is introduced, which is a process of computing attention scores between Q and a set of $(K - V)$ pairs, where Q, K, and V are

---

[24] An attention "head" is named this way to reflect that it is a module that has separate, learnable matrices for producing the query, Q, key, K, and value, V, discussed in Sect. 4.5.1. It is thought of as a specialized *mini-brain* within the AI model to help it selectively focus on certain aspects of the input data.

all computed from *different* inputs. On the basis of SA, the MSA is developed, which is a specific case of the MHA, where the input for the Q, K, and V are all derived from the *same* source, such as the input to a transformer layer. Most of the transformer variants, used in the NLP tasks, adopt the MSA as the attention mechanism. This is because MSA allows the model to attend to different positions in the same input sequence and capture contextual information more effectively, which is particularly important in sequence-to-sequence tasks, e.g., in language translation and generation.

For the MSA computation, it is favorable to conduct liner projection of Q, K, and V through projection matrices, $\hat{W}_i^Q \in \mathbb{R}^{d_q \times d_{\text{proj}}}$, $\hat{W}_i^K \in \mathbb{R}^{d_k \times d_{\text{proj}}}$, and $\hat{W}_i^V \in \mathbb{R}^{d_v \times d_{\text{proj}}}$ to the same hidden space with dimension d_{proj}, which is commonly taken as d_e, i.e., assuming the same feature dimension, d_{proj}, as that of the input, d_e. Suppose h heads are used, for the ith head, conduct a single SA using Eq. 4.53. Finally, concatenating all heads and subsequently projecting the output by the weight matrix, $\hat{W}^O \in \mathbb{R}^{d_e \times d_e}$, the final MSA results are obtained using Eq. 4.54.

$$\text{head}_i = \text{Attention}\left(Q\hat{W}_i^Q, K\hat{W}_i^K, V\hat{W}_i^V\right) \in \mathbb{R}^{N \times d_e/h}, i \in 1, 2, \ldots, h. \qquad (4.53)$$

$$\text{MSA}(Q, K, V) = \text{Concatenate}\left(\text{head}_1, \text{head}_2, \ldots, \text{head}_h\right)\hat{W}^O. \qquad (4.54)$$

4.5.3 Vanilla Transformer

Until now, there have been many variants of transformers developed for different scenarios and pursuing performance enhancement. The vanilla transformer is introduced first in this section. It was initially developed for NLP tasks, e.g., language translation [27]. As illustrated in Fig. 4.20, the key technical features in the vanilla transformer are: (1) positional encoding, PE, (2) an EN-DE structure, and (3) attention mechanisms. These are discussed in the following three sections.

The EN-DE structure is similar to the one used in LSTM (refer to Sect. 4.4.3, which receives a sequential input to encode it into a high dimensional space and then decodes to the output with the designated format. Through the attention mechanisms, e.g., SA and MSA, the transformer can better utilize the relevance between data and the features and then enhance the model's ability to deal with long-range input. Although the SA mechanism considers all inputs, it does not take into account the *position* information of the input. Therefore, PE, discussed below, was proposed to merge the position information of the input into the model.

4.5.3.1 Positional Encoding

In the vanilla transformer, given arbitrary sequential inputs, e.g., words or sentences, they are first embedded into specific forms of vectors or tensors with a feature dimension d_e by

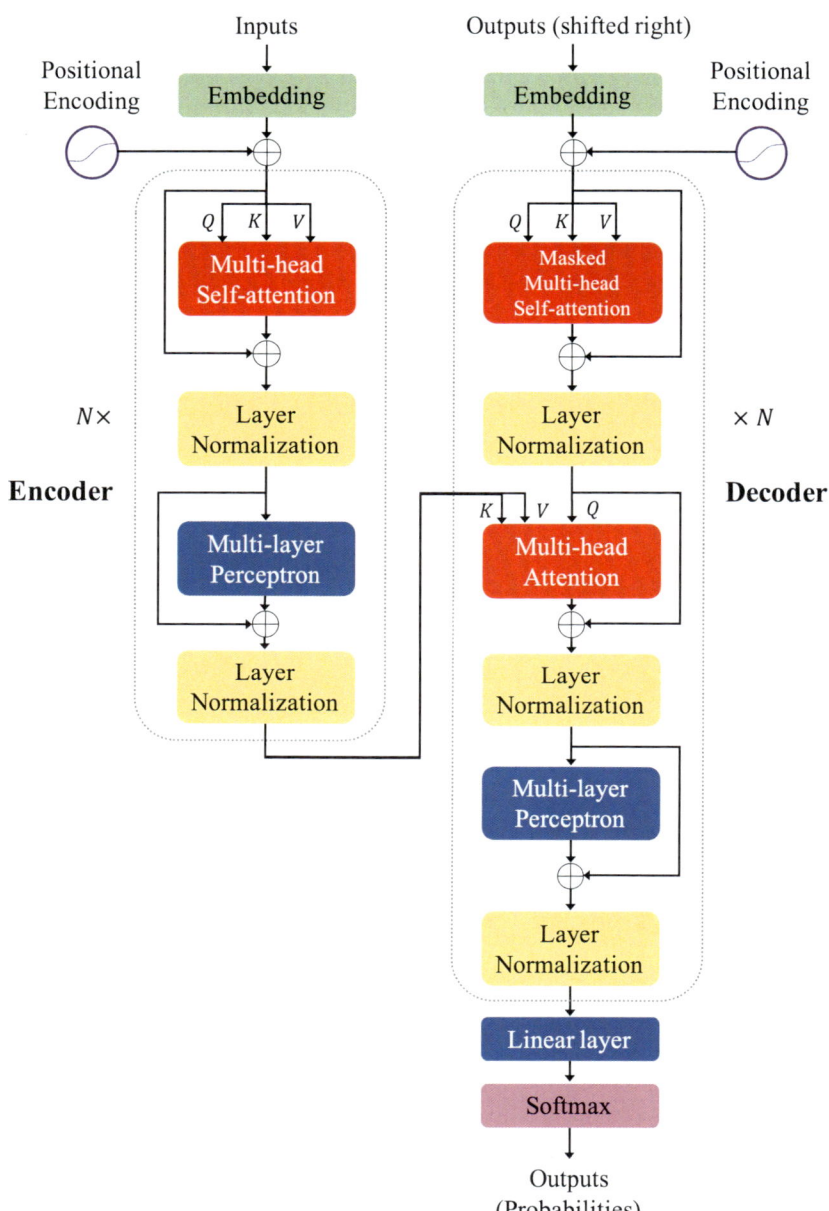

Fig. 4.20 Vanilla transformer

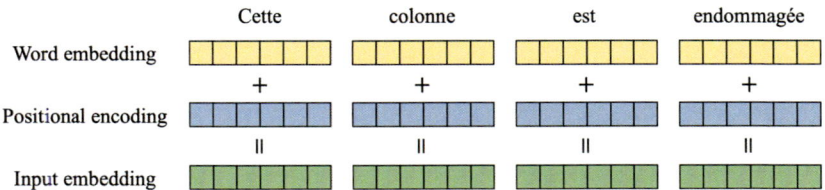

Fig. 4.21 Components of the input embedding for an example of NLP to translate from French to English

embedding[25] models or techniques, e.g., Word2vec[26] [30] or GloVe[27] [31] in NLP. This is also known as *word embedding*. Subsequently, *PE* is introduced to map the position information (e.g., location of the words) into a vector or tensor having the same feature dimension, d_e, as the word embedding. Finally, the produced vectors or tensors by both word embedding and *PE* are combined and then fed into the EN network. This entire process is denoted as *input embedding*. There are many ways to implement *PE*, considering two basic requirements: (1) each position has a unique *PE* form and (2) the relationship between different positions can be captured through certain *affine* transformations. The vanilla transformer adopts "sine" and "cosine" functions, where the angle in their arguments is expressed in radians, with varying frequencies, as follows:

$$PE(p, 2i) = \sin\left(\frac{p}{S^{(2i/d_e)}}\right),$$
$$PE(p, 2i + 1) = \cos\left(\frac{p}{S^{(2i/d_e)}}\right), \tag{4.55}$$

where $PE(p, j)$ is the position function that maps a position in the input sequence to index (p, j) where $j = 2i$ or $2i + 1$ of the *PE* matrix, $p \in \{0, 1, \ldots, L\}$ is the position of the current object (e.g., word) in the entire sequential inputs whose length is $L + 1$, d_e is the dimension of the input embedding, integer $i \in \{0, 1, \ldots, d_e/2\}$ is used for mapping the index of the embedding vector, where $2i$ and $2i + 1$ represent the even and odd dimension indices, respectively, and S is a user-defined scalar, typically taken as $S = 10,000$ in the vanilla transformer [27].

Take an example of language translation from "Cette colonne est endommagée" (French) to "This column is damaged" (English). As illustrated in Fig. 4.21, the (French) input is processed by word embedding word by word and each word is embedded into a vector with a feature dimension taken as $d_e = 6$, where $S = 100$ is considered for this example. Then,

---

[25] Embedding is a concept for a vector representation of the input data, e.g., representing words or phrases as numerical vectors, where each dimension of the vector corresponds to a specific feature or aspect of the word or phrase.

[26] Word2vec is a group of models that are used to produce word embeddings.

[27] GloVe, from Global Vectors, is an unsupervised learning model for distributed word representation related to semantic similarity.

Table 4.1 Position matrix using the *PE* for a NLP example of a (French) sentence

Word sequence	Index, p	Position matrix					
		$i = 0$		$i = 1$		$i = 2$	
Cette	0	$\sin(0) =$ 0.0	$\cos(0) = 1.0$	$\sin(0) = 0.0$	$\cos(0) = 1.0$	$\sin(0) = 0.0$	$\cos(0) = 1.0$
colonne	1	$\sin(1) =$ 0.841	$\cos(1) =$ 0.540	$\sin\left(\frac{1}{4.64}\right) =$ 0.214	$\cos\left(\frac{1}{4.64}\right) =$ 0.977	$\sin\left(\frac{1}{21.54}\right) =$ 0.046	$\cos\left(\frac{1}{21.54}\right) =$ 0.999
est	2	$\sin(2) =$ 0.909	$\cos(2) =$ -0.416	$\sin\left(\frac{2}{4.64}\right) =$ 0.418	$\sin\left(\frac{2}{4.64}\right) =$ 0.909	$\sin\left(\frac{2}{21.54}\right) =$ 0.093	$\cos\left(\frac{2}{21.54}\right) =$ 0.996
endommagée	3	$\sin(3) =$ 0.141	$\cos(3) =$ -0.990	$\sin\left(\frac{3}{4.64}\right) =$ 0.602	$\cos\left(\frac{3}{4.64}\right) =$ 0.798	$\sin\left(\frac{3}{21.54}\right) =$ 0.416	$\cos\left(\frac{3}{21.54}\right) =$ 0.990

a *PE* vector is generated to represent the position information of each word. For example, the word "colonne" is the second word in the sequence, i.e., $p = 1$, where the first word is at the position $p = 0$. Since the embedding feature dimension is taken as $d_e = 6$, $i = 0, 1$ and 2, where $2i = 0, 2$ and 4 and $2i + 1 = 1, 3$ and 5 cover the indices of all 6 positions, i.e., 0 to 5. Based on Eq. 4.55, the calculations of the "sine" and "cosine" functions are repeated for each position and the *PE* form of the word "colonne" is computed as: $\left\{\sin\left(\frac{1}{100^{0/6}}\right), \cos\left(\frac{1}{100^{0/6}}\right), \sin\left(\frac{1}{100^{2/6}}\right), \cos\left(\frac{1}{100^{2/6}}\right), \sin\left(\frac{1}{100^{4/6}}\right), \cos\left(\frac{1}{100^{4/6}}\right)\right\}$, and is simplified into $\left\{\sin\left(\frac{1}{1}\right), \cos\left(\frac{1}{1}\right), \sin\left(\frac{1}{4.64}\right), \cos\left(\frac{1}{4.64}\right), \sin\left(\frac{1}{21.54}\right), \cos\left(\frac{1}{21.54}\right)\right\}$. This is expressed for the whole (French) sentence in Table 4.1, where the *position matrix* is computed for this example.

In the example above, each position corresponds to a different trigonometric function, which encodes a single position into a vector. This scheme for *PE* has several advantages. The use of "sine" and "cosine" functions leads to normalized $PE \in [1, -1]$ with a unique encoding each position due to the different values of these trigonometric functions. Moreover, this formulation of the *PE* quantifies the similarity between different positions enabling the encoding of the relative positions of the words in the sentence.

4.5.3.2 EN-DE Structure

As illustrated in Figs. 4.20 and 4.22, the vanilla transformer adopts the same number of EN and DE networks (a.k.a. blocks) [27], e.g., $N = 6$ as shown in Fig. 4.22. In an EN block, the inputs are computed by the MSA operation first and the corresponding outputs are then combined with the original input by a shortcut connect (the same techniques used in ResNet, refer to Sect. 4.3.3) to help reduce the gradient degradation issue. To reduce the internal covariate shift (discussed in Sect. 4.2.4) and improve the stability and training speed of the model, layer normalization (LN) is performed, which computes the mean and variance of the MSA outputs across each feature dimension and then applies a scale and shift to normalize the distribution, i.e., re-centering it by subtracting the mean and re-scaling it by dividing by the square root of the variance, i.e., std [32]. To increase the nonlinearity and

Fig. 4.22 EN-DE structure in the transformer

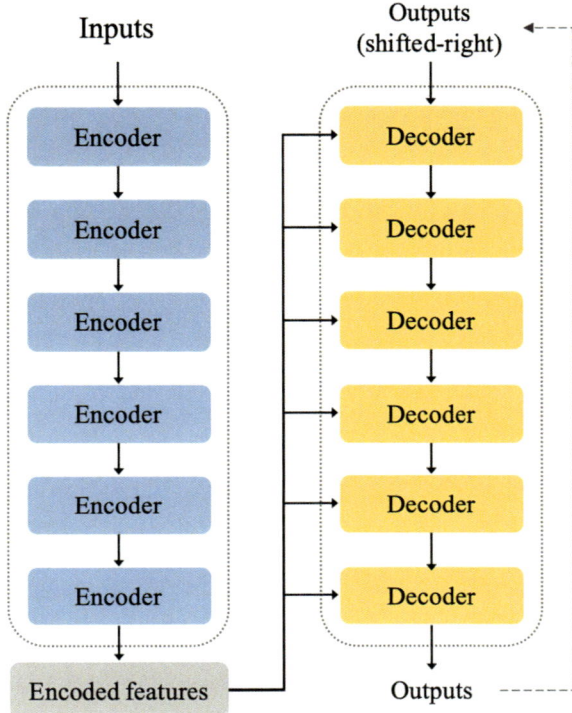

complexity of the network, MLP is followed and accompanied by the shortcut connection. In addition, another round of LN is conducted after the MLP. These steps are then repeated for $N = 6$ times such that the high dimensional encoded features are extracted.

The DE network shares a similar structure to the EN, but it has an additional masked MSA (refer to Sect. 4.5.3.3) with a shortcut connection and it uses MHA instead of MSA, which utilizes information from both the previous layers of the DE and from the extracted encoded features from the EN. Unlike the EN, which has access to all the inputs and performs encoding for the sequential data at once, the DE generates outputs one by one and uses the previous outputs as the new inputs for the next step, which follows an *auto-regressive* manner, as shown in Fig. 4.22. The DE implementation makes use of the *shifting right operation*, refer to Fig. 4.23, through shifting the target sequential data one position to the right before it is fed into the DE [27]. For the same example of language translation in Sect. 4.5.3.1, the training input-output pair is "Cette colonne est endommagée" (French) and "This column is damaged" (English). First, the (French) inputs are encoded into features and fed into the DE along with the special symbol "<start>" to represent the beginning of the sentence in the NLP. Accordingly, the DE generates the first output, "This". Subsequently, "This" along with the previous inputs are fed into the DE, which then generates the second output, "column". The decoding process is repeated until the output gets the "<end>" symbol.

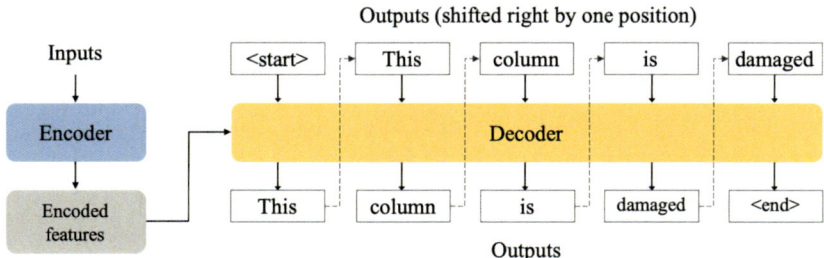

Fig. 4.23 Shifting right operation in the DE

Therefore, as shown in Fig. 4.23, the outputs are "This column is damaged <end>" and the inputs from the previous outputs of the DE are "<start> This column is damaged", which is right shifting the target (English) expression by one position (one word herein) with the padding "<start>" at the beginning of the sentence.

In short, the vanilla transformer is the first model that only relies on the SA to implement the EN-DE structure. In addition, the encoded features from the EN are fed into each DE block separately, which allows the DE to comprehensively consider relevant parts of the input sequence at each step of the decoding process, without being restricted to a fixed context window, as in conventional models, e.g., the one presented in Sect. 4.4.3, refer to Fig. 4.17.

4.5.3.3 Attention Mechanism

Three types of attention mechanisms are adopted in the vanilla transformer: (1) MSA, (2) Masked MSA, and (3) MHA. In the EN network, the MSA is used. As illustrated in Fig. 4.20, after embedding (input embedding is discussed in Sect. 4.5.5.1) and PE, the inputs are fed into the EN blocks. Subsequently, the inputs are copied three times, treated as Q, K, and V, and then attention is computed among them. Because here Q, K, and V are the same, the attention is referred to as SA. In the DE network, similar to the EN, the inputs from the previous outputs are copied three times for the attention computation. The only difference is that after multiplying the matrices Q and K and scaling them, a special mask is applied to the resulting matrix, which sets the negative weights for the input after the current position. This is to ensure that the prediction at the current position only depends on every other previous position in the sentence up to its current position. Accordingly, this still belongs to the SA type but is restricted by a mask. In the middle part of the DE, a general MHA is used because it receives Q, K, and V from different sources and no longer belongs to SA. From [27] and Fig. 4.20, Q is the resulting output from the Masked MSA of the DE and K and V are the resulting outputs from the EN. By computing attention among inputs from the EN and previous DE outputs, more comprehensive and long-term relationships can be explored. For more details, refer to [27].

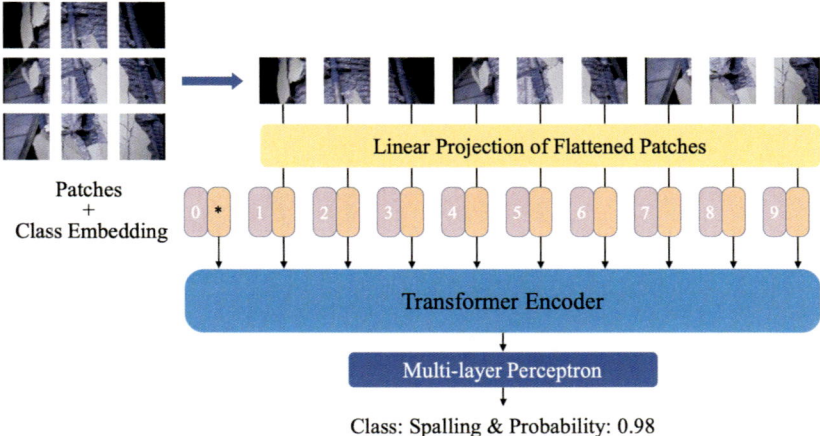

Fig. 4.24 Example of a ViT

4.5.4 Vision Transformer

The transformer achieved successful NLP applications with high computational efficiency and scalability. Moreover, as the model complexity increases, its performance is not saturated, leading to increased interest in adopting the transformer for vision problems. However, applications of transformers remained limited in vision problems until 2018 because directly applying SA to images requires each pixel to globally attend to every other pixel, making the computational complexity quadratic[28] with the number of pixels, which is computationally demanding. Since 2018, researchers started early attempts to apply transformers in vision tasks [33–35], making use of computing attention within neighboring pixel patches or regions instead of globally and accordingly replacing the Conv layers with MHA layers. One of the major milestones is the development of the so-called ViT by the Google research team in 2020 [28], which demonstrated the performance validity of the transformers in vision-based classification problems.

Adopted from [28], the overview of the ViT is illustrated, via a specific example, in Fig. 4.24. Firstly, the input image (e.g., a RC column with concrete cover spalling damage) is split into a few pixel patches and they are flattened, i.e., reshaped into a sequence of flattened (1D row vector) of 2D patches starting with the top left patch and ending with the bottom right patch. Accordingly, a 2D image $x \in \mathbb{R}^{H \times W \times C}$ is first split into N non-overlapping square patches and reshaped into the sequence of flattened 2D patches $x_p \in \mathbb{R}^{N \times P^2 \times C}$, where H and W are the height and width of the original input image, C is the number of channels ($C = 3$ for a RGB image), $P \times P$ is the resolution of each image patch, and $N = H \times W / P^2$ stands for the number of patches. The example in Fig. 4.24 uses $N = 9$

---

[28] An algorithm with quadratic complexity implies that its runtime is directly proportional to the square of the size of the input, i.e., N^2.

Fig. 4.25 Transformer EN in vision transformer

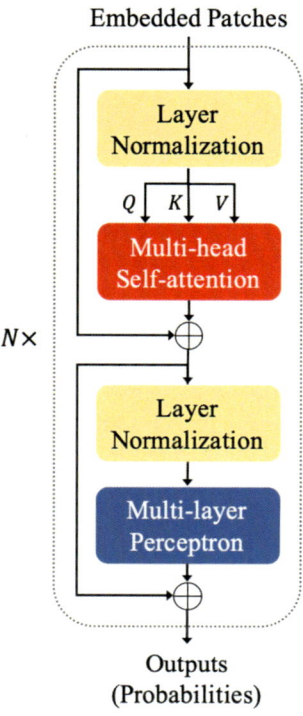

patches. These image patches are treated the same way as sequential tokens[29] (e.g., words in NLP). Subsequently, a linear projection is performed on these patches to map them into a certain dimensional space, D, in a process known as *linear embedding*, refer to Eq. 4.56. In addition, *PE*, i.e., $E_{pos} \in \mathbb{R}^{(N+1) \times D}$, and class embedding, i.e., $E_{cls} \in \mathbb{R}^D$, are conducted and added to the previous embedding to serve as the processed input to the transformer EN, refer to Eq. 4.57. The EN structure used in ViT is illustrated in Fig. 4.25, which is similar to the one designed in the vanilla transformer, Fig. 4.20. It is noted that the "∗" with position index $p = 0$ in the PE function, refer to Eq. 4.55, in Fig. 4.24 represents the *class embedding*. It starts as an extra "learnable blank class" (i.e., token in the context of NLP), where its final output is used as the input into a classification head during pre-training. Similar to the process discussed in Sect. 4.5.3.2, the transformer EN extracts high-dimensional features to be fed into the MLP layers. Finally, through *softmax* activation in MLP, the predictions with probabilities of each class are obtained as the ViT output.

$$E_i = W \cdot x_p^i, \quad i \in \{1, 2, \ldots, N\}, \quad W \in \mathbb{R}^{D \times (P^2 \cdot C)}. \tag{4.56}$$

$$z_i = [E_{cls}; E_1; \ldots; E_N] + E_{pos}. \tag{4.57}$$

---

[29] A token is an instance of a sequence of characters that are grouped together as a useful semantic unit for processing using NLP.

Through experiments [28], it is found that ViT lacks several *inductive biases*[30] available in CNNs, e.g., translational equivariance (refer to Sect. 4.2.3) and locality.[31] Accordingly, its performance is not as good as classical CNN models, e.g., ResNet, on the small- and medium-scale dataset (less than 14 million images). However, if pre-trained ViT on large-scale datasets (14 million to 300 million images) and fine-tuned to small datasets, ViT achieved excellent results approaching or even outperforming the state-of-the-art performance of CNNs on certain benchmark datasets, e.g., ImageNet [14] and CIFAR-10 [36]. To alleviate the dependence of ViT on a large-scale dataset and its inefficient pre-training methods, more advanced vision-based transformers and learning frameworks are developed, e.g., the Swin transformer [29], discussed below.

4.5.5 Swin Transformer

On the basis of ViT, the Swin transformer was developed by a research team from Microsoft Research Asia (MSRA) in 2021, which is one of the state-of-the-art backbone networks for many CV applications [29, 37], e.g., classification, localization, and segmentation. There are two major characteristics of the Swin transformer distinguishing it from the vanilla ViT: (1) hierarchical mechanism and (2) shifted window multi-head self-attention (SW-MSA). These two new characteristics and their advantages for the Swin transformer over the ViT are discussed below.

4.5.5.1 Hierarchical Mechanism

The Swin transformer first constructs hierarchical feature maps starting from small-size patches and gradually merging neighboring patches in deeper layers. This mechanism, illustrated in Fig. 4.26, reduces the feature map size while increasing each patch's receptive field, which helps the network to learn the long-term relationships.

Similar to the settings in ViT, given an image input $x \in H \times W \times C$ and defining an image patch with a pixel resolution of $P \times P$, the Swin transformer first partition the input image into $N = H \times W/P^2$ non-overlapping patches. Each patch, x_p, is concatenated through the third channel of the feature map and the input is then transformed into a feature map of the shape $\left(\frac{H}{P}, \frac{W}{P}, C \times P \times P\right)$. In this manner and as an example, consider an input RGB image ($C = 3$) of size 224×224 and a 4×4 patch size. Thus, the input is partitioned to $\frac{224}{4} \times \frac{224}{4}$, i.e., 56×56 patches, and then transformed to a feature map of the shape (56, 56, 3 × 4 × 4), i.e., (56, 56, 48). These features are then processed through four similar stages, Fig. 4.27, where their differences are: (1) the number of Swin transformer

---

[30] The inductive bias is a set of assumptions or prior knowledge that a ML algorithm or model uses to guide the process of learning from the data. For example, in linear regression, the assumption (inductive bias) is that the output and input are related through a *linear* model.

[31] Locality indicates that adjacent pixels in the image have a certain relationship with each other.

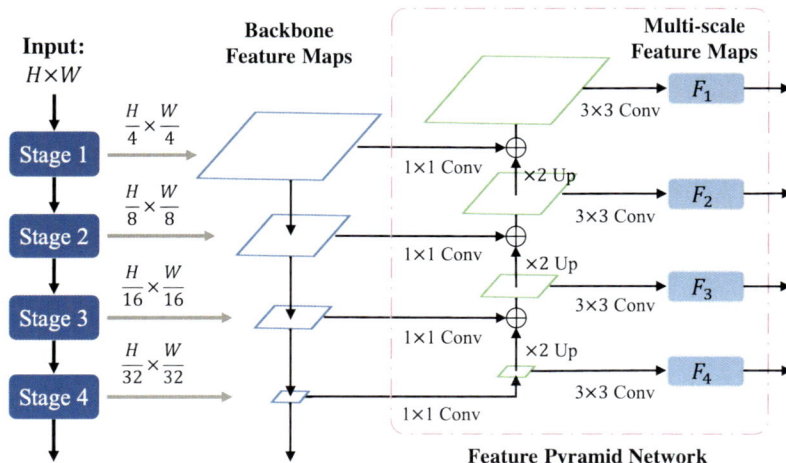

Fig. 4.26 Hierarchical mechanism of the Swin transformer [38]

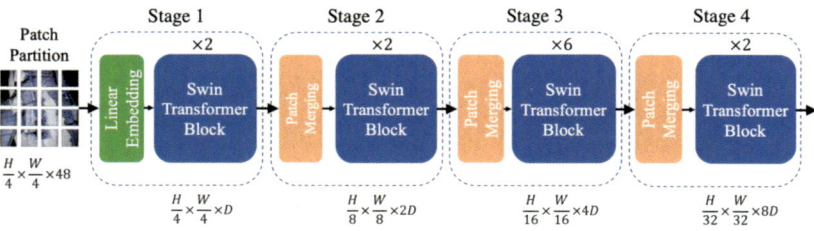

Fig. 4.27 Swin transformer

blocks and (2) the architectures, e.g., the linear embedding layer[32] is used in stage 1 and the patch merging (as discussed below) is placed before the transformer blocks in stages 2 to 4. In that regard, in stage 1, the linear embedding layer is considered to project the features into a pre-determined dimension, D, i.e., $\left(\frac{H}{P}, \frac{W}{P}, D\right)$, where the projected dimension D can be taken, e.g., as 96, which is inherited from the Swin-T model (a special type of Swin transformer) [29], i.e., the features are further projected to a (56, 56, 96) feature space.

In the above example, the features are subsequently fed into the consecutive Swin transformer blocks and repeated for the remaining stages with patch merging operations, discussed in the next paragraph. Each Swin transformer block has the same configuration with LN, shortcut connection, and 2-layer MLP. The exception to this block to block similarity is the used type of SA, i.e., window multi-head self-attention (W-MSA) or SW-MSA, which are introduced in Sect. 4.5.5.2. From [29], the Swin transformer block always repeats even number of times, e.g., 2, 4, and 6. For example, in [38], the authors used 2, 2, 6, and 2 repeat times for stages 1–4, respectively. The example of the connection between the six consecu-

---

[32] The linear embedding layer is a Dense layer with no activation. "Linear" implies that there is no activation (i.e., activation is the identity).

Fig. 4.28 Connection between
6 blocks in stage 3 [38]

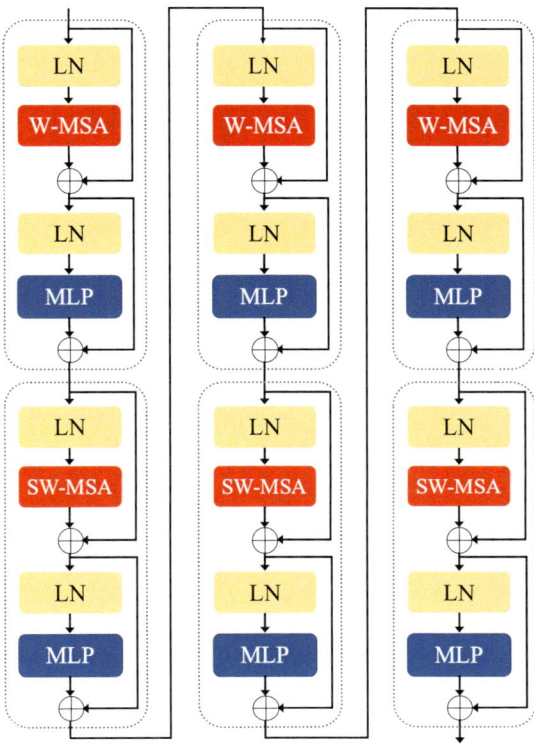

tive transformer blocks in stage 3 (Fig. 4.27) is illustrated in Fig. 4.28, where W-MSA and
SW-MSA are repeatedly used.

To realize a hierarchical representation, the input patches of stages 2–4 are merged through
merging operations to shrink the feature size from the shallow to the deep layers of the
network, i.e., from the input to the output [38]. The patch merging operation down-samples
the input by a merging rate of p by grouping $p \times p$ regions and concatenating the patches
depth-wise to expand the depth by $p \times p$. One example of a 4×4 patch participation to
demonstrate the patch merging using a $p \times p = 2 \times 2$ merging rate is illustrated in Fig. 4.29.
It first separates the patches into four 2×2 groups with the shown indexing, then merges
the sub-parts with the same index into a single group, and finally concatenates these merged
sub-parts along the depth (third) channel. This operation shrinks the feature dimension by
$p = 2$ and expands the depth by $p \times p = 2 \times 2 = 4$. In [38], the merging rate of $p = 2$ was
adopted. However, to avoid overly expanding the depth ($\times 4$ after each merging), a $1 \times 1 \times q$
convolution operation[33] is performed depth-wise, i.e., along the q dimension to further
reduce the depth expansion by a factor of $q = 2$. Therefore, the feature shapes of stages 1–4 in

---

[33] The $1 \times 1 \times q$ convolution layer is used for "cross channels (or feature maps) down-sampling (or
pooling)", i.e., having q 1×1 Conv filters. This reduces the number of channels (i.e., feature maps)
in the depth dimension while introducing nonlinearity.

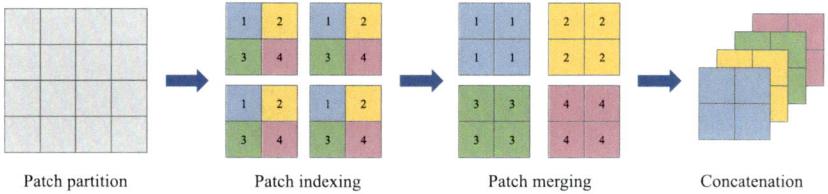

Patch partition   Patch indexing   Patch merging   Concatenation

Fig. 4.29 Patch merging down-sampling example

Fig. 4.26 (note that the example considers a RGB $224 \times 224 \times 3$ input image, i.e., $H = W = 224$ and $C = 3$ with $D = 96$, $p = 2$, and $q = 2$, i.e., considering the 1×1 convolution) are $(56, 56, 96)$, $(28, 28, 192)$, $(14, 14, 384)$, and $(7, 7, 768)$, respectively. Accordingly, through patch merging, the transformer blocks form a *pyramid-like feature representation*, which can utilize both fine and coarse features (from small and large patches) to extract multi-scale semantic and contextual information. This characteristic is useful in the image localization and segmentation tasks.

4.5.5.2 Shifted Window Multi-head Self-attention

To obtain global SA, MSA requires computing the patch relationships between one patch against all other patches, which results in a quadratic complexity, i.e., $\mathcal{O}\left(N^2\right)$, where N is the number of patches.[34] A commonly used image input size is 224×224, which is inefficient, and accordingly impractical, to adopt the standard MSA under such resolution. Therefore, window-based methods, i.e., W-MSA and SW-MSA, are used in the MSA computations, Fig. 4.30, as conducted in [38]. The standard W-MSA partitions the input image or feature maps into multiple equal number of square windows, where each window with a size of M includes a $M \times M$ region of the feature maps, which are obtained from the previous stage of the transformer. Subsequently, the SA is computed within the window, not the whole feature map as shown in Fig. 4.30a.

Similar to the definitions in Sect. 4.5.5.1, $\hat{H}$, $\hat{W}$, and C are the height, width, and channel number of the input window region of the input image or feature map $\hat{X}_w$. Given a $M \times M$ window size, the input is flattened first to $X_w \in \mathbb{R}^{N \times C_w}$, where $N = \hat{H} \cdot \hat{W}/M^2$, $C_w = M^2 \cdot C$. Suppose there are h heads adopted in the MSA computations, X_w is further split into h subsets along its channel dimension, i.e., $X_w = \left\{X_w^1, X_w^2, \ldots, X_w^h\right\}$. For the jth head, $X_w^j \in \mathbb{R}^{N \times C_w/h}$, $j \in \{1, 2, \ldots, h\}$, is embedded to a high dimensional space by a linear model with parameters $W^l \in \mathbb{R}^{C_w/h \times d_e}$ and the projected window feature, A_w^j, is expressed

---

[34] The Big $\mathcal{O}$ notation describes the asymptotic behavior of a function or an algorithm. For example, assume one analyzes an algorithm and the time (or memory), T, it takes to complete a problem of size N is given by $T(N) = c_1 N^2 + c_2 N + c_3$ where c_1, c_2 and c_3 are constants. If one ignores the constant term, c_3, which depends on the used particular hardware to run the algorithm, and also ignores the linear term, $c_2 N$, which grows in a slower manner than the leading term $c_1 N^2$, then one says "$T(N)$ grows at the order of N^2" or $T(N) = \mathcal{O}(N^2)$.

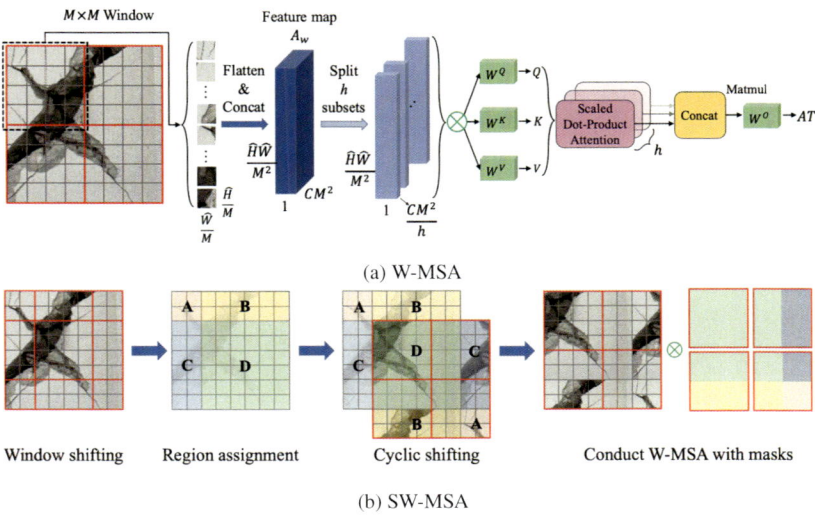

(a) W-MSA

Window shifting   Region assignment   Cyclic shifting   Conduct W-MSA with masks

(b) SW-MSA

Fig. 4.30 Window-based MHA computation mechanisms [38]

as follows:

$$A_w^j = X_w^j W^l \in \mathbb{R}^{N \times d_e}. \tag{4.58}$$

Denote three weight matrices, $W^Q \in \mathbb{R}^{d_e \times d_q}$, $W^K \in \mathbb{R}^{d_e \times d_k}$, and $W^V \in \mathbb{R}^{d_e \times d_v}$, where $d_q = d_k = d_v = d_e/h$ (refer to the classical setting in Sect. 4.5.1 and more details in [27]) and the corresponding Query (Q), Key (K), and Value (V) of the feature, A_w^j, for the jth attention head, are computed as follows:

$$Q^j = A_w^j W^Q \in \mathbb{R}^{N \times d_q}. \tag{4.59}$$

$$K^j = A_w^j W^K \in \mathbb{R}^{N \times d_k}. \tag{4.60}$$

$$V^j = A_w^j W^V \in \mathbb{R}^{N \times d_v}. \tag{4.61}$$

Based on Eqs. 4.52 and 4.53, the result of the jth head attention, $S^j \in \mathbb{R}^{N \times d_v}$, $j \in \{1, 2, \ldots, h\}$, becomes,

$$S^j = \text{Attention}\ (Q^j, K^j, V^j) = softmax \left(\frac{Q^j (K^j)^T}{\sqrt{d_k}} + B \right) V^j, \tag{4.62}$$

where the *softmax* function, Eq. 3.24, converts the input vector of numbers into a vector of probabilities and $B \in \mathbb{R}^{M^2 \times M^2}$ is the relative position bias[35] [29]. Similar to Eq. 4.54, concatenating outputs for all the heads, the attention token, S, is obtained as follows:

---

[35] The learned relative position bias accounts for relative spatial configurations of the visual elements. The relative position along each axis of the feature map is in the range of $[-M + 1, M - 1]$.

$$S = \text{Concatenate}\{S^1, S^2, \ldots, S^h\} \in \mathbb{R}^{N \times d_e}. \tag{4.63}$$

Subsequently, and also following Eq. 4.54, S is projected back to the original dimension by another linear model with parameters $W^o \in \mathbb{R}^{d_e \times C_w}$ and the final patch attention, AT, refer to Fig. 4.30a, is given as follows:

$$AT = SW^o \in \mathbb{R}^{N \times C_w}. \tag{4.64}$$

It is noted that the operation "Matmul" shown in Fig. 4.30a to compute the output AT refers to the "numpy.matmul()" function, part of the "NumPy" library,[36] which returns the matrix product of two 2D arrays.

Using W-MSA significantly reduces the computational complexity and cost [29] compared with the use of the MSA. However, the pixels and patches are non-overlapped between windows, which causes the loss of possible relationships between windows. To address this issue, W-MSA and SW-MSA can be alternately used in the consecutive transformer blocks as illustrated in Fig. 4.28 [38]. Through shifting the window partition, new information from adjacent patches can be accessible in the attention computation, which strengthens the cross-window connections, e.g., in Fig. 4.30b, the $M \times M$ window shifts to the bottom right by $\lfloor M/2 \rfloor$ pixels, where $\lfloor \bullet \rfloor$ indicates the *floor function* of its argument $\bullet$. After window shifting, the number of windows and window configuration are changed. Not to increase the computational burden, an efficient *batch computation approach* proposed in [29] can be adopted. This approach uses *cyclic shifting* of certain patches from the top left to the bottom right (e.g., patches A, B, and C), as clearly illustrated in Fig. 4.30b. The obtained re-arranged image or feature maps have the same number and window configuration used in the previously-discussed W-MSA computations and accordingly the "Attention" can be computed in the same way as conducted in the W-MSA. To avoid mixing up the information from the moved patches with adjacent parts within the window, the *masking mechanism*, refer to Sect. 2.3, can be adopted as illustrated in Fig. 4.30b. With a deeper network from stage 1 to stage 4, both W-MSA and SW-MSA are repeated multiple times. The deeper layers not only have larger receptive fields, but also maintain local information of the data during the forward propagation.

In summary, the Swin transformer establishes a hierarchical feature map that enables the use of structures such as feature pyramid network (FPN) [39] and U-Net [40], refer to Sect. 8.4, to handle multi-scale problems in visual tasks. At the same time, the shifted window adds local priors (i.e., shared parameters) and improves the computational efficiency. Through experiments [29], this combination is found to be effective in performance enhancement for image classification, object detection, and semantic segmentation problems [29, 37], especially for SHM applications, as discussed further in Chap. 12 and Ref. [38].

---

[36] NumPy is a library for the Python programming language with support for large, multi-dimensional arrays and matrices, including a large collection of high-level mathematical functions.

4.6   Exercises

1. Design and draw the architecture of a four-layer feed-forward NN. Describe the forward propagation process and how it calculates the output of the network.
2. Explain backpropagation and how it computes gradients to update the network parameters.
3. Explain the dropout technique and its purpose in reducing over-fitting during training. Discuss how dropout affects the network during both training and testing phases.
4. Suppose the input image has a shape of $128 \times 128 \times 3$, two Conv layers are applied with 32 3 × 3 filters each, compute the shape of the output feature maps in the following two scenarios: (1) with zero padding and (2) without zero padding.
5. Explain the concept of residual learning in ResNet and how it addresses the vanishing gradient problem.
6. Explain the vanishing gradient problem in RNNs and why it occurs during back-propagation. Discuss how the vanishing gradient problem affects learning long-term dependencies in sequences.
7. Discuss situations where using GRU might be preferred over LSTM.
8. Describe the self-attention mechanism and explain how it is computed in a transformer model.
9. Explain how transformers can be adapted for image-based tasks from their original NLP tasks.
10. Discuss the advantages and challenges of using transformers for image classification and other CV tasks for SHM applications.

References

1. I. Goodfellow, Y. Bengio, A. Courville, *Deep Learning* (MIT Press, Cambridge, 2016)
2. K. Hornik, Approximation capabilities of multilayer feedforward networks. Neural Netw. **4**(2), 251–257 (1991)
3. V. Nair, G.E. Hinton, Rectified linear units improve restricted boltzmann machines, in *Proceedings of the 27th International Conference on Machine Learning (ICML-10)* (2010), pp. 807–814
4. A.L. Maas, A.Y. Hannun, A.Y. Ng, Rectifier nonlinearities improve neural network acoustic models. Proc. icml. **30**(1), 3 (2013)
5. D.-A. Clevert, T. Unterthiner, S. Hochreiter, Fast and accurate deep network learning by exponential linear units (ELUs) (2015). arXiv:1511.07289
6. N. Srivastava et al., Dropout: a simple way to prevent neural networks from overfitting. J. Mach. Learn. Res. **15**(1), 1929–1958 (2014)
7. K. Simonyan, A. Zisserman, Very deep convolutional networks for large-scale image recognition (2014). arXiv:1409.1556
8. C. Szegedy et al., Going deeper with convolutions, in *Proceedings of the IEEE Conference on Computer Vision and Pattern Recognition* (2015), pp. 1–9

9. K. He et al., Deep residual learning for image recognition, in *Proceedings of the IEEE Conference on Computer Vision and Pattern Recognition* (2016), pp. 770–778

10. A. Krizhevsky, I. Sutskever, G.E. Hinton, Imagenet classification with deep convolutional neural networks, in *Advances in Neural Information Processing Systems* (2012), pp. 1097–1105

11. J. Nagi et al., Max-pooling convolutional neural networks for visionbased hand gesture recognition, in *2011 IEEE International Conference on Signal and Image Processing Applications (ICSIPA)*. (IEEE, 2011), pp. 342–347

12. S. Ioffe, C. Szegedy, Batch normalization: accelerating deep network training by reducing internal covariate shift (2015). arXiv:1502.03167

13. Y. LeCun et al., Gradient-based learning applied to document recognition. Proc. IEEE **86**(11), 2278–2324 (1998)

14. J. Deng et al., Imagenet: a large-scale hierarchical image database, in *2009 IEEE Conference on Computer Vision and Pattern Recognition* (2009), pp. 248–255

15. C. Szegedy et al., Rethinking the inception architecture for computer vision, in *Proceedings of the IEEE Conference on Computer Vision and Pattern Recognition* (2016), pp. 2818–2826

16. C. Szegedy et al., Inception-v4, inception-resnet and the impact of residual connections on learning, in *Thirty-first AAAI Conference on Artificial Intelligence* (2017)

17. K. He et al., Identity mappings in deep residual networks, in *European Conference on Computer Vision* (Springer, 2016), pp. 630–645

18. S. Xie et al., Aggregated residual transformations for deep neural networks, in *Proceedings of the IEEE Conference on Computer Vision and Pattern Recognition* (2017), pp. 1492–1500

19. Y. LeCun, Y. Bengio, G. Hinton, Deep learning. Nature **521**(7553), 436–444 (2015)

20. S. Hochreiter, J. Schmidhuber, Long short-term memory. Neural Comput. **9**(8), 1735–1780 (1997)

21. K. Cho et al., Learning phrase representations using RNN encoder-decoder for statistical machine translation (2014). arXiv:1406.1078

22. J. Li, M.-T. Luong, D. Jurafsky, A hierarchical neural autoencoder for paragraphs and documents (2015). arXiv:1506.01057

23. A. Graves, A.-R. Mohamed, G. Hinton, Speech recognition with deep recurrent neural networks, in *IEEE International Conference on Acoustics, Speech and Signal Processing* (IEEE, 2013), pp. 6645–6649

24. I. Sutskever, O. Vinyals, Q.V. Le, Sequence to sequence learning with neural networks, in *Advances in Neural Information Processing Systems* 27 (2014)

25. J.Y.-H. Ng et al., Beyond short snippets: deep networks for video classification, in *Proceedings of the IEEE Conference on Computer Vision and Pattern Recognition* (2015), pp. 4694–4702

26. D. Bahdanau, K. Cho, Y. Bengio, Neural machine translation by jointly learning to align and translate (2014). arXiv:1409.0473

27. A. Vaswani et al., Attention is all you need, in *Advances in Neural Information Processing Systems* 30 (2017)

28. A. Dosovitskiy et al., An image is worth 16x16 words: transformers for image recognition at scale (2020). arXiv:2010.11929

29. Z. Liu et al., Swin transformer: Hierarchical vision transformer using shifted windows, in *Proceedings of the IEEE/CVF International Conference on Computer Vision* (2021), pp. 10012–10022

30. T. Mikolov et al., Distributed representations of words and phrases and their compositionality, in *Advances in Neural Information Processing Systems* 26 (2013)

31. J. Pennington, R. Socher, C.D. Manning, Glove: global vectors for word representation, in *Proceedings of the 2014 Conference on Empirical Methods in Natural Language Processing (EMNLP)* (2014), pp. 1532–1543

32. J.L. Ba, J.R. Kiros, G.E. Hinton, Layer Normalization (2016). arXiv:1607.06450
33. P. Ramachandran et al., Stand-alone self-attention in vision models, in *Advances in Neural Information Processing Systems* 32 (2019)
34. H. Hu et al., Local relation networks for image recognition, in *Proceedings of the IEEE/CVF International Conference on Computer Vision* (2019), pp. 3464–3473
35. J.-B. Cordonnier, A. Loukas, M. Jaggi, On the relationship between self-attention and convolutional layers (2019). arXiv:1911.03584
36. A. Krizhevsky, Learning multiple layers of features from tiny images. Technical Report TR-2009 (2009)
37. X. Xu et al., An improved swin transformer-based model for remote sensing object detection and instance segmentation. Remote Sens. **13**(23), 4779 (2021)
38. Y. Gao et al., Multi-attribute multi-task transformer framework for vision-based structural health monitoring. Computer-Aided Civil and Infrastructure Engineering (2023)
39. T.-Y. Lin et al., Feature pyramid networks for object detection, in *Proceedings of the IEEE Conference on Computer Vision and Pattern Recognition* (2017), pp. 2117–2125
40. O. Ronneberger, P. Fischer, T. Brox, U-net: convolutional networks for biomedical image segmentation, in *Medical Image Computing and Computer-Assisted Intervention-MICCAI 2015: 18th International Conference, Munich, Germany, October 5-9, 2015, Proceedings, Part III 18* (Springer, 2015), pp. 234–241

Part II
Introducing AI to Vision-Based SHM

Data plays an important role in the whole AI's life cycle. Before delving into applications, Chap. 5 initiates a discussion on several fundamental aspects of structural vision data collection. These include types of vision data, procedures for preparing structural image data, and the establishment of a benchmark framework and dataset for structural image detection of multi-tasks, known as PEER Hub ImageNet (PHI-Net or ϕ-Net). Moreover, to satisfy various practical engineering scenarios and demands, several variants of the ϕ-Net are investigated.

To substantiate the feasibility of integrating AI into vision-based SHM, key SHM applications are explored in Chaps. 6–8. These applications focus on structural image classification, structural damage localization, and structural damage segmentation, respectively.

In Chap. 6, the concept of structural image-based transfer learning (TL) is introduced. This mechanism aims to overcome recognition challenges and potential data shortages, ultimately boosting the AI model effectiveness. Firstly, preliminary classification experiments are conducted to investigate two strategies in TL, namely, feature extractor and fine-tuning. Subsequent extensive numerical experiments on eight sub-datasets of the ϕ-Net are carried out. Furthermore, hierarchical transfer learning (HTL) is explored to deal with these vision tasks in a sequential manner. Finally, a case study on a post-disaster damage assessment is presented.

Chapter 7 builds upon the knowledge from the classification tasks and broadens the scope to include structural damage localization. Unlike the discrete class labels in the classification tasks, specific localization evaluation metrics such as Intersection over Union (IoU) and Average Precision (AP) are introduced. Subsequently, several typical localization models are presented, including the single-stage YOLO series and the two-stage R-CNN series. Additionally, two case studies are presented focusing on the following two tasks: (1) concrete spalling detection and (2) bolt loosening detection.

Chapter 8 provides an overview of how the AI technologies can effectively address the structural damage segmentation task. Similar to the localization tasks, the segmentation-specified IoU and AP are introduced as evaluation metrics. Two classical segmentation models are presented, namely, fully convolutional network (FCN) and U-Net. To demonstrate the practicality of employing these models for damage segmentation, one case study of recognizing concrete spalling areas is investigated in detail.

Structural Vision Data Collection and Dataset   5

5.1   Structural Vision Data

5.1.1   Structural Vision Data Types

Images and videos are the two most commonly used data types in vision-based SHM. The image represents the instantaneous state in the structure and the video provides continuous changes of the state of the structure, which is composed of a sequence of frames.[1] In this book, the authors focus on image data treating them as the foundation of the whole vision-based SHM endeavor.

An image has different storage formats, typical examples are listed as follows:

- Joint photographic experts group (JPEG): It is a common image format used with the advantages of small storage capacity and fast loading speed, e.g., ImageNet [1], CelebA[2] [2], and MS COCO[3] [3]. Although a JPEG file usually takes less space for storage, it is a lossy (irreversible) compression format, which is prone to blurring and noise. In vision-based SHM, it is widely used for capturing images of structural damage or defects.
- Portable network graphics (PNG): It is a popular image format, which retains the original image information in a losslessly (reversible) compressed format. As a trade-off, a PNG file takes more storage than a JPEG one. In vision-based SHM, using PNG can capture images of structural components or damage with sharp edges or precise details.
- Tagged image file format (TIFF): It is a more flexible format that can store both lossless and lossy image data, making it suitable for a wide range of applications. A TIFF file can

---

[1] A frame refers to a single still image that is displayed on the screen at any given time.

[2] CelebFaces attributes dataset (CelebA), https://mmlab.ie.cuhk.edu.hk/projects/CelebA.html.

[3] Microsoft common objects in context (MS COCO), https://paperswithcode.com/dataset/coco.

provide higher quality and more information on structural images than a PNG one, but it takes more storage space.

There are many other storage formats, e.g., Bitmap (BMP) (raster, i.e., a grid of coordinates on a display space, graphics image file format used to store bitmap digital images) and RAW (minimally processed data from the image sensor, e.g., from a digital camera). The choice of the format for vision-based SHM depends on the specific application and the requirements for image quality, resolution, and file size. Nowadays, JPEG and PNG are the two most used formats in vision-based SHM.

5.1.2   Issues Related to Structural Vision Data

As pointed out in Sects. 1.3.1 and 1.3.2, most past studies in vision-based SHM faced the issue of lacking a large number of structural images, especially labeled ones, and thus conducted experiments based on a small-scale or low-variety image dataset. Zhang et al. [4] used 2,000 3D asphalt pavement surface images for pavement crack detection and Liang [5] collected 1,164 bridge-related images to conduct post-disaster inspection of RC bridge systems. However, to increase the data, many past studies focusing on pixel level images adopted cropping methods to crop raw high-resolution images into many sub-images. Cha et al. [6] collected 332 raw images taken from buildings with 277 training images ($4{,}928 \times 3{,}264$ pixels) and 55 test images ($5{,}888 \times 3{,}584$ pixels) for validation. All training images were further cropped to 40,000 sub-images (256×256 pixels). Xu et al. [7] collected only 12 images ($3{,}264 \times 4{,}928$ pixels) from a steel box girder of a bridge taken by a consumer-grade camera, then reshaped and cropped them to 240,448 sub-images (24×24 pixels). Similarly, Xu et al. [8] gathered 350 raw images and augmented their small dataset to 67,200 sub-images (64×64 pixels). Dorafshan et al. [9] obtained 18,000 sub-images (256×256 pixels) via cropping 100 raw images ($2{,}592 \times 4{,}608$ pixels), which are for concrete panels to simulate a RC bridge. Although cropping increases the number of image data, it brings drawbacks in training the DL models. Limited raw images only contain a few cases, similar scenarios, and specific types of structural components. After cropping, the high similarity in texture, lighting condition, etc., between training and test datasets can lead to good experimental results but lack generalization to more unseen scenarios. Thus, any high accuracy obtained on such test datasets may not be reliable and not truly reflect the generalization of the DL model where the model just memorizes similar data. Moreover, using cropping does not introduce new information with an inefficient use of computational resources where the learning of the DL models can directly and more efficiently utilize the full-size images instead of the sub-parts.

From the perspective of the dataset scale, the amount of data after cropping in the previous studies is still much smaller than that of ImageNet [1] (14 million), MNIST [10] (70,000), or CIFAR-10 [11] (70,000). For Structural Engineering applications, there exist online-accessible relatively large image collections like DesignSafe,[4] or self-taken and collected datasets. However, most of these datasets lack data variety and do not have well-defined labels, requiring additional efforts. As a result, increasing the dataset scale, along with its variety, is necessary. In addition, the datasets used in previous studies are inconsistent e.g., different studies may apply different label definitions. This is mainly attributed to the lack of shareable and benchmark datasets between researchers, which is an impediment to comparing results from different research groups for validation and further improvement of their models and algorithms.

A large-scale open-source well-labeled benchmark dataset of images with different attributes of structures and their components is needed. Until 2018, such a dataset was non-existent for general vision-based SHM. However, some datasets existed mainly for concrete crack detection, namely, CrackForest [12], Tomorrow's Road Infrastructure Monitoring and Management (TRIMM) [13], and structural defects network (SDNET2018) [14], developed in 2018. CrakForest and TRIMM include images from pavement, where vision patterns of cracks are similar to those from structural members but the cause of crack formation is different due to different loading and boundary conditions. In the SDNET2018 dataset, other than pavement images, it also contains images of walls and bridge decks. However, the number of such raw images in the SDNET2018 dataset is limited to only 54 bridge decks and 72 walls. In general, these datasets are for pixel level crack detection and only contain limited concrete component surface areas. Thus, in the decision-making procedure of SHM, more information related to the structural component type, structural damage severity, etc., is needed, which can only be extracted from images of much larger structural vision views.

5.2   PEER Hub ImageNet (PHI-Net or ϕ-Net)

Inspired by the establishment of the large-scale high-variety dataset of ImageNet [1], it was proposed to construct a Structural ImageNet, namely, PHI-Net or ϕ-Net [15], which contains images relevant to Structural Engineering, such as buildings and bridges, with both damaged and undamaged status. Based on past experiences from reconnaissance efforts [16–18], several structural attributes affecting most of the past structural assessment efforts are considered in ϕ-Net, i.e., scale of the target object, vision pattern, severity of the damage, type of structural component, etc. The constructed ϕ-Net framework and dataset can directly contribute to the development of automatic detection and recognition tasks in the vision-based SHM.

---

[4] A cyberinfrastructure, https://www.designsafe-ci.org, developed by the Natural Hazards Engineering Research Infrastructure (NHERI) program to enable research and educational advancements aimed at preventing natural hazard events from becoming societal disasters.

5.2.1   Automatic Detection Framework of ϕ-Net

The scale of the target object is directly related to the distance from the camera to the target. Pre-clustering the images based on such distance, i.e., close, mid-range, and far, related to the scene level (explained in Sect. 5.2.2.1), reduces the problem complexity and subsequent detection can be performed more accurately. Vision patterns and severity of damage can intuitively inform the engineers on how severe the structure is damaged with predefined engineering criteria based on the scene levels. For images taken from a close or mid-range distance, such images are more related to material loss and deterioration, i.e., concrete cover spalling, steel reinforcing bar buckling, etc. On the contrary, for images taken from a far distance, damage severity is related, in this case, to the failure status, i.e., complete or partial collapse of a building or a bridge structure. The type of structural component can also provide information for decision-making since the different structural components and their damage conditions play different roles in the overall integrity of the structural system.

Based on domain knowledge, structural vision tasks have certain relationships with each other, i.e., a certain hierarchy exists. A developed framework with a hierarchy tree structure is depicted in Fig. 5.1, where the grey boxes represent the detection tasks for the corresponding attributes, white boxes within dashed lines are possible labels of the attributes to be chosen for each task, and ellipses in boxes represent other choices or conditions, which represent the possibility for further expansions (scalability) based on newly collected data or future research interests. In the detection procedure, one starts from a root node where recognition tasks are conducted layer by layer and node (gray box) by node until another leaf node is reached. The output label of each node describes a structural attribute. Each structural image may have multiple attributes, e.g., one image can be categorized as pixel level, concrete, damaged state, etc. Thus, in the terminology of CV, this is a *multi-attribute classification problem*. However, in a pilot investigation, intended mainly for benchmarking[5] purposes, these attributes are treated independently.

From the perspective of Structural Engineering, the structural system is complex. While evaluating the health condition of the structure, information about both global and local details should be considered. For example, a structural elevation image, which belongs to the structural level, may reveal the global outside conditions of the structure, e.g., collapse mode and general damage state. Subsequently, indoor object level images, collected from key locations of the interior (or closer view) of the structure based on the system-level plan, can provide detailed information for the structural components (e.g., column, beam, or foundation), material type (e.g., steel, wood, or concrete), and damage level and type. When only focusing on small regions of the component, pixel level images can be used for local damage severity evaluation, e.g., crack width determination, or checking concrete cover spalling and reinforcing bar buckling conditions. Even though the attribute information of

---

[5] Benchmarking in AI, specifically ML and DL, is the development of a benchmark model (sometimes called a state-of-the-art model) to be used to compare performance of other models. A benchmark is basically the best known model on a given dataset for a given problem.

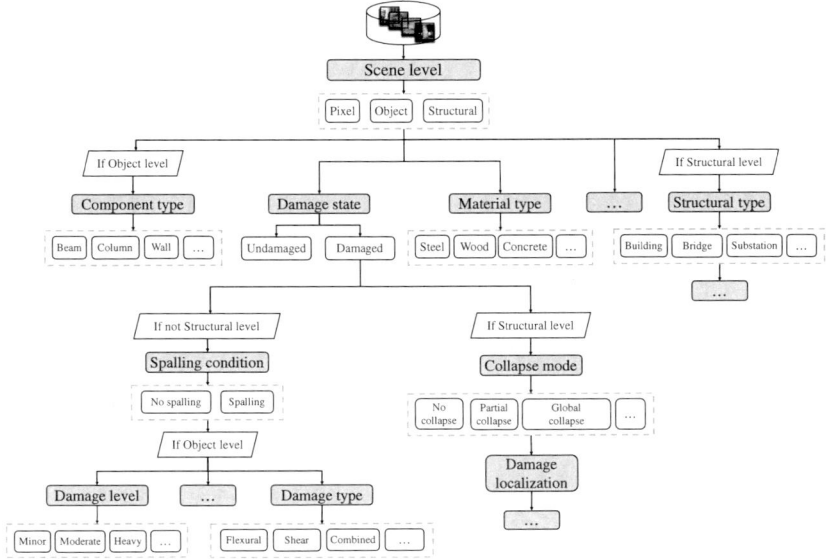

Fig. 5.1 Automatic detection framework of the φ-Net

a single image extracted from the φ-Net framework is limited, if we consider a large set of images taken from both inside and outside of the target structure, e.g., a building, multiple attributes recognized by the DL models can be very informative. This can be the case at least for initial *rapid assessment* to be followed by more detailed, time-consuming, and costly investigations, only if deemed necessary. Therefore, assembling information to describe certain structural attributes from all collected images can finally lead to a comprehensive view of the structural health condition, which is consistent with post-hazard reconnaissance strategies, typically conducted nowadays by human experts.

The hierarchy relationship described in Fig. 5.1 is just one reasonable choice among multiple possible combinations. The φ-Net framework is generic and it can be easily adjusted to fit most structural images, i.e., the branches and leaf nodes can be expanded and node sequences can also be adjusted based on the analyst's research or application demands. For example, in Sect. 5.6, several branches are combined and the sequence of the nodes is re-ordered to construct an extended hierarchy tree structure for more general scenarios, namely, next generation φ-Net (φ-NeXt).

5.2.2   Benchmark Detection Tasks

In the original φ-Net [15], the following eight classification problems are designed as benchmark tasks based on key attributes: (1) scene level, (2) damage state, (3) spalling condition (material loss), (4) material type, (5) collapse mode, (6) component type, (7) damage level, and (8) damage type. Based on Structural Engineering judgment, Tasks 1–4 are easier than Tasks 5–8. Moreover, from Fig. 5.1, the hierarchy tree is expandable with more branches and deeper depth, if desired, i.e., boxes with ellipses are possible detection tasks of other attributes for future considerations.

5.2.2.1 Task 1: Scene Level

From Fig. 5.1, the scene level classification acts as an initial task and subsequent tasks are pursued according to their different levels. Because images collected from reconnaissance efforts broadly vary in terms of distance, camera angle, camera type, emphasized target, etc., the scene level helps to decrease the doubt of the scale issues in the image and reduces the complexity of subsequent tasks. By filtering out the irrelevant data and dividing the images more granularly into different levels, the classifiers for the following tasks can achieve an improved recognition performance. In the ϕ-Net dataset, the benchmark scene level task is defined with three classes: *pixel level (P)*, *object level (O)*, and *structural level (S)*. An image taken from a very close distance or only containing part of the component belongs to P, Fig. 5.2a, an image taken from a mid-range and involving major targets such as single or multiple components belongs to O, Fig. 5.2b, and an image containing most parts of the structure or identifying its outline belongs to S, Fig. 5.2c.

5.2.2.2 Task 2: Damage State

In the vision-based SHM, damage state is one of the most important indicators of structural health. In this task, the definition is straightforward where any observable damage pattern on the structural surface implies a *damaged state (DS)*, Fig. 5.3a, otherwise it is an *undamaged state (US)*, Fig. 5.3b.

5.2.2.3 Task 3: Spalling Condition (Material Loss)

Spalling is usually defined as flakes of material that are broken off of a component or pertain to the loss of the cover of the reinforcing bars cage in a structural component. It often takes place in RC and reinforced masonry structures, which may lead to severe consequences due to the reductions in the cross-sectional area and moment of inertia, in conjunction with increased susceptibility of corrosion in the reinforcing bars. Spalling can be induced by expansion forces produced chemically due to the corrosion or mechanically due to loads or cracks. Most spalling images in ϕ-Net are collected from post-earthquake reconnaissance. In this task, the spalling condition is a binary classification task with two classes: *spalling (SP)*, Fig. 5.4a, and *non-spalling (NS)*, Fig. 5.4b.

(a) Pixel level (P)

(b) Object level (O)

(c) Structural level (S)

Fig. 5.2 Sample images used in the scene level task in ϕ-Net

(a) Damaged state (DS)

(b) Undamaged state (US)

Fig. 5.3 Sample images used in the damage state task in ϕ-Net

5.2.2.4 Task 4: Material Type

Structural material properties have a significant impact on the structural response. Thus, identifying the type of material used for a structural component is important for automated vision-based SHM. The commonly used materials are steel, concrete, masonry, and wood. One of the difficulties in this task is that the structural components are usually covered with non-structural finishes, e.g., plaster, making it difficult to accurately identify the material type from surface images. However, several structures and components are made from exposed steel members such as braces and plates and can be easily recognized by the shape and surface texture. Therefore, in this task, simplifications are made for the time being to only classify between *steel (ST)*, Fig. 5.5a, and *other materials (OM)*, Fig. 5.5b.

5.2.2.5 Task 5: Collapse Mode

Due to different photographic distances, camera angles, etc., it is reasonable to evaluate the severity of damage based on different scene levels. From the view of structural-level images, which contain the global information of the structure, the collapse mode is defined as *non-collapse (NC)*, *partial collapse (PC)*, and *global collapse (GC)*. NC, Fig. 5.6a, includes

(a) Spalling (SP)

(b) Non-spalling (NS)

Fig. 5.4 Sample images used in the spalling condition task in ϕ-Net

undamaged or slightly damaged patterns while the structure is intact. PC, Fig. 5.6b, corresponds to only a part of the structure (fewer components) collapsed while the remaining parts are still intact. GC, Fig. 5.6c, represents catastrophic damage in the structure (including a significant number of its components) or evidence of permanent global excessive deformation.

5.2.2.6 Task 6: Component Type

In the structural system, vertical components, i.e., columns and shear walls, provide lateral stiffness to resist lateral forces induced, e.g., by earthquakes. On the other hand, horizontal components, i.e., beams, diaphragms, and joints, transfer the horizontal loads to the vertical components. These components have different importance levels, especially in seismic design where engineers adopt concepts like "strong column weak beam" and "strong joint weak component" in buildings, see for example [19]. Thus, identifying the type of a structural component is informative for SHM. From the hierarchical relationship in the ϕ-Net, the component type identification is a subsequent task at the object level with four classes: *beam* (B), Fig. 5.7a, *column* (C), Fig. 5.7b, *wall* (W), Fig. 5.7c, and *other compo-*

(a) Steel (ST)

(b) Other materials (OM)

Fig. 5.5 Sample images used in the material type task in ϕ-Net

nents (OC), Fig. 5.7d. It is challenging to identify whether a wall is a structural shear wall or a non-structural infill wall (e.g., a partition bounded by a structural frame) from the images. Therefore, for simplicity, both types are grouped herein into one category in ϕ-Net, even though they have pronounced differences in terms of the provided lateral stiffness and strength to the structural system [32]. For a component other than beam, column, or wall, e.g., joint, stair, window, or brace, it is treated as OC.

5.2.2.7 Task 7: Damage Level

This task refers specifically to the damage level evaluation from images at the pixel and object levels, to complement the collapse mode task at the structural level. It is defined with four classes: *no damage (ND)*, which is the same as the class "US" in Task 2, Fig. 5.3b, *minor damage (MiD)*, *moderate damage (MoD)*, and *heavy damage (HvD)*. MiD, Fig. 5.8a, implies that there are only small and thin cracks or a few small spalling areas on the cover of the structural components. With further development of damage, cracks become wider and spalling areas propagate without failure, indicating MoD, Fig. 5.8b. HvD, Fig. 5.8c, means that the damage area is large and approaching the failure state of the structural component.

(a) Non-collapse (NC)

(b) Partial collapse (PC)

(c) Global collapse (GC)

Fig. 5.6 Sample images used in the collapse mode task in φ-Net

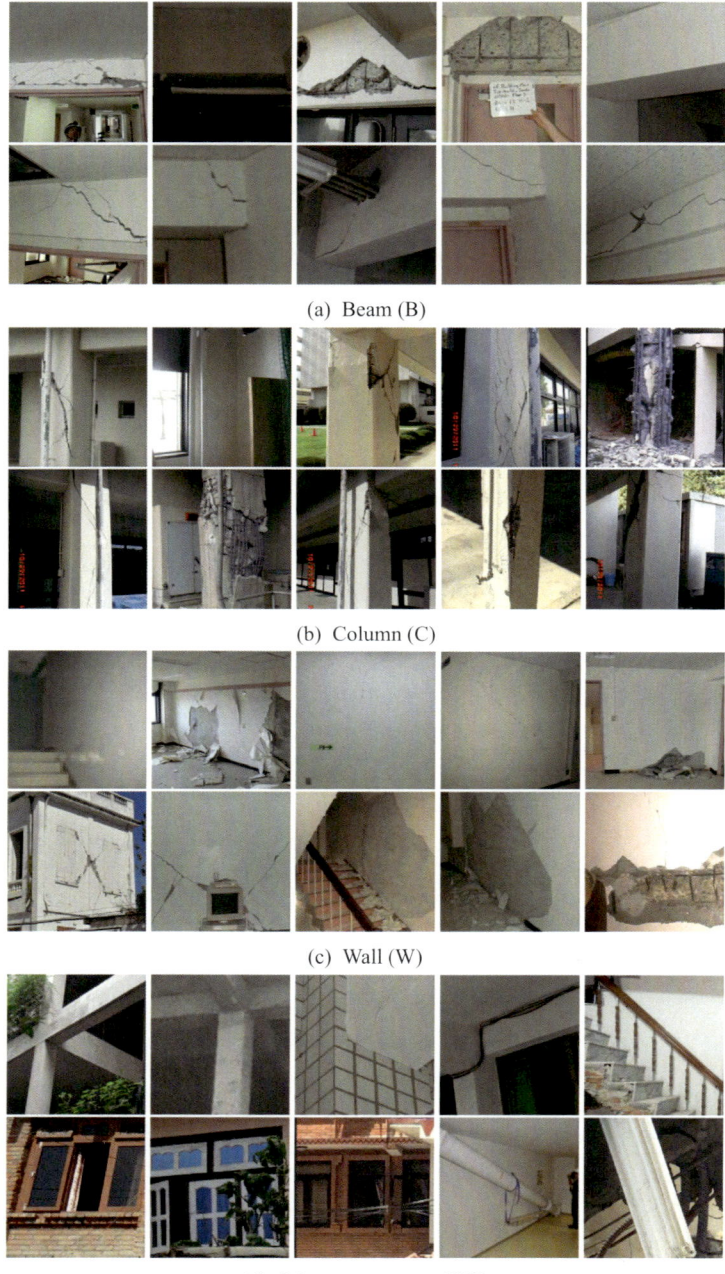

(a) Beam (B)

(b) Column (C)

(c) Wall (W)

(d) Other components (OC)

Fig. 5.7 Sample images used in the component type task in ϕ-Net

(a) Minor damage (MiD)

(b) Moderate damage (MoD)

(c) Heavy damage (HvD)

Fig. 5.8 Sample images used in the damage level task in ϕ-Net

5.2.2.8 Task 8: Damage Type

Not restricted to a specific damage representation, e.g., reinforcing bar buckling, concrete spalling, etc., in this task of ϕ-Net, damage type refers to a complex and abstract semantic vision pattern, defined as *flexural damage (FD)*, *shear damage (SD)*, and *combined damage (CD)*. These damage types have direct implications on the mechanical properties, structural behavior, and also design guidelines, e.g., seismic design. Analogous to the failure types in [19] for seismic loading on RC structures, definitions are as follows: (1) if most cracks occur in the horizontal or the vertical direction of a component with vertical (e.g., column) or horizontal (e.g., beam) edges, respectively, at the component mid-height (e.g., column), mid-span (e.g., beam), or at the component fixed ends (i.e., restrained from rotation), it is a flexural-type damage, which is more *ductile*, Fig. 5.9a, (2) if most cracks occur in a diagonal direction, or form a "X" or a "V" pattern, it is a shear-type damage, which is more *brittle*, Fig. 5.9b, and (3) if the distribution and orientation of the cracks are irregular and mixed, or accompanied with heavy spalling, it is a combined-type damage, Fig. 5.9c.

5.3 Data Preparation

The procedure used to establish the ϕ-Net dataset consisted of data collection, preprocessing, labeling, and splitting. These four steps are discussed in the following sections.

5.3.1 Data Collection

To construct a large-scale image dataset, a large number of images are needed, e.g., ImageNet contains over 14 million images [1]. Researchers in Structural Engineering typically have large numbers of photographs taken from reconnaissance efforts or laboratory experimental investigations, which are good resources for DL training. Moreover, there exist online resources, i.e., databases and search engines storing raw structural images of pre- and post-disaster status but without labels. Besides self-taken and donated images, raw images in the ϕ-Net were collected from NISEE library archive,[6] DesignSafe, EERI (Earthquake Engineering Research Institute) learning from earthquakes (LFE) archive,[7] Baidu Images,[8] and Google Images.[9] Until October 2023, over 100,000 structural images with variant quality were collected, where a large amount was taken from reconnaissance efforts after recent earthquakes, e.g., the 2017 Mexico City earthquake. Multiple worldwide sources contributed to the variety of the data in the current ϕ-Net dataset.

---

[6] https://nisee.berkeley.edu/elibrary.

[7] https://www.eeri.org/projects/learning-from-earthquakes-lfe.

[8] https://image.baidu.com.

[9] https://images.google.com.

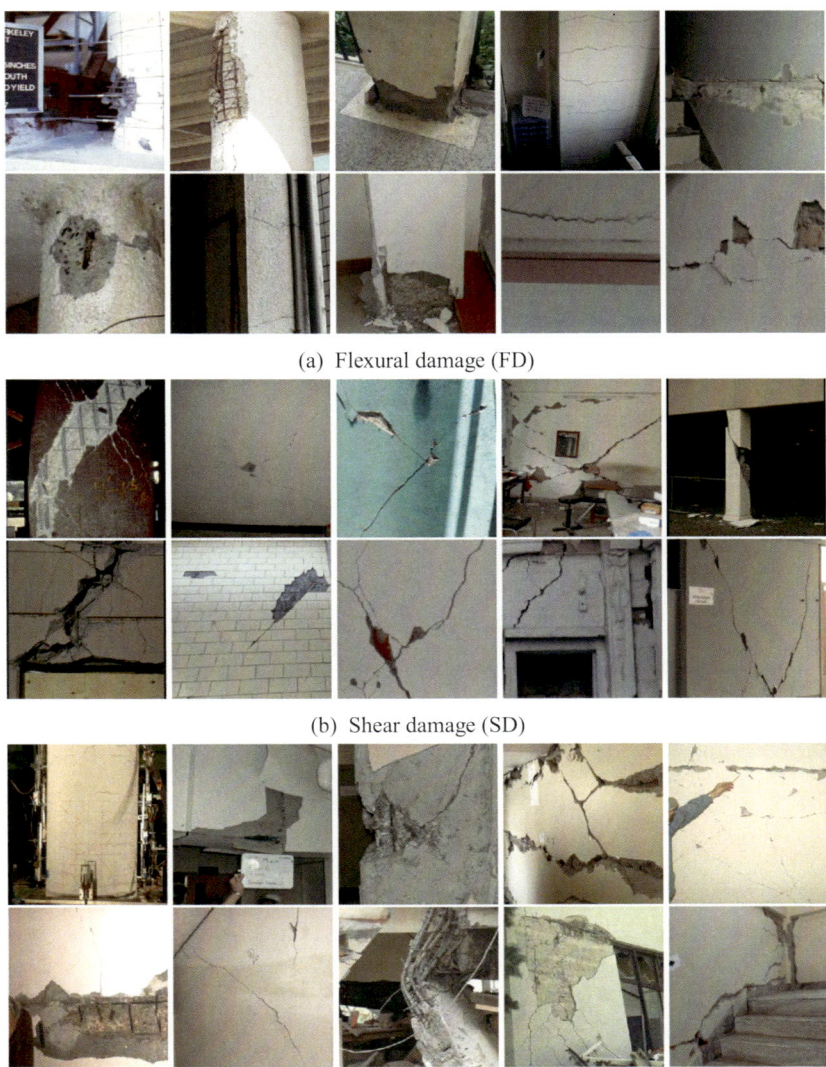

(a) Flexural damage (FD)

(b) Shear damage (SD)

(c) Combined damage (CD)

Fig. 5.9 Sample images used in the damage type task in ϕ-Net

5.3.2   Data Preprocessing

Since the collected images vary broadly with different qualities, image preprocessing was conducted before labeling following these four steps:

1. Manually eliminate low-resolution images (lower than 448×448 pixels) and noisy images containing too many irrelevant objects, e.g., vehicles, trees, people, furniture, etc., where the major content of the image is unrelated to damage or other structural attributes from a human expert perspective.
2. To further increase the dataset size, especially for the pixel and object levels, sub-parts, e.g., beam, column, and wall, were cropped from some high-resolution structural images, refer to Fig. 5.10.
3. To avoid significant distortions to the image features due to stretching or rescaling used in training,[10] cropping was applied again to make the aspect ratio of the images roughly less than 2 (ideally, it should be 1) for the sake of training.
4. To avoid low quality and low resolution, cropped images with a resolution lower than 224×224 pixels were eliminated.

5.3.3   Data Labeling

As discussed above, eight benchmark tasks are independently treated in ϕ-Net [15]. However, in the labeling procedure, the order is followed according to the framework in Fig. 5.1, starting from the labeling attribute at the scene level and then progressing toward different tasks. The labeling procedure is crowd-sourced via an online labeling tool,[11] which provides a user-friendly graphical interface for annotation where users need to only click the button corresponding to the desired label to choose, Fig. 5.11. In this labeling procedure, to avoid bias, a majority voting mechanism is adopted, where multiple experts may label one image and the final label is the one with the majority voting. Moreover, a reference voting mechanism is introduced as shown in the right part of Fig. 5.11, which represents the percentages of voting for each choice based on the opinions of other experts who labeled the same image. The reference voting mechanism is intended to help in judging some ambiguous images. In some cases, there may exist an image unrelated to the currently proposed attributes or the user may not be able to make a decision on a certain image. Thus, a "skip" button is added to bypass such an image. Once the user completes labeling one task, the selected choice is shown at the bottom right corner, then the user can advance to the next task of the above mentioned hierarchical relationship by clicking "next".

---

[10] For example, while feeding arbitrary images into the vanilla VGG-16, refer to Sect. 4.3.1, the images need to be reshaped to the network default input size 224×224. This reshaping causes stretching and rescaling.

[11] https://apps.peer.berkeley.edu/spo.

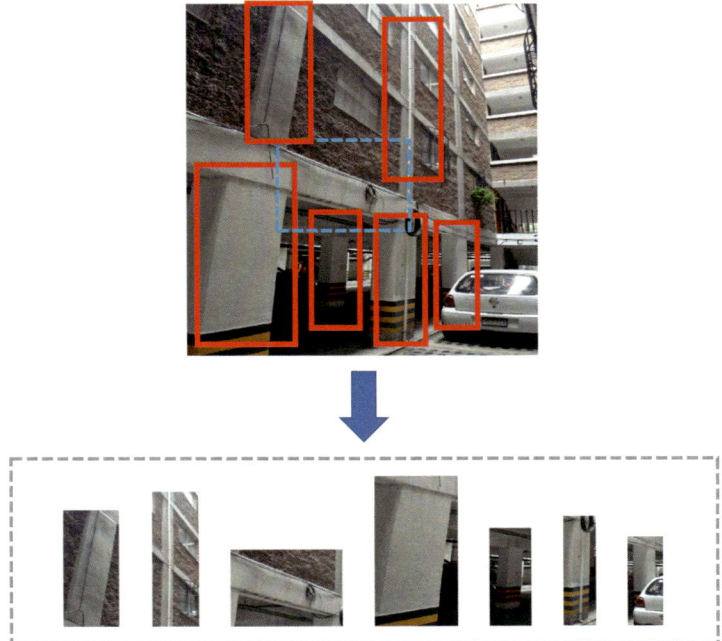

Fig. 5.10 Sub-parts are cropped from a high-resolution structural level image

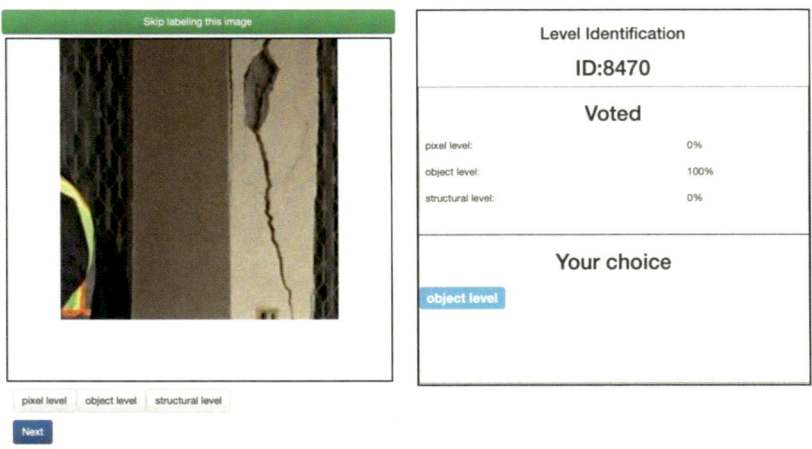

Fig. 5.11 Graphical user interface of the online labeling tool

In the labeling process of the ϕ-Net, 20 volunteers with a Structural Engineering background were recruited. This is in addition to receiving contributions from worldwide researchers and experts with relevant background knowledge. In order to avoid bias in the current ϕ-Net dataset, each added labeled image to the database was labeled with the majority voting among at least three experts and was discarded if it had less than three votes. Finally, **36,413** labeled images are now available in the open-source ϕ-Net database [15].

5.3.4 Data Splitting

As a benchmark dataset for training and validating ML and DL models, splitting the raw dataset into deterministic training and test sets is required. However, the ϕ-Net is a multi-attribute dataset where each image contains multiple labels according to the relevant attributes. Thus, simply splitting raw images by a fixed training : test split ratio may lead to different training : test ratios of labels in each task and its attributes. This is not the case in common classification benchmark datasets, e.g., MNIST [10] and CIFAR-10 [11]. As an example, in the damage state (Task 2), only 13,271 images have valid labels (DS and US) and the remaining 23,142 images are unlabeled for this task. Simply adopting as an example "training/test test = 9/1" across the entire image set may lead to the undesirable situation where all or most of the images with valid labels are in the training set and no such images appear in the test set for this particular task. Therefore, a new split algorithm is developed where the training/test ratio is replaced by a range instead of a deterministic fixed number for each attribute. Usually, a fixed split ratio for a single attribute dataset is from 4/1 to 9/1 and herein the split ratio range for the multi-attribute and the relatively small ϕ-Net dataset, e.g., compared with ImageNet [1], is chosen from 8/1 to 9/1.

Algorithm 5.1 shows the pseudo-code of the multi-attribute split algorithm. Firstly, according to the number of labeled images in the whole dataset ($N_{total} = 36,413$), the number of training images ($N_{training} = \alpha \cdot N_{total} = 32,407$) is randomly sampled without replacement from 1 to 36,413, where the trial split ratio factor $\alpha \approx 0.89$ is assigned for a better convergence performance after multiple numerical trials. The remaining 4,006 images are accordingly assigned as a test dataset. Subsequently, the training/test ratios are checked for each attribute. If there existed any conflicts where the ratio is out of the predefined range of 8/1 to 9/1, the sampled images are rejected and the *recursive loop* is run again until one reaches a satisfactory split. Since eight attributes are treated independently, eight task-oriented sub-datasets, one for each attribute with training and test sets, are further separated, Tables 5.1, 5.2, 5.3, 5.4, 5.5, 5.6, 5.7 and 5.8.

Algorithm 5.1 Multi-attribute split algorithm for ϕ-Net

1: **Required**: *function* Generate_Training_Index(x) to randomly generate x training indices without replacement;

2: *function* Generate_Test_Index(N, y, *training_index*) to generate y test indices, where $N = x + y$;

3: *function* Compute_Split(task_id, training_index, test_index) to determine the ratio of a specific task.

4: **Define**: trial split ratio, α, total number of data, N_{total}, lower (Bound_L) and upper (Bound_U) bounds of the ratio.

5: **Function**: Multi_Attribute_Split(α, N_{total}, Bound_L, Bound_U)

6:   Set done = False

7:   $N_{training} \leftarrow \alpha \cdot N_{total}$

8:   $N_{test} \leftarrow (1 - \alpha) \cdot N_{total}$

9:   **while** not done **do**

10:     training_index $\leftarrow$ Generate_Training_Index($N_{training}$)

11:     test_index $\leftarrow$ Generate_Test_Index(N_{total}, N_{test}, training_index)

12:     # *Loop over all attributes or tasks*

13:     **for** task_id $\in$ [task_1, task_2, ..., task_8] **do**

14:       ratio $\leftarrow$ Compute_Split(task_id, training_index, test_index)

15:       # *Check if violating the boundary assumptions*

16:       **if** ratio < Bound_L or ratio > Bound_U **then**

17:         # *Use recursion*

18:         Multi_attribute_Split(α, N_{total}, Bound_L, Bound_U)

19:       **end if**

20:     **end for**

21:     done $\leftarrow$ True

22:   **end while**

23:   **return** training_index, text_index

Table 5.1 Number of images in the scene level (Task 1) of ϕ-Net

Set	Pixel level	Object level	Structural level	Total
Training	7,690	8,111	8,508	24,309
Test	965	962	1,070	2,997
Total	9,073	8,655	9,578	27,306

Table 5.2 Number of images in the damage state (Task 2) of ϕ-Net

Set	Damaged state	Undamaged state	Total
Training	6,282	5,529	11,811
Test	745	715	1,460
Total	7,027	6,244	13,271

Table 5.3 Number of images in the spalling condition (Task 3) of ϕ-Net

Set	Non-spalling	Spalling	Total
Training	4,294	2,604	6,898
Test	527	310	837
Total	4,821	2,914	7,735

Table 5.4 Number of images in the material type (Task 4) of ϕ-Net

Set	Steel	Other materials	Total
Training	1,806	6,506	8,312
Test	209	770	979
Total	2,015	7,276	9,291

Table 5.5 Number of images in the collapse mode (Task 5) of ϕ-Net

Set	Non-collapse	Partial collapse	Global collapse	Total
Training	322	379	525	1,226
Test	39	40	67	146
Total	361	419	592	1,372

Table 5.6 Number of images in the component type (Task 6) of ϕ-Net

Set	Beam	Column	Wall	Others	Total
Training	511	1,618	2,268	358	4,755
Test	60	205	265	49	579
Total	571	1,823	2,533	407	5,334

Table 5.7 Number of images in the damage level (Task 7) of ϕ-Net

Set	No damage	Minor	Moderate	Heavy	Total
Training	1,551	869	799	919	4,138
Test	207	93	104	94	498
Total	1,758	962	903	1,013	4,636

Table 5.8 Number of images in the damage type (Task 8) of ϕ-Net

Set	No damage	Flexural	Shear	Combined	Total
Training	1,598	476	826	1,193	4,093
Test	215	46	99	132	492
Total	1,813	522	925	1,325	4,585

Typically, for ML and DL studies, it is suggested to maintain a similar label distribution (percentage of each label with respect to all labels) among the training, test, and full datasets. Therefore, the distributions among the eight tasks in the ϕ-Net dataset are examined as listed in Tables 5.9, 5.10, 5.11, 5.12, 5.13, 5.14, 5.15 and 5.16. All results indicate that the ratios among these three datasets are nearly consistent.

Table 5.9 Label distribution in training, test, and full dataset in Task 1 of ϕ-Net

Set	Pixel level (%)	Object level (%)	Structural level (%)
Training	31.6	33.4	35.0
Test	32.2	32.1	35.7
Full	31.7	33.2	35.1

Table 5.10 Label distribution in training, test, and full dataset in Task 2 of ϕ-Net

Set	Damaged state (%)	Undamaged state (%)
Training	53.2	46.8
Test	51.0	49.0
Full	53.0	47.0

Table 5.11 Label distribution in training, test, and full dataset in Task 3 of ϕ-Net

Set	Non-spalling (%)	Spalling (%)
Training	62.2	37.8
Test	63.0	37.0
Full	62.3	37.7

Table 5.12 Label distribution in training, test, and full dataset in Task 4 of ϕ-Net

Set	Steel (%)	Other materials (%)
Training	21.7	78.3
Test	21.3	78.7
Full	21.7	78.3

Table 5.13 Label distribution in training, test, and full dataset in Task 5 of ϕ-Net

Set	Non-collapse (%)	Partial collapse (%)	Global collapse (%)
Training	26.3	30.9	42.8
Test	26.7	27.4	45.9
Full	26.3	30.5	43.1

Table 5.14 Label distribution in training, test, and full dataset in Task 6 of ϕ-Net

Set	Beam (%)	Column (%)	Wall (%)	Others (%)
Training	10.7	34.0	47.7	7.5
Test	10.4	35.4	45.8	8.5
Full	10.7	34.2	47.5	7.6

Table 5.15 Label distribution in training, test, and full dataset in Task 7 of ϕ-Net

Set	No damage (%)	Minor (%)	Moderate (%)	Heavy (%)
Training	37.5	21.0	19.3	22.2
Test	41.6	18.7	20.9	18.9
Full	37.9	20.8	19.5	21.9

Table 5.16 Label distribution in training, test, and full dataset in Task 8 of ϕ-Net

Set	No damage (%)	Flexural (%)	Shear (%)	Combined (%)
Training	39.0	11.6	20.2	29.1
Test	43.7	9.3	20.1	26.8
Full	39.5	11.4	20.2	28.9

5.4 ϕ-Net Dataset Statistics

The ϕ-Net dataset, discussed in Sects. 5.2 and 5.3, is composed of **36,413** effectively labeled images. It can be either treated as a one multi-attribute (i.e., multi-label) dataset or as consisting of eight independent sub-datasets corresponding to their eight tasks (and their attributes). The statistics of the labels are shown in Fig. 5.12. These eight datasets with both training and test sub-sets are open-sourced online[12] and researchers can download the data according to instructions for non-commercial usage. All relevant files are compressed in the form of zip files and they share the same data storage architecture containing **X_train.npy** (i.e., training data), **y_train.npy** (i.e., training labels), **X_test.npy** (i.e., test data), and **y_test.npy** (i.e., test labels). All the images are converted to NumPy array and can be easily loaded using several available computer platforms, e.g., Python or MatLab.

The direct usage of the ϕ-Net dataset is to train and validate analysts' self-developed algorithms and models through comparisons with the reported values in this section. Besides the benchmarking purpose, the open-sourced 36,413 images can facilitate future augmentation of self-collected or other open-source datasets from previous studies with or without labels. For example, the ϕ-Net pixel level concrete images with labels "DS" and "US" can be added

---

[12] https://apps.peer.berkeley.edu/phi-net.

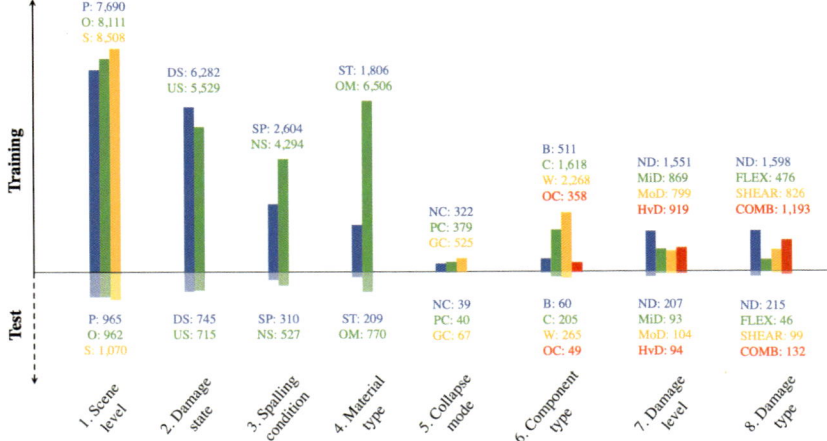

Fig. 5.12 Statistics of the labeled data in the ϕ-Net dataset

into SDNET2018 [9] to increase the variety of the crack images from multiple structural components and different loading conditions. For the dataset used in [5], the proposed local, component, and system levels are consistent with their respective counterparts, namely, pixel, object, and structural levels of the ϕ-Net. Thus, the ϕ-Net images can be combined with the dataset from [5]. In addition, the ϕ-Net dataset can be extended to other CV tasks, e.g., damage localization and segmentation, with additional relabeling efforts by the analysts, as discussed in Sect. 5.6. In conclusion, complementing the ϕ-Net data increases both the number and variety of the data and it is expected to improve the DL modeling accuracy and robustness in similar future studies. It is worth noting that this image dataset is not only limited to the scope of DL but can also be beneficial to other data-driven studies using general ML or traditional methods, e.g., [20, 21].

The developed framework for the original ϕ-Net and its associated dataset only focused on the structural surface images and cannot reflect the invisible inner damage. Research activities related to fusing new data and information from Lidar [22], 3D point cloud [23], vibration-based methods using time series signals [24, 25], etc., are recommended for future endeavors.

5.5   Extensions of the ϕ-Net Dataset Hierarchy

The flowchart of the ϕ-Net framework, shown in Fig. 5.1, indicates the hierarchical relationships between the different tasks. These relationships and their hierarchy typically depend on the Structural Engineering domain knowledge and past experiences. One possible hierarchical framework is proposed herein, namely, hierarchy ϕ-Net, as presented in Fig. 5.13, which combines several branches for efficiency in the computations and introduces con-

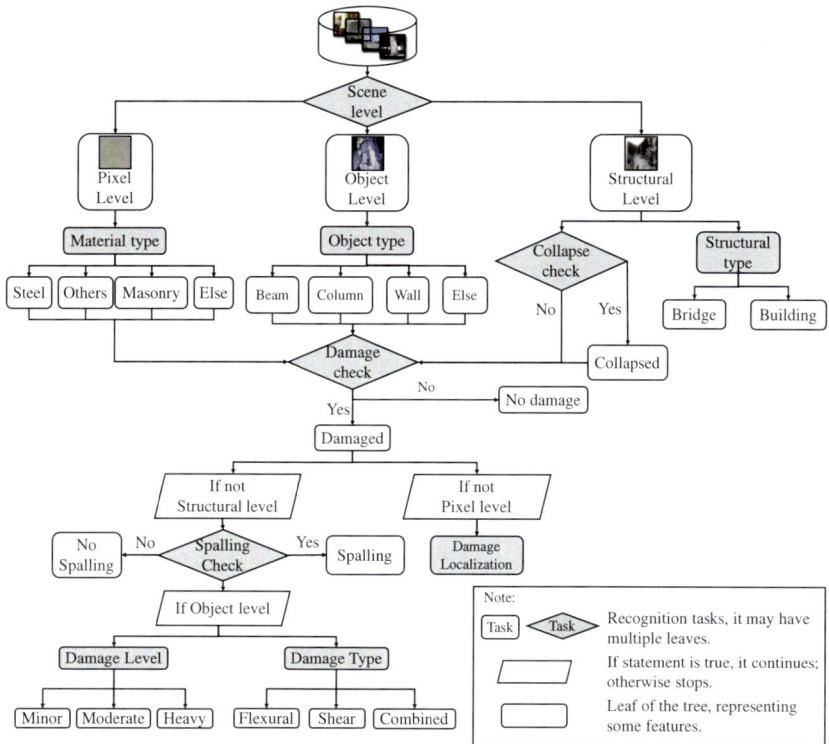

Fig. 5.13 Hierarchy of the extended ϕ-Net framework

ditional nodes (shown as diamond shapes) for tree-branch selections. The grey nodes in Fig. 5.13 represent detection tasks, where all previously-discussed eight benchmark tasks are included. Therefore, the ϕ-Net dataset is also suitable for this extended framework. Hierarchy ϕ-Net has a clearer order and stronger correlation between the root and the leaf tasks than the original framework (Fig. 5.1) by the additional emphasis on the two types of diamond nodes (i.e., scene level and damage state tasks) to make them play the most important roles in the whole framework. The remaining nodes are now connected and follow a clear sequence. Thus, it is expected that the hierarchy ϕ-Net to be beneficial for future studies. However, the order and selection of nodes (i.e., tasks) can be further improved by exploring more hierarchical relationships in future improvements and expansions of the ϕ-Net dataset.

From Fig. 5.13, the hierarchy ϕ-Net has certain depths representing the complexity of the whole framework and the related problems. In some cases, the user may only focus on simpler tasks, e.g., scene level and damage state, instead of using the whole framework. In such cases, a hierarchical pipeline with shallow depth can be applied as shown in Fig. 5.14. When only using the information until the "depth-2" layer, it is denoted as depth-2 representation, where the root and leaf tasks are scene level and damage state classifications, respectively. From

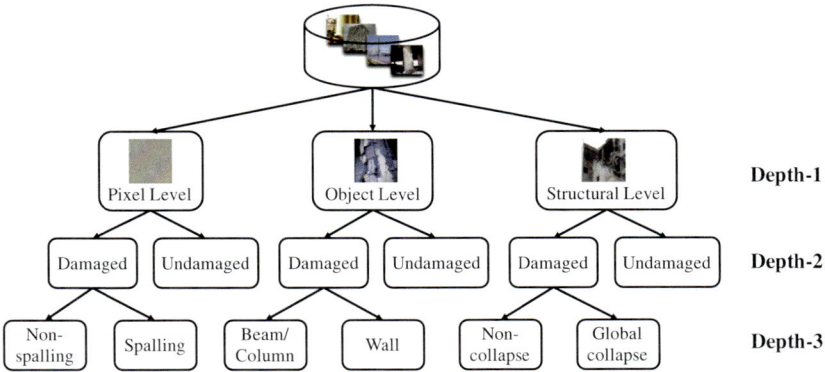

Fig. 5.14 Shallow depth representation of the ϕ-Net

the nodes in the depth-2 layer, the pipeline partitions the damaged and undamaged states into 6 sub-categories, known as leaf nodes, where DS or US can be further separated and sorted as follows:

1. From the pixel level with the damaged state (PD);
2. From the object level with the damaged state (OD);
3. From the structural level with the damaged state (SD);
4. From the pixel level with the undamaged state (PU);
5. From the object level with the undamaged state (OU); and
6. From the structural level with the undamaged state (SU).

From the ϕ-Net, more information, as prior knowledge, can be added by further expansions of the leaf nodes in depth-2 into a deeper layer. Thus, these nodes become intermediate ones. This is denoted as "depth-3" representation, if expanded to one more layer. The leaf nodes in the depth-3 layer represent sub-categories in finer classification tasks, e.g., component type, spalling condition, and collapse check. Accordingly, depth-3 representation has lower variations within the sub-classes than depth-2, but it may lose information by excluding some sub-classes leading to possibly less diversity. Moreover, for the CV-based SHM problems, leaf nodes in depth-3 are still thought of as being complex due to different lighting conditions, shapes, textures, surrounding objects, damage patterns, etc. Thus, dealing with images in the subsets of the leaf nodes is still challenging compared with several well-studied CV applications, e.g., handwriting digits [10] and face recognition problems [26].

5.6   Multi-task Extension to ϕ-NeXt

Analogous to the CV domain, besides the image classification task, object detection [5, 27] and segmentation [28, 29] are two important applications in vision-based SHM. Technically, these two tasks sometimes can be treated as two downstream tasks from the classification task, e.g., using pre-trained parameters from the classification task to fine-tune the model for localization and segmentation [30, 31]. However, most of the previous studies considered these tasks independently ignoring the inter-task relationships and not accounting for the possibility of shared features among different data sources. Similar to the multi-attribute settings in the original ϕ-Net, the results of the localization and segmentation tasks can be re-defined as two special attributes, i.e., the *object location* is represented by coordinates of the bounding box of the target object and the pixel-wise labels *annotate the class of each pixel* of the image. Finally, by adding these two attributes to the multiple classification attributes, the vision-based SHM can be comprehensively reformed and unified as a *multi-attribute multi-task problem*.

Based on the ϕ-Net, a new multi-attribute multi-task version of the automatic SHM framework is proposed, namely, the next generation ϕ-Net (i.e., ϕ-NeXt). Firstly, the same eight classification tasks defined in Sect. 5.2.2 are maintained as the fundamental ones. Subsequently, damage localization and segmentation tasks are included as downstream tasks, which are denoted as Tasks 9 and 10, respectively. These two new tasks are described as follows:

- **Task 9: Damage localization**. It localizes the damage patterns by bounding boxes. For different damage types, they are annotated by different bounding boxes as shown in Fig. 2.3c.
- **Task 10: Damage segmentation**. It quantifies the damage by finding the whole damaged area, where each pixel has its own label, and regions of pixels with the same label are grouped and segmented as one object (i.e., class).

Finally, the ten tasks (the original eight in ϕ-Net and the two new ones, discussed above) are reorganized into a similar but simplified framework as shown in Fig. 5.15. Similar to the original ϕ-Net, the number of tasks of ϕ-NeXt can be expanded upon based on the SHM demands and more typical tasks, e.g., locating and segmenting damage patterns of reinforcing bar exposure, steel corrosion, and masonry crushing, are recommended as future extensions. The developed ϕ-NeXt dataset is further discussed in Chap. 12.

In summary, the ϕ-NeXt expands the original ϕ-Net into a multi-attribute multi-task version, which describes the characteristics of the collected structural image data in a more comprehensive manner covering the classification, localization, and segmentation tasks. Such framework together with its dataset is helpful in utilizing the relationships between tasks and different data sources leading to a more accurate and robust AI model performance. This comprehensive framework ultimately aims to enhance the resiliency of the AI-based SHM system. More relevant topics are discussed in detail in Chap. 12.

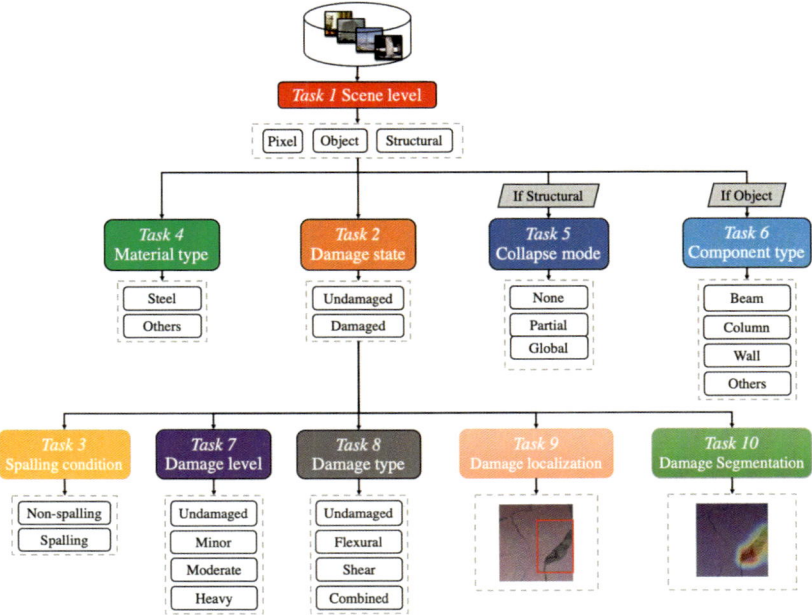

Fig. 5.15 ϕ-NeXt framework

5.7 Exercises

1. The ϕ-Net follows a hierarchical tree structure with a flexible sequence of nodes that can be tailored according to specific requirements. Using the visual representations in Figs. 5.1 and 5.13 as a guide, consider an engineering scenario and plot a new and simplified version of ϕ-Net hierarchical tree by arranging nodes differently to meet the specific demands of the considered scenario.
2. What are common operations and techniques involved in the data preparation (e.g., preprocessing) stage? Provide a brief discussion regarding their purposes and benefits.
3. Download the 8 datasets for the 8 classification benchmark tasks from the ϕ-Net official website (https://apps.peer.berkeley.edu/phi-net). Subsequently, generate bar charts illustrating the distribution of each label within the training and test sets for every task. Discuss the data statistics and elucidate their potential impact on the performance of the trained DL models.
4. Find and present information on 3–5 publicly available (i.e., open-sourced) datasets related to SHM, providing an overview of their key characteristics, discussing similarities with and differences from ϕ-Net and ϕ-NeXt.

References

1. J. Deng et al., Imagenet: "A large-scale hierarchical image database", in *2009 IEEE Conference on Computer Vision and Pattern Recognition* (2009), pp. 248–255
2. Z. Liu et al., Deep learning face attributes in the wild, in *Proceedings of the IEEE International Conference on Computer Vision* (2015), pp. 3730–3738
3. T.-Y. Lin et al., Microsoft coco: common objects in context, in *Computer Vision-ECCV 2014: 13th European Conference, Zurich, Switzerland, September 6–12, 2014, Proceedings, Part V 13* (Springer, 2014), pp. 740–755
4. A. Zhang et al., Automated pixel-level pavement crack detection on 3D asphalt surfaces using a deep-learning network. Comput.-Aided Civil Infrast. Eng. **32**(10), 805–819 (2017)
5. X. Liang, Image-based post-disaster inspection of reinforced concrete bridge systems using deep learning with Bayesian optimization. Comput.-Aided Civil Infrast. Eng. **34**(5), 415–430 (2019)
6. Y.-J. Cha, W. Choi, O. Büyüköztürk, Deep learning-based crack damage detection using convolutional neural networks. Comput.- Aided Civil Infrast. Eng. **32**(5), 361–378 (2017)
7. Y. Xu et al., Identification framework for cracks on a steel structure surface by a restricted Boltzmann machines algorithm based on consumer-grade camera images. Struct. Control Health Monit. **25**(2), 1031–1045 (2018)
8. Y. Xu et al., Surface fatigue crack identification in steel box girder of bridges by a deep fusion convolutional neural network based on consumergrade camera images. Struct. Health Monit. **18**(3), 653–674 (2019)
9. S. Dorafshan, R.J. Thomas, M. Maguire, Comparison of deep convolutional neural networks and edge detectors for image-based crack detection in concrete. Constr. Build. Mater. **186**, 1031–1045 (2018)
10. Y.L. Cun et al., Gradient-based learning applied to document recognition. Proc. IEEE **86**(11), 2278–2324 (1998)
11. A. Krizhevsky, Learning multiple layers of features from tiny images. Technical Report TR-2009 (2009)
12. Y. Shi et al., Automatic road crack detection using random structured forests. IEEE Trans. Intell. Transp. Syst. **17**(12), 3434–3445 (2016)
13. R. Amhaz et al., Automatic crack detection on two-dimensional pavement images: an algorithm based on minimal path selection. IEEE Trans. Intell. Transp. Syst. **17**(10), 2718–2729 (2016)
14. M. Maguire, S. Dorafshan, R.J. Thomas, SDNET2018: a concrete crack image dataset for machine learning applications (2018)
15. Y. Gao, K.M. Mosalam, PEER Hub ImageNet: a large-scale multiattribute benchmark data set of structural images. J. Struct. Eng. **146**(10), 04020198 (2020)
16. C. Koch et al., A review on computer vision based defect detection and condition assessment of concrete and asphalt civil infrastructure. Adv. Eng. Inf. **29**(2), 196–210 (2015)
17. B. Li, K.M. Mosalam, Seismic performance of reinforcedconcrete stairways during the 2008 Wenchuan earthquake. J. Perform. Constr. Facil. **27**(6), 721–730 (2013)
18. H. Sezen et al., Performance of reinforced concrete buildings during the August 17, 1999 Kocaeli, Turkey earthquake, and seismic design and construction practise in Turkey. Eng. Struct. **25**(1), 103–114 (2003)
19. J. Moehle, *Seismic Design of Reinforced Concrete Buildings* (McGraw Hill Professional, 2014)
20. T. Khuc, F.N. Catbas, Structural identification using computer vision-based bridge health monitoring. J. Struct. Eng. **144**(2), 04017202 (2018)
21. Y. Yang, S. Nagarajaiah, Dynamic imaging: real-time detection of local structural damage with blind separation of low-rank background and sparse innovation. J. Struct. Eng. **142**(2), 04015144 (2016)

22. L. Dong, J. Shan, A comprehensive review of earthquake-induced building damage detection with remote sensing techniques. ISPRS J. Photogramm. Remote Sens. **84**, 85–99 (2013)
23. N. Charron et al., Automated bridge inspection using mobile ground robotics. J. Struct. Eng. **145**(11), 04019137 (2019)
24. J.P. Santos et al., Early damage detection based on pattern recognition and data fusion. J. Struct. Eng. **143**(2), 04016162 (2017)
25. Y. Gao et al., Auto-regressive integrated moving-average machine learning for damage identification of steel frames. Appl. Sci. **11**(13), 6084 (2021)
26. A. Radford, L. Metz, S. Chintala, Unsupervised representation learning with deep convolutional generative adversarial networks (2015). arXiv:1511.06434
27. X. Yang et al., Deep learning-based bolt loosening detection for wind turbine towers. Struct. Control Health Monit. **29**(6), e2943 (2022)
28. S.O. Sajedi, X. Liang, Uncertainty-assisted deep vision structural health monitoring. Comput.-Aided Civil Infrast. Eng. **36**(2), 126–142 (2021)
29. Y. Zheng et al., Multistage semisupervised active learning framework for crack identification, segmentation, and measurement of bridges. Comput.-Aided Civil Infrast. Eng. **37**(9), 1089–1108 (2022)
30. K. He et al., Mask r-cnn, in *Proceedings of the IEEE International Conference on Computer Vision* (2017), pp. 2961–2969
31. Z. Liu et al., Swin transformer: hierarchical vision transformer using shifted windows, in *Proceedings of the IEEE/CVF International Conference on Computer Vision* (2021), pp. 10012–10022
32. K.M. Mosalam, S. Günay, Progressive collapse analysis of reinforced concrete frames with unreinforced masonry infill walls considering in-plane/out-of-plane interaction. Earthq. Spectra **31**(2), 921–943 (2015). https://doi.org/10.1193/062113EQS165M

Structural Image Classification

6

As mentioned in Chap. 2, structural image classification is one of the most fundamental tasks in vision-based SHM. However, the introduction of AI technologies into the field of SHM is not straightforward. Therefore, in this chapter, the feasibility of applying AI methods in vision-based SHM is explored. This is mainly evaluated by the accuracy and efficiency of the trained AI models. ML and DL can achieve very accurate or promising results through training on a big dataset, but the performance may degrade on a smaller-scale dataset. However, the terms "big" and "small" of the scale of the dataset are ambiguous and subjective, because there is no explicit definition or specific threshold value to clearly separate them. It is also inappropriate and sometimes impossible to infer the results simply by examining the scale of the collected dataset without validation experiments. This issue is demonstrated for the DL performance related to the eight ϕ-Net benchmark vision classification tasks, introduced in Sect. 5.2.2, where the amounts of the labeled data in the eight ϕ-Net sub-datasets vary from 1,372 to 27,306 images. Compared with traditional studies in vision-based SHM, these datasets are considered somewhat "big". However, when compared to common CV benchmark datasets, e.g., ImageNet [1], MNIST [2], and CIFAR-10 [3], these eight datasets are "small" or even "very small". Therefore, to validate the feasibility of using DL methods in vision-based SHM tasks with these sub-datasets, comprehensive benchmarking experiments are required.

Conducting comprehensive numerical experiments on the eight sub-datasets of the ϕ-Net is highly resource-consuming. Before doing so, a preliminary study is performed first on a smaller-scale dataset for validating the effectiveness and applicability of the DL methods in vision-based SHM, where the datasets are manually and carefully selected from the original ϕ-Net sub-datasets. To alleviate the over-fitting issues (refer to Sect. 3.5) during the training while dealing with such small-scale datasets, two classical DL strategies, namely, DA and TL, are introduced. In addition, two strategies or patterns of TL, namely, feature extractor and fine-tuning, are explored. It is noted that most current DL-based image classification applications are based on the TL strategies instead of training from scratch with random initialization of the DL model parameters.

6.1   Data Augmentation

The data deficiency is usually thought of as one of the most troublesome issues in vision-based SHM. From experience in the CV domain, DA is a widely-used strategy to mitigate this deficiency in DL applications.

In DA, the number of raw input samples is increased via a series of affine transformations or other preprocessing operations, e.g., rotation, zoom, translation, flip, scale, whitening, and adding noise. Three typical examples are illustrated in Fig. 6.1. In general, DA of image data has two types: (i) *online* where the CNN model is trained with a processed batch of images via several forms of transformation (e.g., rotation and translation) at each training iteration and (ii) *offline* where new images are generated from transformation of the raw images and added into the original training set before the training process starts and then the CNN model is trained using this fixed and augmented set of images.

6.2   Transfer Learning Mechanism

In some engineering fields, AI, especially DL, has already achieved great success in recent years, but it heavily relies on large amounts of data, especially labeled (i.e., annotated) datasets. In real applications, sometimes useful "*big data*" is very rare or expensive to collect. Usually, the worst case is that only limited data can be collected. In order to mitigate the great dependency of the DL accuracy on the available data size and to maximize the usage of existing datasets, TL became a very effective tool in addition to the DA approaches, discussed in Sect. 6.1. Important definitions [4, 19], related to the subsequent discussion, are stated as follows:

(a) Rotation   (b) Zoom   (c) Flip

Fig. 6.1 Examples of DA

Definition 6.1 (*Domain*) A *domain*, $\mathcal{D}$, consists of two parts, namely, a feature space, $\mathcal{X}$, and a marginal distribution, $P(X)$, i.e., $\mathcal{D} = \{\mathcal{X}, P(X)\}$, with X representing the set of instances, i.e., $X = \{x_1, x_2, \ldots, x_n\}$, where $x_i \in \mathcal{X}$.

Definition 6.2 (*Task*) A *task*, $\mathcal{T}$, consists of a label space, $\mathcal{Y}$, and a prediction function, $f(\cdot)$, i.e., $\mathcal{T} = \{\mathcal{Y}, f(\cdot)\}$. Moreover, Y is the set of labels corresponding to X, i.e., $Y = \{y_1, y_2, \ldots, y_n\}$, where $y_i \in \mathcal{Y}$, and y_i is commonly expressed as $y_i = f(x_i)$.

Definition 6.3 (*Transfer Learning*) Given a *source domain*,[1] $\mathcal{D}_S$, and its learning task, $\mathcal{T}_S$, a *target domain*,[2] $\mathcal{D}_T$, and its *target task*, $\mathcal{T}_T$, the objective of the TL is to improve the prediction function, $f(\cdot)$, in learning the $\mathcal{T}_T$ in $\mathcal{D}_T$ using the knowledge from the $\mathcal{D}_S$ and its $\mathcal{T}_S$.

According to different conditions between the source and target domains and tasks, TL can be classified into three sub-types: inductive, transductive, and unsupervised [4]. In the *inductive TL*, $\mathcal{T}_T$ is different from $\mathcal{T}_S$, irrespective of whether $\mathcal{D}_S$ and $\mathcal{D}_T$ are the same or different. It usually requires labeled data available in $\mathcal{D}_T$. The inductive TL involves pre-training a model on the dataset in $\mathcal{D}_S$ and then fine-tuning this pre-trained model on the dataset in $\mathcal{D}_T$ with fewer labeled data. If no labeled data in $\mathcal{D}_S$ are available, a self-taught learning setting[3] can be adopted [5]. In the *transductive TL*, labeled data are available only in $\mathcal{D}_S$. There are two cases: (1) the feature spaces between domains are different, i.e., $\mathcal{X}_S \neq \mathcal{X}_T$, or (2) the feature spaces of the two domains are the same, but they have different marginal probability distributions of the data, i.e., $P(X_S) \neq P(X_T)$. In the *unsupervised TL*, $\mathcal{T}_T$ is different from but related to $\mathcal{T}_S$ and there are no labeled data in either $\mathcal{D}_S$ or $\mathcal{D}_T$. While linking to the vision-based SHM, for example, the eight benchmark detection tasks in the ϕ-Net belong to the inductive TL, which requires a few labeled data in $\mathcal{D}_T$.

There are four common approaches that can be applied according to the requirements of the knowledge transfer from $\mathcal{D}_S$ to $\mathcal{D}_T$. These are listed as follows (more details can be found in [4]):

1. **Instance transfer**: Re-weight some labeled data in $\mathcal{D}_S$ for use in $\mathcal{D}_T$.
2. **Feature representation transfer**: Find better feature representations to reduce the difference between domains.
3. **Parameter transfer**: Find shared parameters or priors between domains.
4. **Relational knowledge transfer**: Build mapping of relational knowledge between domains.

---

[1] Source domain refers to the original task and data distribution that the model is trained on.

[2] Target domain refers to the new task and data distribution that the model will be used for.

[3] As discussed in [5], an example of self-taught learning consists of two stages. First, learn a representation using only unlabeled data. Then, apply this representation to the labeled data and use it for the classification task. Once the representation has been learned in the first stage, it can then be applied repeatedly to different classification tasks.

6.2.1 TL in Vision Classification Tasks

In recent years, significant efforts have taken place in applying TL to visual classification problems, especially in deep CNNs. Usually, one major issue in training deep CNNs is that, with the increasing depth of the network, the model becomes very complex requiring excessive time and computational resources. In addition, with the deficiency of training data, over-fitting and difficulties in training[4] emerge as other problematic issues. Thus, as discussed above, TL can improve learning which model to use and greatly improve the efficiency of the training process.

Compared with general TL approaches, the details of using TL in CNNs are somewhat different. In the CNN, the parameters in the *shallow layers* represent low-level features, such as color, texture, and edges[5] of objects, while parameters in the *deep layers* attempt to capture more complicated and abstract high-level features [6]. Therefore, the major objective of TL in CNNs is to make use of parameters in a well-trained model from the dataset in the source domain to improve the training using the dataset in the target domain. If the two datasets are similar, some low-level features of the CNN model are expected to be similar, which can be shared, and the high-level features can be tuned by TL. Thus, usually in the CNN, TL plays the roles of feature representation transfer and parameter transfer [7].

The original training dataset for pre-trained models, e.g., VGGNet, InceptionNet, and ResNet, is the ImageNet [1], which includes thousands of images related to buildings, bridges, pillars, walls, etc., belonging to the civil (e.g., structural) engineering field. Therefore, for benchmark detection tasks in the domain of Structural Engineering, the source and target domains are assumed to be related. Moreover, from the perspective of task objectives, for example, the component type classification between beam, column, wall, and others is a similar but easier task than that of the ImageNet classification problem, since it reduces the 1,000 classes [1, 8] to only 4 classes. According to the domain knowledge of Structural Engineering and the characteristics of the parameter transfer, low-level features in the CNN might be useful to be adapted to new tasks, such as *texture* might be beneficial to spalling condition identification and *pixel change in directions* might provide information for damage type though understanding the cracking morphology. Moreover, mid- to high-level features can be learned by retraining some parts of the network. This is illustrated in Fig. 6.2 using a hypothetical CNN made of five Conv blocks, where the first three are retained (i.e., their parameters are kept frozen) via TL, while the last two are retrained. In addition, the mechanism of TL can be explained through the feature transfer. As mentioned above, the CNN's parameters represent some features and training on the source domain plays the role of feature learning. Accordingly, retraining on the target domain is actually transferring some features from the source domain to the target domain and thus learning

---

[4] Training difficulties include training accuracy saturation, where increasing the size of the training set cannot improve the limitations of the model, e.g., the accuracy cannot be improved when using a *linear model* to classify data that is *separable in only a nonlinear way*.

[5] Edge is a boundary or transition region with a sharp change of pixels within neighboring areas.

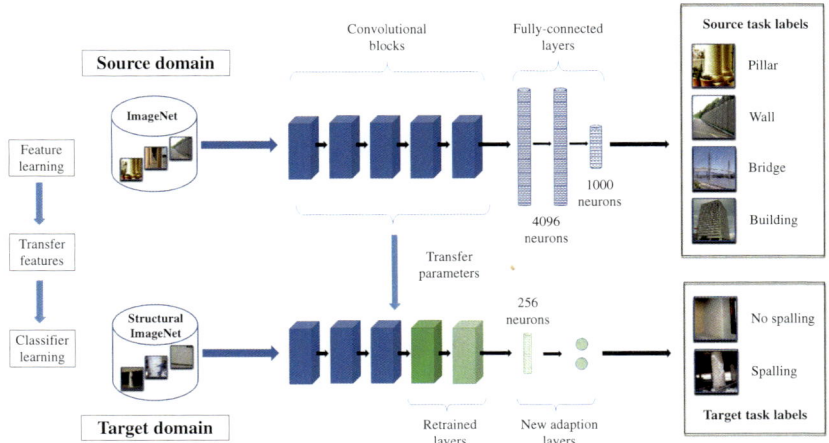

Fig. 6.2 TL illustration for the spalling condition detection

the classifier for the target domain. During the implementation, these pre-trained parameters can relax the procedure of parameter initialization, which might help deal with the training accuracy saturation issues occurring when using small datasets for training.

In general, there are two different strategies or patterns while conducting TL in deep CNNs: (i) *feature extractor* and (ii) *fine-tuning*. The configurations and procedures for training a VGGNet [9], as an example, using the feature extractor and the fine-tuning strategies are illustrated in Fig. 6.3. These two strategies are discussed in detail in the next two sub-sections.

6.2.2   Feature Extractor

The feature extractor (sometimes referred to in the literature as "off-the-shelf CNN" [10] or "bottleneck feature for CNN" [11]), as its name reflects, extracts the features from the data, i.e., the images. In this case, the Conv operations are only performed once in the feed-forward procedure. The outputs of the final layer before the FC-layers are taken as the bottleneck feature (i.e., the features to be extracted by the present model through the TL approach) of the training data and then multiple classifiers can be applied in the classification problem, refer to the example of the feature extractor in VGGNet, Figs. 6.3a and 6.4.

The advantage of implementing the feature extractor in TL is that it greatly decreases the training time and number of epochs due to already extracted features. Instead of training a large number of Conv layers for multiple epochs, only features are fed into a shallow NN stacked by the FC-layers or even linear classifiers such as a SVM, which is faster to train. Although the Conv operation is computationally expensive and still needs a long time to extract the features by passing the input (e.g., images) through the CNN layers with pre-

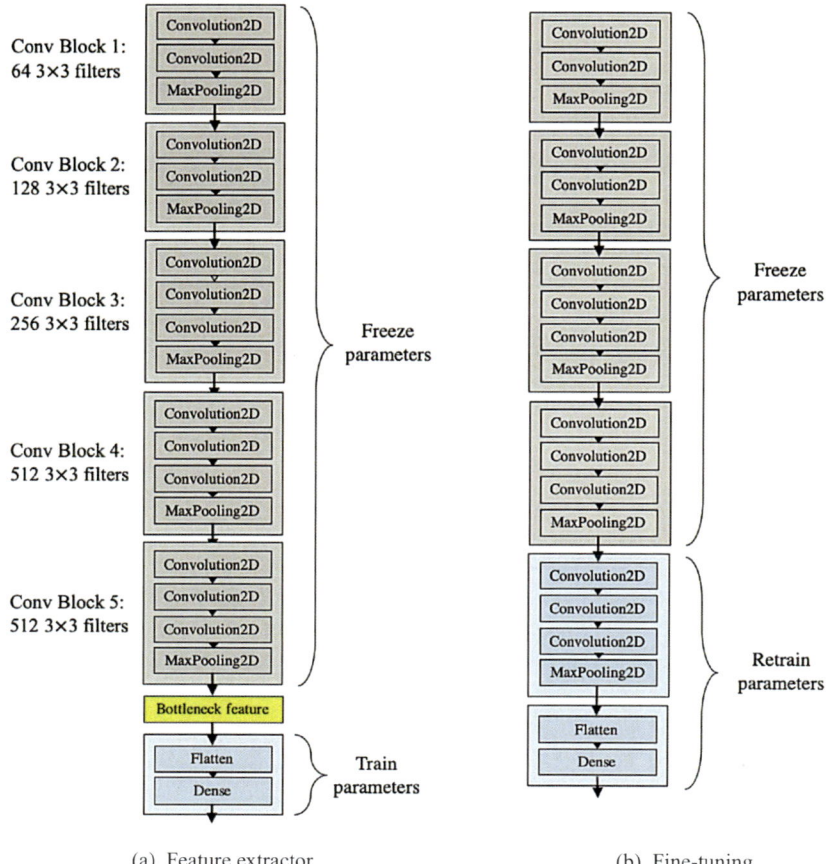

(a) Feature extractor   (b) Fine-tuning

Fig. 6.3 TL strategies applied to VGG-16

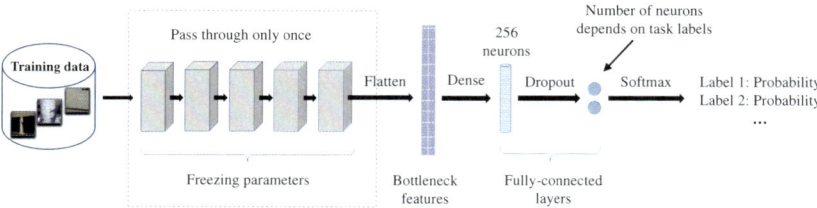

Fig. 6.4 Procedure of the feature extractor of the TL

trained filter parameters and weights, this procedure is only performed once, refer to Fig. 6.4. Thus, all feature extractor procedures require much less computing time than other methods not involving TL. However, one shortcoming of the use of the feature extractor is that online DA would not work for this procedure since the features are only extracted once, but the offline DA can be applied to increase the amount of data, which leads to the undesirable increases of the data storage demands and extraction time of the features, compared with the online DA.

6.2.3 Fine-Tuning

Compared with the feature extractor, fine-tuning retrains some parts of the CNN. The fine-tuning retrains the Conv blocks instead of just retraining a few Conv layers, refer to Figs. 6.3b and 6.5. As mentioned above, Conv blocks in the shallow layers represent low-level features, which might be similar for both the source and target domains, and the objective of fine-tuning the last several Conv blocks is to adjust the mid- to high-level features to the target domain.

In fine-tuning, DA tricks can be applied in the training to avoid the over-fitting problem. The trade-off is that the new augmented (i.e., generated) images go through a forward propagation at each training epoch, which is time-consuming due to the expensive convolution computations. However, the parameters in some parts of the Conv layers are fixed where their backward calculations are skipped and only parameters in the unfixed layers are adjusted by the gradient descent algorithm in the backpropagation process. Thus, the use of fine-tuning is still less time-consuming than training a regular CNN from scratch. Moreover, as in many DL applications, the GPU-based parallel computing alleviates the excessive computational time demand.

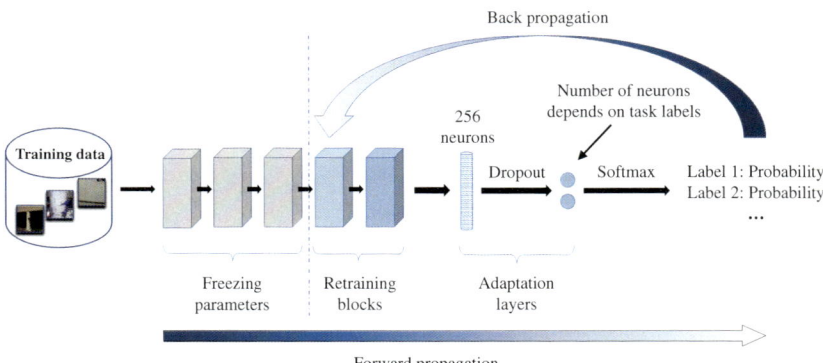

Fig. 6.5 Procedure of the fine-tuning of the TL

Theoretically, fine-tuning is treated as one type of better parameter initialization (refer to Sect. 4.1.2) approaches, where the parameters from pre-training on the source dataset usually have better initialization than the random Gaussian one. This is usually reflected in the observations that fine-tuning takes fewer epochs to converge and has more stable performance during training. More details are introduced and discussed in Sect. 9.6.1.

6.3   TL from ImageNet to a Small-Scale Structural Dataset

Typically, a high-resolution image contains more information with large sizes of pixel matrices. However, the large image occupies more memory and requires a large *network input size*, which is limited by the computer hardware. It is to be noted that the network input size is typically defined with 4 variables as (N, W, H, C). It is an internal property of the network itself, where N is the number of data (i.e., images) fed to the network and W, H, and C correspond to the respective designated image width, height, and the number of channels, e.g., 3 for RGB images. On the contrary, the small image costs less computation resources but it might lose some information and possibly produce misleading results. In the case of vision-based SHM, low-resolution images may make it difficult to identify minor damages, which can lead to unreliable results for damage evaluation. Therefore, the process of vision-based SHM is usually a trade-off between computational efficiency and recognition accuracy. The selection of the network input size is part of the network design. The common input size of a gray-scale image is $W \times H = 224 \times 224$ for the input layer used in most popular deep CNN architectures, e.g., VGG-16, VGG-19 [9], and ResNet-50 [12]. If the images are either larger or smaller than 224×224, they should be re-scaled or cropped to exactly fit the 224×224 size to be able to use these existing networks. Thus, it is important to investigate what is an appropriate network input size for deep CNNs in vision-based SHM problems.

While conducting TL with the fine-tuning strategy, the ratio of the retrainable parts of the deep CNN is a hyper-parameter to be determined before the onset of the network training. Even though in many CV applications, a deep CNN is retrained thoroughly by fine-tuning the parameters in all Conv layers during each epoch, sometimes this strategy is inapplicable due to limited computational resources or may lead to severe over-fitting issues due to limited amount of training data. Thus, only retraining the partial network reduces the number of trainable parameters and decreases the model complexity, which may alleviate the over-fitting problem. However, only retraining a small portion of the network may also lead to under-fitting issues (refer to Sect. 3.5), since some basic features in the early shallower layers may differ from the source domain to the target domain, which also need to be fine-tuned, in this case, similar to the later deeper layers. Therefore, the effect of the ratio of the network part to be retrained is an important aspect to be carefully investigated as it may significantly affect the trained network performance when using TL.

There exist many network architectures, as introduced in Sect. 4.3, that can be applied in DL, such as VGGNet, InceptionNet, and ResNet. Moreover, different networks may be more suitable for different problems and tasks. Thus, the architecture of the utilized network is another important hyper-parameter to be accurately evaluated. In this chapter, exploring the feasibility of deep CNN models for structural image classification is introduced. For this purpose, the following three important questions are first studied through experiments on small-scale datasets, where the DL models are pre-trained from ImageNet [1] for the general low-level features.

1. What is a suitable *input size* of the network?
2. What is a good *retrain ratio* while fine-tuning the pre-trained model?
3. What is a better *network configuration* or architectural design?

The implementation of these experiments was conducted on the TensorFlow platform.[6] The used computational hardware was Alienware 15R3 with a single GPU: Nvidia Geforce GTX 1060 and its Central Processing Unit (CPU): Inter(R) Core i7-7700HQ @2.80 GHz with RAM: 16.0 GB.

6.3.1 ϕ-Net-2000

Since conducting comprehensive verification experiments on the entire datasets of the ϕ-Net is highly resource-consuming, a preliminary investigation using a mini-version of ϕ-Net is explored in this section. Specifically, the scope is only limited to four classification tasks (component type, spalling condition, damage level, and damage type). Moreover, simplifications are introduced in the class labels for the component type and the damage level tasks. These modifications are listed as follows:

- In the component type task, the original labels "beam" and "column" are combined to a broad class "beam/column" and the component label "other component" in the ϕ-Net is not considered herein. Thus, this task is reduced to a binary classification: *beam/column* versus *wall*.
- In the damage level task, the original labels "moderate damage" and "heavy damage" are combined into a broad class "moderate to heavy damage". Thus, this task is reduced to a three-class classification: *undamaged state*, *minor damage*, and *moderate to heavy damage*.

---

[6] TensorFlow is an end-to-end ML platform. For details, refer to https://www.tensorflow.org/.

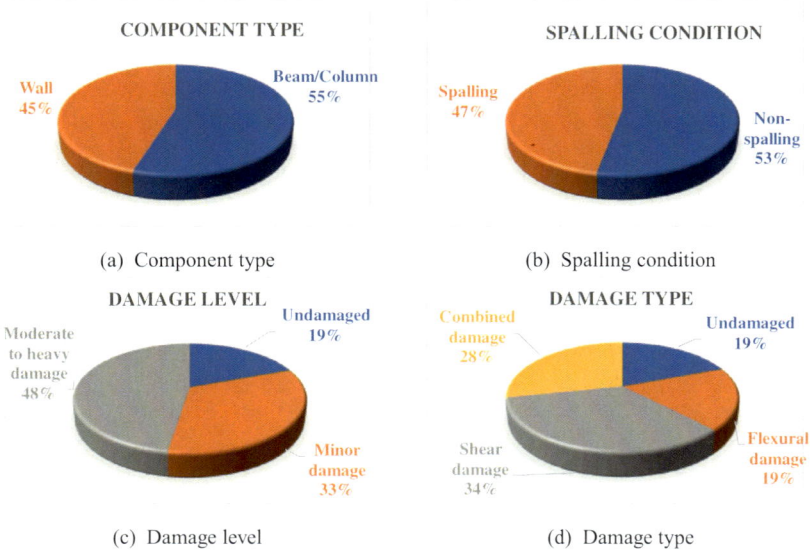

(a) Component type (b) Spalling condition

(c) Damage level (d) Damage type

Fig. 6.6 Label ratios in the ϕ-Net-2000 dataset

In the TL experiments with a small-scale dataset, 2,000 images from the ϕ-Net dataset are selected, pre-processed, and relabeled manually based on the above modifications, to form the required small-scale dataset (ϕ-Net-2000). It is noted that all 2,000 images have four labels with respect to the considered four detection tasks. The label statistics for each task are shown in Fig. 6.6. For the shown two binary tasks, the label ratios are 55/45% and 53/47% and these are judged to be balanced datasets. For the other two multi-class tasks, since there are no dominating labels but only fewer data for the undamaged cases (e.g., no crack and no damage) and considering the fact that a multi-class labeled dataset is generally expensive to collect, the shown distributions in Fig. 6.6 are acceptable. Moreover, although it may not be desirable to detect an undamaged structure as damaged, it is unacceptable and dangerous to do the converse, that is, detect a damaged structure as undamaged. Therefore, using such ratios (with less percentage of the undamaged labels) for the multi-class detection tasks is reasonable from a practical (i.e., conservative) engineering perspective.

For supervised learning classification problems, the labeled data is split into two sets, one for training the model and the other for testing its generality. For the ϕ-Net-2000 (labeled images = 2,000), the ratio of the training and test sets is empirically set to 4:1, i.e., 1,600 images for training and 400 ones for test. To avoid loss of generality for the prediction model, the label ratios (Fig. 6.6) in both training and test sets are kept the same, i.e., all tasks in both sets have the same data distributions.

Table 6.1 Details of the network input size experiment

Large	Medium	VGG	Tiny
448×448	256×256	224×224	128×128

6.3.2   Input Size

For simplicity, only the feature extractor approach is applied to investigate the influence of the network input size. In order to determine the appropriate input size, four network input (i.e., image) size[7] categories are designed and re-scaled for computer experiments, namely, *VGG size* (the size used in the prototype VGGNet), *large size*, *medium size*, and *tiny size*, refer to Table 6.1 for details of these empirical choices. For example, when using the tiny size in the network input layer, all images are re-scaled from their raw image size to the fixed size of 128×128 to match this network input size.

6.3.2.1 Network Configuration

In the experiments of the network input size, VGGNet variants are introduced, which are modified to accommodate three additional input sizes (large, medium, and tiny), as listed in Table 6.1. From the perspective of CV, structural images including damage patterns are much different from natural objects in ImageNet [1]. In TL, usually network modifications are employed due to different image statistics between the source and the target domains, such as the type of objects and typical viewpoints [13]. As a result, several modifications are made to replace the original FC-layers with the so-called adaptation FC-layers (denoted, for short, as the *adaptation layers*) in the target domain including the following: (1) different neurons and different numbers of FC-layers are applied after the Conv blocks and (2) the shape of the *softmax* layer (the final FC-layer uses *softmax* as the activation function) is altered depending on the number of classes to be identified. The dimensions of the adaptation layers usually depend on experience and cross-validation (refer to Sect. 3.6).

Compared with training using the dataset of the prototype VGG-16, i.e., with over 1.3 million images from the ILSVRC-2012 dataset, one subset of the ImageNet [1, 9], the ϕ-Net-2000 dataset only has 2,000 well-defined and properly labeled images. Thus, in the absence of "big data" and to avoid severe over-fitting issues, fewer numbers of neurons and fewer FC-layers are placed after the last pooling layer. Preliminary computations with 256, 1024, and 4096 neurons in the first adaptation layer resulted in insignificant differences. However, over-fitting took place in cases with large numbers of neurons. Thus, considering the cost of computations, an adaptation layer with 256 neurons is a reasonable choice. In the final design, two adaptation FC-layers are used where the first has 256 neurons and the second is a *softmax* layer, which is adjusted according to the number of classes to be

---

[7] For consistency, the network input size is taken as the used image size.

Table 6.2 VGG-type network configurations used with the ϕ-Net-2000 dataset

Network architecture			Output shape of models with different input sizes			
Block	Layer type	Filter size (#)	VGG size	Large size	Medium size	Tiny size
Input	Input/Image	–	$(N, 224, 224, 3)$	$(N, 448, 448, 3)$	$(N, 256, 256, 3)$	$(N, 128, 128, 3)$
Conv Block 1	Convolutional	3×3 (64)	$(N, 224, 224, 64)$	$(N, 448, 448, 64)$	$(N, 256, 256, 64)$	$(N, 128, 128, 64)$
	Convolutional	3×3 (64)	$(N, 224, 224, 64)$	$(N, 448, 448, 64)$	$(N, 256, 256, 64)$	$(N, 128, 128, 64)$
	Max Pooling	–	$(N, 112, 112, 64)$	$(N, 224, 224, 64)$	$(N, 128, 128, 64)$	$(N, 64, 64, 64)$
Conv Block 2	Convolutional	3×3 (128)	$(N, 112, 112, 128)$	$(N, 224, 224, 128)$	$(N, 128, 128, 128)$	$(N, 64, 64, 128)$
	Convolutional	3×3 (128)	$(N, 112, 112, 128)$	$(N, 224, 224, 128)$	$(N, 128, 128, 128)$	$(N, 64, 64, 128)$
	Max Pooling	–	$(N, 56, 56, 128)$	$(N, 128, 128, 128)$	$(N, 64, 64, 128)$	$(N, 32, 32, 128)$
Conv Block 3	Convolutional	3×3 (256)	$(N, 56, 56, 256)$	$(N, 128, 128, 256)$	$(N, 64, 64, 256)$	$(N, 32, 32, 256)$
	Convolutional	3×3 (256)	$(N, 56, 56, 256)$	$(N, 128, 128, 256)$	$(N, 64, 64, 256)$	$(N, 32, 32, 256)$
	Convolutional	3×3 (256)	$(N, 56, 56, 256)$	$(N, 128, 128, 256)$	$(N, 64, 64, 256)$	$(N, 32, 32, 256)$
	Max Pooling	–	$(N, 28, 28, 256)$	$(N, 56, 56, 256)$	$(N, 32, 32, 256)$	$(N, 16, 16, 256)$
Conv Block 4	Convolutional	3×3 (512)	$(N, 28, 28, 512)$	$(N, 56, 56, 512)$	$(N, 32, 32, 512)$	$(N, 16, 16, 512)$
	Convolutional	3×3 (512)	$(N, 28, 28, 512)$	$(N, 56, 56, 512)$	$(N, 32, 32, 512)$	$(N, 16, 16, 512)$
	Convolutional	3×3 (512)	$(N, 28, 28, 512)$	$(N, 56, 56, 512)$	$(N, 32, 32, 512)$	$(N, 16, 16, 512)$
	Max Pooling	–	$(N, 14, 14, 512)$	$(N, 28, 28, 512)$	$(N, 16, 16, 512)$	$(N, 8, 8, 512)$
Conv Block 5	Convolutional	3×3 (512)	$(N, 14, 14, 512)$	$(N, 28, 28, 512)$	$(N, 16, 16, 512)$	$(N, 8, 8, 512)$
	Convolutional	3×3 (512)	$(N, 14, 14, 512)$	$(N, 28, 28, 512)$	$(N, 16, 16, 512)$	$(N, 8, 8, 512)$
	Convolutional	3×3 (512)	$(N, 14, 14, 512)$	$(N, 28, 28, 512)$	$(N, 16, 16, 512)$	$(N, 8, 8, 512)$
	Max Pooling	–	$(N, 7, 7, 512)$	$(N, 14, 14, 512)$	$(N, 8, 8, 512)$	$(N, 4, 4, 512)$
FC-layer	Flatten	–	$(N, 25{,}088)$	$(N, 100{,}352)$	$(N, 32{,}768)$	$(N, 8{,}192)$
	Dense	–	$(N, 256)$	$(N, 256)$	$(N, 256)$	$(N, 256)$
	Dense	–	$(N, 2/2/3/4)$	$(N, 2/2/3/4)$	$(N, 2/2/3/4)$	$(N, 2/2/3/4)$

identified in the classification task at hand. Compared with the prototype of VGG-16, these new configurations make the network more computationally efficient and compatible with the adopted TL to circumvent the data deficiency issue.

The detailed configurations including the filter sizes and output shape for each layer of these VGG-type networks (variants of the VGG-16) adapted for the ϕ-Net-2000 dataset are listed in Table 6.2, where N denotes the number of data input to the network, i.e., 2,000 colored images with an initial input depth of 3 for the RGB channels. The notion of the outputs $2/2/3/4$ corresponds to the number of classes in the four adopted classification tasks, i.e., for the respective tasks in Fig. 6.6a–d.

6.3.2.2 Training Configuration

As mentioned above, the network input size experiment is conducted only using the feature extractor by performing the convolution computations once, refer to Fig. 6.4. For computational efficiency, no DA tricks are used in the feature extractor. Instead, dropout is applied where the dropout rate is typically taken as about 0.5. However, considering that the

VGG-type model is pre-trained on a large dataset with over 14 million images and, on the other hand, the training dataset in this study has only 1,600 images, a heavier dropout ratio of 0.7 is used to avoid over-fitting.

The training is carried out by optimizing the categorical cross-entropy loss, Eq. 3.6, using *Adam optimization* [14] based on the backpropagation algorithm, Eqs. 4.7 and 4.8, which only requires first-order gradients with little memory requirements. Three key parameters in Adam optimization are β_1, β_2, and ϵ.[8] They are respectively assigned the empirical values 0.9, 0.999, and 10^{-8}. To avoid making a rapid convergence toward the pre-trained parameters, the learning rate η (Eq. 4.8) is set to 10^{-4}. Considering the potential limitations in the available computational resources in practical situations, the mini-batch approach is applied and the batch size is taken as 32. The number of training epochs is set to 25 and is allowed to increase if the results do not converge (such as for the damage level and damage type tasks). For the feature extractor, the initialization of the parameters is conducted only in the FC-layers. The weight terms W are sampled from a Gaussian distribution with zero mean and 10^{-2} variance.[9] The adopted evaluation criteria are the overall accuracy, Eq. 3.12, and the CM, Fig. 3.1.

6.3.2.3 Results and Analysis

Figure 6.7a–d plot the histories of the accuracy of both training and test processes for all four detection tasks. It is noted that in damage level and damage type tasks, more training epochs are needed for convergence, requiring an upper limit of 50 epochs instead of the specified 25 used for the other two binary tasks. The first observation is that computations converge quickly at the very beginning, which is due to the fact that extracted features are fixed and the training process is only working on different nonlinear combinations of the features to reduce the loss with a small size FC-layers. However, with poor or non-optimal features, it is also possible to converge to local minima. It is usually difficult for the adopted algorithm to escape these local minima and such a situation should be carefully handled. For example, preventing the loss function from getting stuck in local minima is provided by the momentum value in Adam optimization algorithm. Conceptually, this takes place by providing a basic impulse to the loss function in a specific direction, which helps the function avoid narrow or small local minima.

The model using the "tiny" input size performed worse than the other three due to lower resolution, causing a lack of information. This phenomenon can also be attributed to insuf-

---

[8] β_1 is a tuning hyper-parameter for the exponentially decaying rate for the momentum term. β_2 is another tuning hyper-parameter for the exponentially decaying rate for the velocity term. ϵ is not a tuning hyper-parameter and its very tiny value avoids any division by zero error in the implementation. For more details, refer to [14].

[9] For normal random variable $W = \sigma_W Z + \mu_W$ where μ_W is its mean and σ_W is its std or σ_W^2 is its variance. The usual notation is that $W \sim N\left(\mu_W, \sigma_W^2\right)$. Note that Z is the standard (having zero mean and unit variance) normal (Gaussian) random variable, i.e., $Z \sim N(0, 1)$.

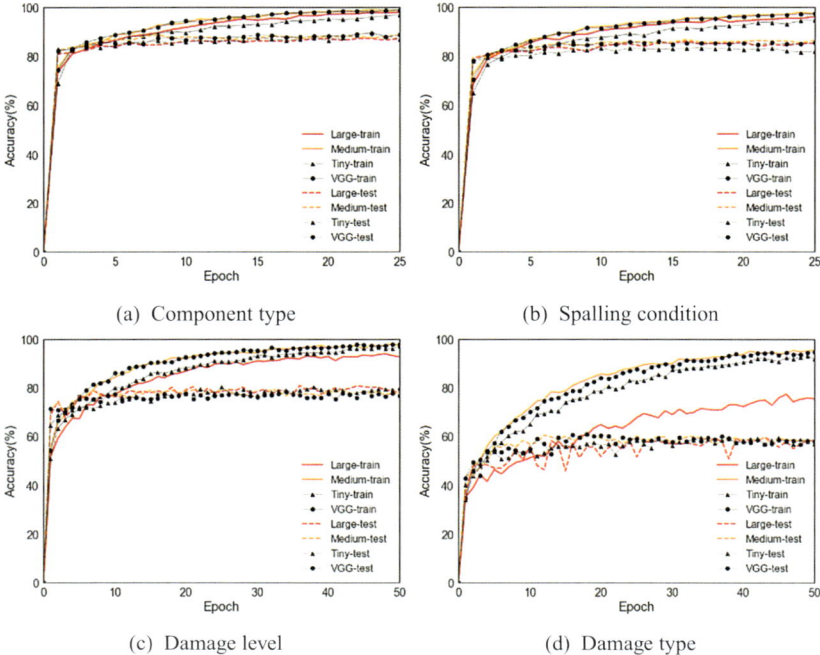

(a) Component type   (b) Spalling condition

(c) Damage level   (d) Damage type

Fig. 6.7 Accuracy histories for the four recognition tasks using feature extractor

ficient features, which makes it harder to achieve good training accuracy with fewer training epochs, let alone lower test accuracy. On the contrary, the models using the "large", "medium", or "VGG" sizes performed well and also similarly for the binary tasks. However, inputting an image with higher resolution with the "large" input size did not perform better than the "medium" or "VGG" sizes for the multi-class tasks. This observation is attributed to the fact that more parameters need to be trained with a large model caused by a large input. The VGG size seems to be the relatively optimal one amongst the four considered sizes for all tasks. Therefore, it is concluded that the VGG-type network receiving an image with a size of 224×224 can provide enough information to handle the considered four tasks, with which the CNN model can achieve good and stable results. The model with too large (i.e., costly) or too small (i.e., less informative) input size would probably perform worse. Table 6.3 lists the best accuracy for the four detection tasks from the VGG size input. Clearly, the results for the binary classifications are very promising considering the short computational time and the possibility of using a small dataset. Meanwhile, even though there are severe over-fitting problems in the multi-class tasks, compared with the lower bound accuracy values of 33.3 and 25.0% using random guesses for the respective tasks with three and four classes, the obtained accuracy values of 77.0% for the damage level task (3 classes) and 57.7% for the damage type task (4 classes) are reasonably acceptable considering the small scale of the used dataset.

Table 6.3 Best accuracy results (%) in the network input size experiment

Task	# of classes	Training accuracy	Test accuracy
Component type	2	99.1	88.8
Spalling condition	2	97.1	85.4
Damage level	3	97.5	77.0
Damage type	4	94.2	57.7

In general, the above results reflect some degree of over-fitting. For the two binary tasks, it is observed that there is about 10% over-fitting (obtained as the difference in accuracy between the training and test). On the other hand, over-fitting gap values are around 20% and 37% for the three-class and four-class tasks, respectively. In all tasks, the more difficult the task, the more severe the over-fitting. Since training accuracy in all tasks can reach over 94%, one concludes that the main reason for the over-fitting is due to less training data because the used CNN architectures already hold enough complexity to classify the problem accurately. Thus, the CNN loses the generality to extend the learned parameters from a small training dataset to the unseen test dataset. In other words, this is equivalent to the generality issue of the extracted features, attributed to the fixed Conv blocks. Therefore, even though low-level features of the VGGNet might be common and shareable, due to the difference between the source and the target domains, the high-level features from the source domain could be incompatible with those of the target domain. Thus, there are three recommended ways to improve the performance of the model to be pursued in future studies: (1) increase the training dataset, (2) use the same architecture for feature extraction but in addition fine-tune some parts of the CNN, and (3) change to a better architecture. As mentioned above, additional labeled training data (option 1) are usually expensive to obtain. Therefore, fine-tuning (option 2) and implementing other network architectures (option 3) are more feasible choices.

In *predictive analytics*, the overall accuracy for prediction alone may not be very reliable if the dataset is imbalanced or there are more than two classes in the dataset. On the other hand, the CM, introduced in Sect. 3.4.2, is a better way to evaluate the results. Therefore, normalized CM (note that the normalization of the CM entries is performed with the number of predictions of each class making the sum of each row $= 1.0$ with possible slight round-off errors) is computed for each classifier of the VGG size input experiment, which presents the probability of correct and incorrect predictions with values broken down for each class. From the CMs shown in Fig. 6.8, both binary tasks performed quite well with high correct prediction accuracy and low misclassification error. However, in the spalling condition task, 20% of "spalling" images are misclassified as "non-spalling", a finding that should be carefully assessed and possibly attributed to some deficiency in the detection of the change of texture (indicative of spalling condition) in these misclassified images. In the damage level task, "undamaged" and "moderate to heavy damage" states achieved better results

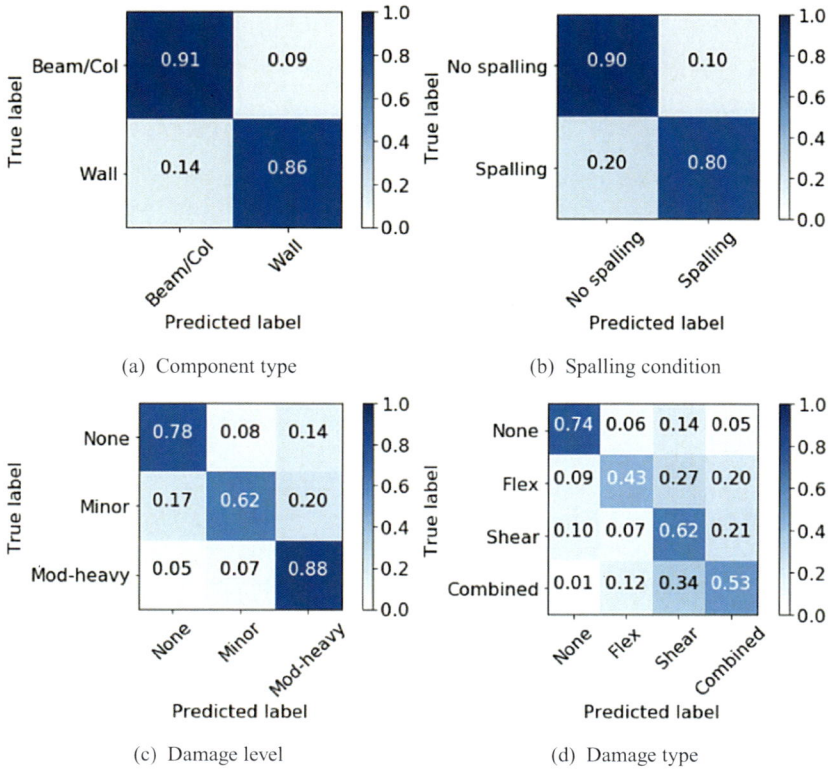

(a) Component type   (b) Spalling condition

(c) Damage level   (d) Damage type

Fig. 6.8 Normalized CMs of the test prediction using feature extractor

than the "minor damage" state, which can be explained by the fact that the existence of an edge is more obvious in the two extreme classes compared with the intermediate class. Finally, in the damage type task, the prediction of the "undamaged" state still behaved well but its accuracy slightly decreased due to the increase in the number of classes from 3 to 4. Moreover, because the "flexural damage", "shear damage", and "combined damage" classes have some similarities, which may confuse the classifier, they are more prone to misclassification.

Regarding the computational efficiency, even though similar CNN architectures are used in the four considered tasks (Table 6.3) for the TL using feature extraction, the number of neurons in the FC-layers differs due to different network input size and number of classes in each task. Because of the dimension shrinkage of the FC-layers with fewer classes (maximum number of classes in the considered four tasks is only 4), the difference in the parameter numbers for different tasks (more precisely, for different number of classes to be classified) does not significantly affect the total number of parameters, resulting in less impact on the computational time for these four tasks, but the input size does. It is understandable that in the feature extractor, where the Conv blocks are used for feature extraction, the larger

Table 6.4 Average computational times for the considered four tasks in seconds for different network input sizes (numbers in parentheses are the ratios with respect to the VGG size)

Procedure	Large	Medium	VGG	Tiny
Feature extraction	219.5 (3.3)	83.5 (1.3)	65.9 (1.0)	35.2 (0.5)
Training per epoch	3.98 (3.2)	1.52 (1.2)	1.26 (1.0)	0.63 (0.5)

the input is, the more the features to be extracted leading to a large increase in the number of neurons in the first FC-layer. For example, the number of parameters in the "large" size is nearly 3 times that in the "VGG" size and 23 times that in the "tiny" size. Thus, as listed in Table 6.4 for both feature extraction time and the training time per epoch following the feature extraction phase, the "large" size indeed costs significantly more computational time (more than 6 times) than that of the "tiny" size, with a significant portion of the total computational time consumed in the Conv operations.

In summary, the detection results through the feature extractor are quite promising even with some misclassification. The results indicate the great generalization of the VGG-type architecture and the recognition accuracy can clearly be improved by increasing the training data size or by using more parameters in the CNN architecture through fine-tuning. A significant portion of the computational time is consumed in the feature extraction procedure. However, it is only performed once for TL with feature extractor compared with training a vanilla CNN where the feature extraction is performed for all epochs, leading to a very fast training for the classifiers (Table 6.4). Thus, in large SHM projects, the feature extractor can be used in preliminary analysis before running costly computations on the complete CNN.

6.3.3   Retrain Ratio

Continuing with the VGG-type networks using an input size of $W \times H = 224 \times 224$, more characteristics of the TL are explored in this section. Instead of fine-tuning by layers, the model, consisting of five Conv blocks (Table 6.2), is retrained via these Conv blocks. Moreover, the experiment is conducted by gradual comparisons, such as starting from only retraining the last block and retaining the first four, then retraining the last two blocks and retaining the first three, and so on. Considering the small dataset, retraining more Conv blocks leads to a closer resemblance to the case of training the whole network from scratch, which is more susceptible to over-fitting. Meanwhile, since the parameters of part of the CNN are adjusted by backpropagation, the online DA is applied to reduce the over-fitting. This is beneficial to reduce the storage demand required for the offline DA used in the feature extractor. Refer to Sect. 6.1 for related discussion about the online and offline DA.

6.3.3.1 Training Configuration

The same training and test datasets, as for the previous network input size experiment using the feature extractor, are used for the present fine-tuning experiment. Different from the feature extractor approach, for each epoch, the training process goes through the parameters in the few deeper "unlocked" Conv blocks and updates these blocks once. An online DA approach, coupled with a mini-batch, refer to Sect. 4.1.3, is applied. During training, whenever a new batch of data arrives, all images in that batch are passed into a *transformation unit*, which contains random choices for zoom in/out, flip horizontally or vertically, rotate with random angles within five degrees, and translate in random directions (up/down/left/right). After such an online DA process, the input image is now a newly transformed image. These online DA tricks can generate more new data with more training epochs, which proved to be an effective approach to dealing with the over-fitting issues. Moreover, the online DA approach can save significant data storage memory requirements compared with the offline one, because the generated images only exist during the training process. After each training epoch, the generated images with this DA are replaced by newly generated ones for the next epoch, i.e., new data are created with an increasing number of training epochs to reduce the over-fitting of the model. However, these *generated data* are highly correlated to the original ones, without introducing new information. Therefore, for better and more reliable performance, more *real data* are required to be collected and more advanced DA methods should be developed.

In this experiment, some parameters, e.g., W in Eq. 6.1, are re-tuned in the Conv blocks. To avoid ruining the parameters during the retraining from those of the pre-trained model, the learning schema differs from that of the feature extractor experiment. Herein, the recommended optimization method is to use the SGD *with momentum* considering a small learning rate [15]. Aiming to minimize the loss function L (taken as the categorical cross-entropy Loss, Eq. 3.6), SGD updates the model parameters W by moving in the direction opposite (notice the negative term in Eq. 6.1) to the gradient of the loss function L with respect to W, i.e., $\nabla_W L$. In that regard, the moving step size is controlled by the learning rate, η. Moreover, another momentum hyper-parameter, μ, is added in the update step of the W parameters, Eq. 6.1. Analogous to the momentum in physics (defined as the product of an object's mass and its velocity.), this second momentum term works as a friction mechanism to avoid a rapid parameter change and helps the model achieve a better convergence with the momentum variable μ acting as the coefficient of friction applied to the change of the parameters W, i.e., ΔW in Eq. 6.1.

$$W \leftarrow W - \eta \cdot \nabla_W L + \mu \cdot \Delta W. \tag{6.1}$$

6.3.3.2 Results and Analysis

The histories of the loss are important indicators to evaluate the convergence and to learn about the performance of the model. As shown in Fig. 6.9a–d, with more parameters, the fine-tuning model converges after more rounds of training than that of the feature extractor (compare the maximum number of epochs on the horizontal axes of the plots in Fig. 6.9, i.e., 100, to those of Fig. 6.7, i.e., 25 and 50). It is noted that in Fig. 6.9a for component type and Fig. 6.9c for damage level, the test accuracy is slightly higher than the training accuracy for the case of retraining two blocks, which is explained by the effect of the dropout. When the dropout is applied, during training some neurons are randomly dropped and during testing no neurons are dropped, which means that the trained network is less complex than the complete one used for the test. This is analogous to *making network learning hard to make testing easier*. Considering the over-fitting issues, fine-tuning one Conv block performs well (the difference between training and test accuracy is small) which implies that no over-fitting occurs and slightly over-fitting takes place when fine-tuning two blocks (larger difference between training and test accuracy, especially for the four-class classification task, Fig. 6.9d). The over-fitting issues in retraining two blocks are due to the lack of data since more parameters need to be determined. Thus, there is no need to retrain more than two blocks, e.g., three or more blocks in this case, which is expected to have more severe over-fitting problems.

The loss is usually not used as a dominating metric (e.g., to estimate the overall accuracy) to evaluate the goodness of the model and a few discussions are available about the value of the loss in previous studies. However, the histories and trends of the loss can still be used for understanding and monitoring the supervised learning process. If the loss is decreasing, it shows that the network is learning as intended. For the two binary tasks (component type, Fig. 6.9a, and spalling condition, Fig. 6.9b) with retraining one block, both training and test losses are continuing to decrease with roughly 36% and 26% linear reduction rates (i.e., estimated slop of the loss plots), respectively. However, the losses are still not sufficiently close to zero even after 100 epochs, which indicates that there still exists room for model performance improvement. On the other hand, for the two multi-class tasks (damage level, Fig. 6.9c, and damage type, Fig. 6.9d) with retraining one block, following an initial big drop, both training and test loss plots become flat where the linear reduction rates of the losses decrease to 20% and 7%, respectively, while also the losses are still high (close to 1.0 and 1.3 after 100 epochs for the damage level and damage type tasks, respectively). This observation indicates that it is hard to obtain more improvement with further training based on the current parameter complexity, when retraining only one block. While retraining two blocks, the training loss plots behave very well for all tasks where the losses are much lower, between 0.1 and 0.2 for all tasks. However, the test losses do not decrease and instead oscillate after about 20 epochs with their plots becoming jagged. This is an evidence of *over-fitting*, indicating that the network has *enough complexity* to learn some features but *lacks the generality* to extend these features to new unseen data due to the small size dataset used for training. In general, even with some degree of over-fitting, the models with retraining two blocks have lower losses than those with only retraining one block (ignoring some very minor exceptions during small intervals of oscillations near the 80th epoch for the spalling condition task), implying a better performance. In addition, with the possible exception of

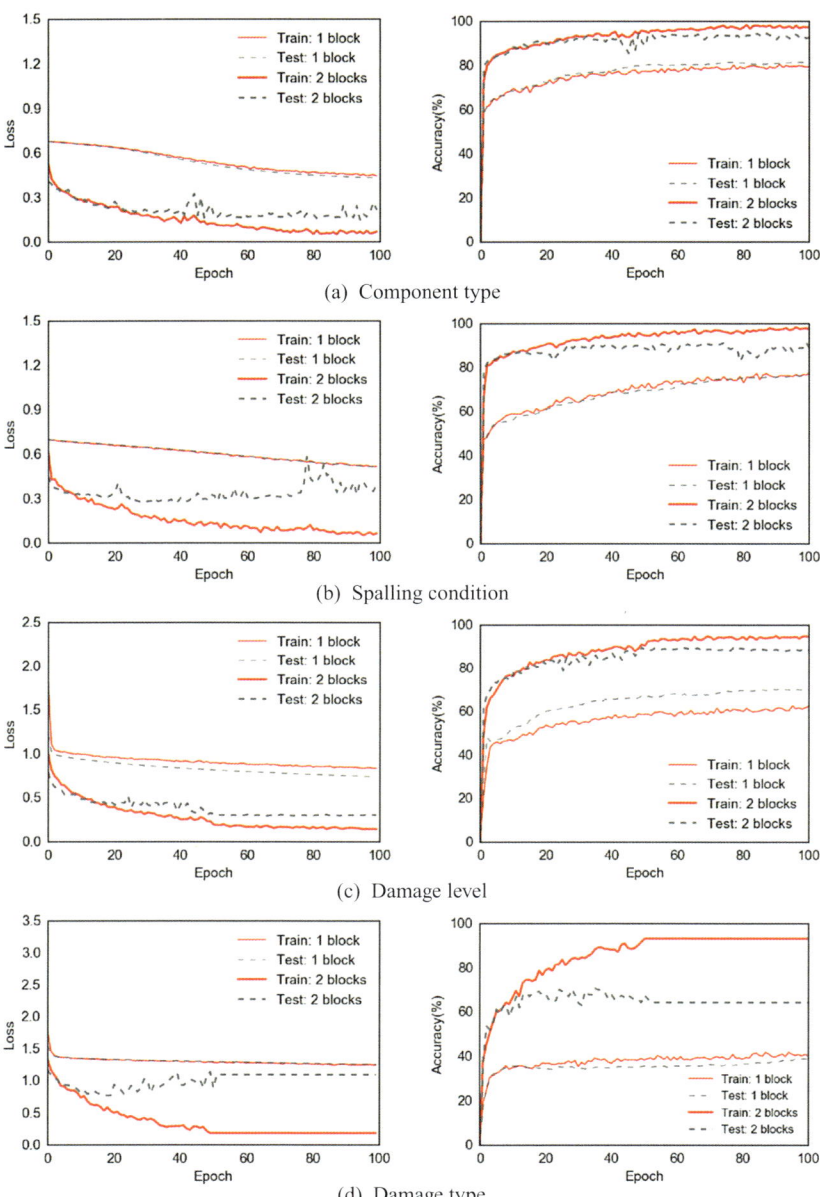

(a) Component type

(b) Spalling condition

(c) Damage level

(d) Damage type

Fig. 6.9 Loss and accuracy histories of the four recognition tasks using fine-tuning

the damage type task and to a lesser extent the spalling condition task, the accuracy results indicate that some degree of over-fitting is still acceptable for retraining two blocks.

The accuracy is the most important indicator to evaluate the performance of the DL models. Its plots in Fig. 6.9 for both training and test show significant improvement from retraining one to retraining two Conv blocks for all four detection tasks. This is especially true for the damage level detection task, which clearly shows the benefits of retraining more parts of the network to efficiently learn features of the damage level. Regardless of the training accuracy, the best values of the test accuracy with retraining two blocks for the component type, the spalling condition, and the damage level reach about 90%, which is very promising considering the use of such a small dataset of the ϕ-Net-2000. Even for the damage type task that corresponds to a slightly less than 70% test accuracy with retraining two blocks, i.e., a 30% increase over the case of retraining only one block, the result is much higher than the 25% accuracy for the *random guess* of this four-class classification problem.

From the above analysis, retraining one block leads to under-fitting but stable performance. On the contrary, retraining two blocks leads to over-fitting but great improvement in accuracy. Therefore, it is believed that for the ϕ-Net-2000 dataset and tasks, more important parameters representing the structural features exist in the second to last Conv block. By increasing the dataset size beyond 2,000 images, it is expected that future studies may fine-tune more Conv blocks to achieve even better performance for the DL models.

The model with retraining two Conv blocks has better performance than that with only retraining one block and the best prediction accuracy is listed in Table 6.5. Compared with the results obtained by applying the feature extractor, the fine-tuning approach reduced the over-fitting by making use of the DA and adjusting the mid- to high-level features, including greatly improving the recognition accuracy. For the binary tasks, the test accuracy by fine-tuning two blocks is improved by $94.5 - 88.8 = 5.7\%$ to $91.5 - 85.4 = 6.1\%$. More interestingly, the test accuracy for the damage level improved by $89.7 - 77.0 = 12.7\%$ with only 5.6% over-fitting (i.e., difference between training accuracy of 95.3% and test accuracy of 89.7%), which shows the good performance of the fine-tuning TL strategy when dealing with multi-class classification problems. Even though for the hardest damage type task (4 classes), the model with retraining two blocks improves the test accuracy by $68.8 - 57.7 = 11.1\%$ and reduced over-fitting from $94.2 - 57.7 = 36.5\%$ (Table 6.3) to $91.8 - 68.8 = 23.0\%$ (Table 6.5). Moreover, the normalized CMs are computed for the fine-tuning experiment, Fig. 6.10. Compared with the feature extractor, Fig. 6.8, the recognition accuracy for the "beam/column" and "non-spalling" classes increased from 0.91 to 0.99 and from 0.90 to 0.97, respectively, but the misclassifications for "wall" and "spalling" still have the same values, i.e., 0.14 and 0.20, respectively, which indicates the need for more training data. For the multi-class tasks, the accuracy has been significantly improved, from 0.78 to 0.88 for "none", from 0.62 to 0.82 for "minor", and from 0.88 to 0.94 for "moderate to heavy" damage level classes. However, for the damage type, although the accuracy is higher for all four classes, there still exists around 20% misclassification errors among the three types of damage, which should be evaluated carefully in future studies.

Table 6.5 Best accuracy results (%) in the retrain ratio experiment (for comparison, the numbers in parentheses are from Table 6.3 for the feature extractor TL)

Task	# of Classes	Training accuracy	Test accuracy
Component type	2	98.4 (99.1)	94.5 (88.8)
Spalling condition	2	98.5 (97.1)	91.5 (85.4)
Damage level	3	95.3 (97.5)	89.7 (77.0)
Damage type	4	91.8 (94.2)	68.8 (57.7)

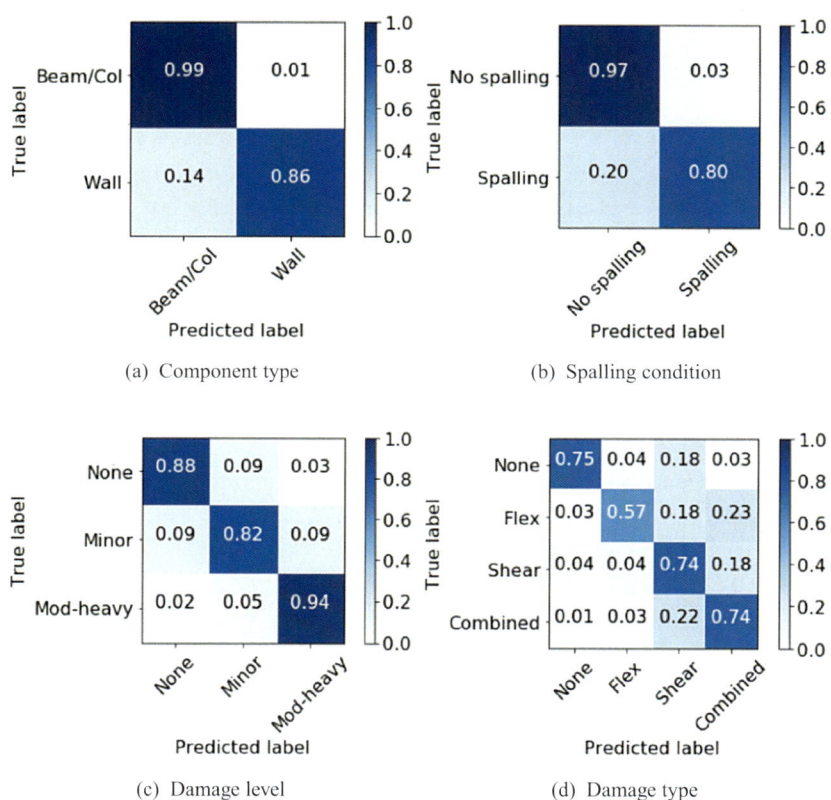

(a) Component type   (b) Spalling condition

(c) Damage level   (d) Damage type

Fig. 6.10 Normalized CMs of the test predictions by fine-tuning two Conv blocks

The computational efficiency of the training is manifested by the number of parameters the model backpropagates. From Table 6.6, it is obvious that, except for the adaptation layers, most parameters are in the last several Conv blocks. The last two blocks plus the adaptation layers contain nearly 90% of the parameters of the whole network and the last block plus the adaptation layers contain 64% of the parameters. However, the training time does not vary linearly with the number of parameters where even though only 10% of the parameters in the

Table 6.6 Trainable parameters for different retrain ratios in the backpropagation

Retrained blocks	Component type	Spalling condition	Damage level	Damage type
FC-layers	6,423,298	6,423,298	6,423,555	6,423,812
One block + FC-layers	13,502,722	13,502,722	13,502,979	13,503,236
Two blocks + FC-layers	19,402,498	19,402,498	19,402,755	19,403,012
All	21,137,986	21,137,986	21,138,243	21,138,500

Table 6.7 Computational time in seconds for different retrain ratios

Procedure	Retrain one block	Retrain two blocks	Retrain all
Training per epoch	128	133	160

first three frozen Conv blocks, more computation time than 10% is needed to backpropagate into these first three blocks, i.e., $\frac{160-133}{160} \times 100 \approx 17\%$, Table 6.7.

In summary, by retraining more parameters and invoking the DA, the classifier behaves better than the one based on the feature extractor in all aspects except for the training time. This is illustrated by comparing Tables 6.4, 6.5, 6.6 and 6.7, where the training time per epoch for the case of retraining two blocks (133 s) is significantly larger by two orders of magnitude (a factor of about 106) than that for the case of the feature extraction using the VGG input size (1.26 s). Therefore, a trade-off between performance accuracy and computational cost should be considered when selecting between the two strategies of TL for deep CNNs, i.e., fine-tuning by retraining the deeper blocks versus feature extraction.

6.3.4   Network Configuration

As introduced in Sect. 4.3, there are many well-known network architectures achieving promising performance of DL models in many domains. An investigation of the influence of using two different network architectures, namely, VGGNet and ResNet, is conducted and discussed in this section. In addition, ResNet-50 and ResNet-101, as two ResNet variants, are examined with feature extraction and fine-tuning TL. Comparisons are made first between these two different depths (50 and 101 layers) of the ResNet. Subsequently, the best ResNet amongst these two depths and the VGG-type network are compared and discussed.

6.3.4.1 Training Configuration

Similar to the VGG-type networks, discussed in Sects. 6.3.2 and 6.3.3, the FC-layers after the original ResNet GAP layer are replaced by two adaption layers with 256-neurons FC-layer and a *softmax* layer, whose final output equals the number of classes of the task at hand. The remaining configurations of the network architectures in both ResNet-50 and ResNet-101 are taken the same as the ones in [12] with zero padding added to the input images as deemed necessary. As mentioned in Sect. 4.3.3, the first residual unit in each Conv block is set as the Conv unit to match the dimension between the Conv layers and the remaining residual units, using the identity mapping, shown in Fig. 4.9a. The major difference between ResNet-50 and ResNet-101 is that the latter has more identity residual units in Conv block # 4 which consists of 3 layers (applied 23 times as identified using "*" in Table 6.8, instead of 6 times for the former, leading to additional $(23 - 6) \times 3 = 51$ Conv layers for a total of $50 + 51$ layers in ResNet-101), which might help with extracting mid- to high-level features. The detailed configurations are listed in Table 6.8, where N represents the amount of data, i.e., 2,000 in the ϕ-Net-2000.

While implementing the feature extractor, all parameters are fixed in the Conv blocks and the output tensor before the FC-layer (Fig. 6.4) is extracted as features. The weights are initialized from a Gaussian distribution with zero mean and 10^{-2} variance (biases are initialized with zero values) and they are updated to optimize the categorical cross-entropy loss, Eq. 3.6, using Adam optimization algorithm [14] based on the backpropagation algorithm, Eqs. 4.7 and 4.8, where the three key parameters, β_1, β_2, and ϵ are, as before, assigned with the empirical values of $0.9, 0.999$, and 10^{-8}, respectively. The learning rate, η, is set to 10^{-3} and manually changed to 10^{-4}, for better convergence, after 200 epochs of training.

Table 6.8 Configurations of ResNet-50 and ResNet-101 [filter size and (# filters) and "# block applications"]

Block	Output shape	ResNet-50{101*}
Input	$(N, 224, 224, 3)$	–
Conv Block 1	$(N, 112, 112, 64)$	$7 \times 7(64)$, stride 2
Conv Block 2	$(N, 56, 56, 256)$	Max Pooling, filter 3×3, stride 2
		$[1 \times 1(64); 3 \times 3(64); 1 \times 1(256)] \times$ "3"
Conv Block 3	$(N, 28, 28, 512)$	$[1 \times 1(128); 3 \times 3(128); 1 \times 1(512)] \times$ "4"
Conv Block 4	$(N, 14, 14, 1024)$	$[1 \times 1(256); 3 \times 3(256); 1 \times 1(1024)] \times$ "6{23*}"
Conv Block 5	$(N, 7, 7, 2048)$	$[1 \times 1(512); 3 \times 3(512); 1 \times 1(2048)] \times$ "3"
GAP	$(N, 2048)$	–
FC-layer	$(N, 256)$	Activated by *ReLU*
FC-layer	$(N, 2/3/4)$	Activated by *softmax*

Since the difference between ResNet-50 and ResNet-101 is the number of identity residual units in block 4, while adopting fine-tuning in this experiment, only either Conv block 5 or Conv blocks 4 and 5 are tuned for both ResNet-50 and ResNet-101 to assess the possible over-fitting due to the small size of the training dataset. Moreover, the categorical cross-entropy loss is optimized using the SGD with momentum variable $\mu = 0.9$, in an *exponentially decaying learning schedule*, as follows:

$$\eta_t = \eta_0 \cdot e^{-d \cdot k}, \tag{6.2}$$

where the chosen values of the hyper-parameters are: the initial decaying value $\eta_0 = 10^{-4}$ and the decaying rate $d = 10^{-6}$. It is noted that the "learning schedule" is usually defined as the strategy or algorithm to adjust the learning rate during the training process according to the iteration number, k. To alleviate the negative influence of over-fitting, the online DA is applied during the training, whose setting is the same as that in the retrain ratio experiment, Sect. 6.3.3.

6.3.4.2 Results and Analysis

The computation results of ResNet-50 and ResNet-101 using feature extractor with an input size of 224×224 are shown in Fig. 6.11. It is observed that all tasks converge fast and their performance does not significantly improve even with adjusting to a lower learning rate, η, after about 200 training epochs. This observation is attributed to the fixed features extracted by the Conv blocks. Another observation is that the ResNet-50 and ResNet-101 share similar performance, where the prediction accuracy of the two binary tasks reaches nearly 80% and that of the damage level task reaches about 66% without over-fitting. However, for the damage type task, the best accuracy values of both ResNet-50 and ResNet-101 are unsatisfactory being about 40%, which is not much higher than the 25% of the random guess of this four-class classification task. This issue is attributed to less information and inadequate features extracted by the fixed Conv blocks. Compared with the VGG-type networks (discussed in Sects. 6.3.2 and 6.3.3), the accuracy of both ResNets is lower because their pre-trained Conv blocks may have a weaker generalization than those of the VGG-type networks and the extracted features by these blocks of the source domain may not be well-aligned with the new recognition tasks in the target domain. However, the performance of the ResNets is expected to be improved via fine-tuning.

Regarding the computational efficiency, the feature extractor consumes a small computational time, about 28 and 48 s for its process in the ResNet-50 and ResNet-101, respectively, and only 0.2 s per training epoch for both networks. Thus, training the ResNets is much faster (i.e., $1.26/0.2 \approx 6$ times faster) than the VGG-type networks (Table 6.4). Since the performance of the feature extractor for ResNet-50 is comparable to that of ResNet-101, ResNet-50 with feature extractor can be considered for preliminary analysis related to the tasks of the component type identification, the spalling check, and the damage level eval-

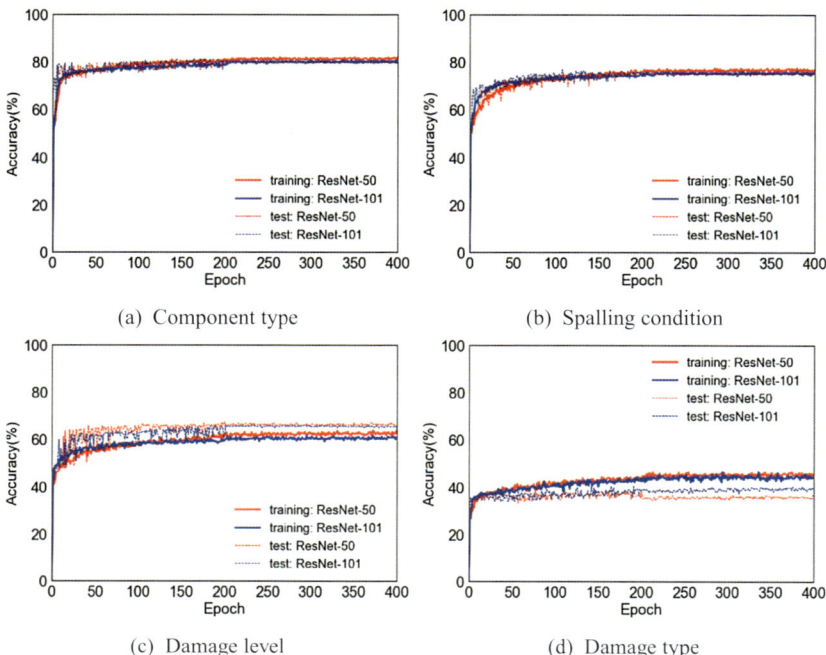

Fig. 6.11 Accuracy histories of ResNet-50 and ResNet-101 using feature extractor

uation. On the other hand, for the damage type task, more studies should be conducted to mitigate the misclassification risks.

While adopting the fine-tuning strategy, the prediction performance for both ResNets is greatly improved and the results are shown in Fig. 6.12. Slight differences between the two networks are observed. The test accuracy of the first three tasks reaches over 90% and it is over 70% with about 20% over-fitting for the damage type task. This indicates that for the current small dataset, ResNet-50 holds sufficient complexity for the recognition tasks in addition to its computational efficiency (about 82 and 160 s per training epoch for ResNet-50 and ResNet-101, respectively). Moreover, the plateau in the test accuracy history, within the first 10 epochs, is attributed to incompatible initial parameter values and a large learning rate. With online DA and decaying learning rate, the model escapes this plateau within about 5–10 additional training epochs and finally achieves very promising results for all four tasks.

In general, ResNet-101 using fine-tuning achieves the best performance in terms of test accuracy in all four tasks among all considered models and the best results of all models are listed in Table 6.9. Comparing its results with the previous VGGNet cases, ResNet-101 outperforms the VGG-type network in the last three tasks and has a good improvement in the hardest task (i.e., damage type), where it improves its test accuracy by $74.7 - 68.8 = 5.7\%$. For the component type task, ResNet-101 is only $94.5 - 94.0 = 0.5\%$ lower than VGGNet in the test accuracy, which is an acceptable discrepancy for this very high accuracy. In addition,

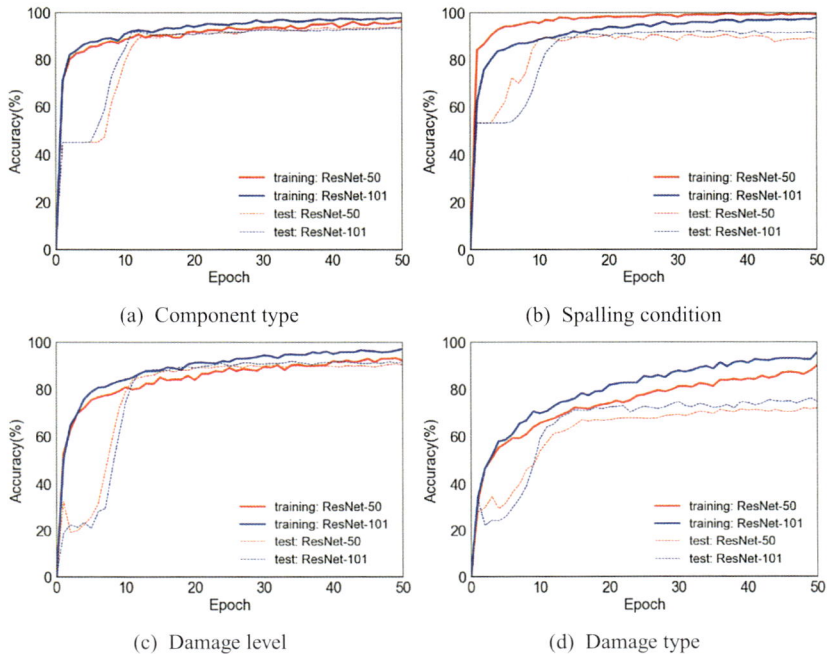

(a) Component type   (b) Spalling condition

(c) Damage level   (d) Damage type

Fig. 6.12 Accuracy histories of ResNet-50 and ResNet-101 using fine-tuning

Table 6.9 Best test accuracy results (%) of VGGNet, ResNet-50, and ResNet-101

Task	# of Classes	VGGNet (Table 6.5)	ResNet-50	ResNet-101
Component type	2	94.5	93.5	94.0
Spalling condition	2	91.5	88.5	92.0
Damage level	3	89.7	90.5	91.5
Damage type	4	68.8	72.0	74.5

ResNet-50 achieves equivalently accurate results compared with those of the VGGNet, where it has higher test accuracy in the harder multi-class tasks (i.e., damage level and damage type), and it has slightly lower test accuracy in the easier binary tasks. Regarding the computational efficiency, the time used in fine-tuning two Conv blocks in ResNet-101 (160 s per training epoch) is the same as that for retraining the whole VGGNet (refer to Table 6.7). However, due to more trainable parameters in ResNet-101, such a time increment is not large and still acceptable. On the other hand, ResNet-50 needs only 82 s per training epoch, which is much less than that of the VGGNet computational time when retraining two blocks (i.e., 133 s per training epoch from Table 6.7). Thus, the family of ResNets shows a competitive performance in computational efficiency and classification accuracy, where the fine-tuned ResNet-101 is the optimal model for the considered four recognition tasks using the ϕ-Net-2000 dataset.

6.3.5 Summary and Discussion

Three major experiments using TL on the small-scale ϕ-Net-2000 dataset are conducted and compared with respect to: (1) network input size, (2) retrain ratio, and (3) network configuration. Conclusions from these experiments are listed as follows:

- The optimal network input size is 224×224, which is used in training the prototypes VGGNet and ResNet. Other sizes would perform worse due to either lack of information or computational complexity and inefficiency.
- Some parameters in the final two Conv blocks represent structural damage features. In the retrain ratio experiment, the VGG-type network with retraining two Conv blocks achieves a very promising accuracy of about 90% for the binary and the three-class tasks. Due to the complexity of the four-class task and the use of a small training dataset, about 69% test accuracy (with a 23% over-fitting gap value) is obtained, which is much higher than what is expected from the lower bound random guess for a four-class classification problem, i.e., 25%.
- The computations of both ResNet-50 and ResNet-101 are fast under feature extractor but with similar, yet limited, accuracy especially for tasks with numbers of classes larger than 2. However, their performance can be significantly improved through fine-tuning, where ResNet-101 has higher test accuracy than that of ResNet-50. When compared with the best variant of the VGGNet, the fine-tuned ResNet-101 outperforms it in most detection tasks and it is considered a better model. Moreover, ResNet-50 is thought to be at a similar accuracy level to the best variant of VGGNet, but it is more computationally efficient.

In conclusion, TL is an effective training approach to mitigate data deficiency in vision-based SHM tasks. Its feature extractor strategy acts as a preliminary and computationally efficient approach with a reasonable recognition accuracy on the target dataset. On the other hand, the fine-tuning strategy of TL can be considered as an improvement tool beyond the feature extractor if good computation resources are available. All experiments in this section indicate the favorable generalization properties and transferability of classical deep CNNs (i.e., VGGNet and ResNet) pre-trained from ImageNet [1] to structural image datasets, which validate the effectiveness and feasibility of deep CNNs in vision-based SHM, especially when effective TL strategies are utilized. In addition, under the limitation of small-scale datasets, the findings in this section reveal a high potential for more complex tasks with larger datasets. These observations are the motivation for conducting more complex benchmark tasks using the complete ϕ-Net dataset in the next section.

6.4   TL from ImageNet to ϕ-Net

Based on the ϕ-Net dataset discussed in Chap. 5 and the computer experiments in Sect. 6.3 using ϕ-Net-2000, a comprehensive investigation applying DL in general tasks of vision-based SHM is explored in this section. The purpose of this investigation is to provide benchmarks and reference performance via conducting comprehensive computer experiments on all eight benchmark classification tasks (refer to Sect. 5.2.2) using the full ϕ-Net dataset. Based on and extending the study in Sect. 6.3, three well-known deep CNN models are used, namely, VGG-16, VGG-19 [16], and ResNet-50 [12] with overall accuracy, defined by Eq. 3.12, considered as the evaluation metric. Moreover, both DA and TL are adopted for better performance. In all, four comparison cases are performed for each task and model as listed below:

1. Train the model from scratch, denoted as the baseline model (BL model).
2. Train the model from scratch + DA (BL-DA model).
3. Fine-tune using TL a pre-trained model by ImageNet [1] (TL model).
4. Fine-tune using TL a pre-trained model by ImageNet [1] + DA (TL-DA model).

6.4.1   Experimental Setup

Based on the conclusions in the network input size experiment in Sect. 6.3.2, the input size for all models is taken the same as 224×224. However, several network architecture modifications are made to accommodate the difference between the ImageNet [1] and ϕ-Net datasets. These are summarized as follows:

- For VGG-16 and VGG-19 models, FC-layers after the Conv blocks are replaced by three adaptation layers. The nonlinear *ReLU* activation function is used in the first two layers together with 256 neurons in the first layer and 128 neurons in the second layer. The number of neurons in the third adaptation layer, which is a *softmax* layer, is adjusted to satisfy the number of classes in each task.
- For the ResNet-50 model, FC-layers after the Conv blocks are replaced by two adaptation layers, where the first layer has 1024 neurons and is activated by *ReLU*. Similar to the VGG-16 and VGG-19 models, the second adaptation layer, which is a *softmax* layer, is modified according to the number of classes in each task.

For a fair comparison, the training configurations for each task are kept fixed. From a preliminary convergence analysis, these settings are: 50 training epochs, 64 batch size, test accuracy computed at the end of each training epoch, and SGD applied along with a piece-wise decaying learning rate with a learning rate schedule of 10^{-2} (first 20 epochs), 10^{-3} (21–30 epochs), 10^{-4} (31–40 epochs), 10^{-5} (41–45 epochs), and 10^{-6} (the remaining epochs). The intuition behind this learning rate schedule is that in the early stages of training, a larger learning rate is helpful to avoid having the model trapped in local minima and the model is expected to somewhat freely explore the search space for the optimal solution. After training for the early epochs, the learning rate decreases, where such a smaller learning rate stabilizes the model and avoids large oscillations. Finally, to avoid over-fitting, dropout is applied in the adaptation layers with a dropout rate, refer to Sect. 4.1.4, of 0.5.

In models (3) and (4) using TL, the model parameters in the Conv layers are loaded from the ImageNet [1] pre-trained models and the parameters in the following adaptation layers are initialized randomly. On the other hand, in models (2) and (4) with DA, transformations are performed from random combinations of six possibilities: (a) horizontal translation within 10% of the total image width, (b) vertical translation within 10% of the total image height, (c) rotation within 5 degrees, (d) zoom in less than 120% of the original image size, (e) zoom out less than 80% of the original image size, and (f) horizontal flip.

It should be noted that due to the random factors involved in training, such as the parameter initialization, the input sequence of data and their batches, and the use of SGD and dropout, the network produces varying results across different runs. In particular, poorly initialized parameters may result in an un-trainable scenario, where the loss remains stuck at the beginning. Therefore, it is recommended to train the network multiple times and select the best model.

Considering the high computational demands for all the designated tasks and their respective comparison groups (i.e., the different models considered in the comparison) and also taking into account that the primary objective of this benchmark study is not to achieve the highest accuracy, but rather to explore the feasibility of the AI model in SHM, each case is performed only once, unless the training loss remains unchanged during the initial epochs, which are set as 10 herein. The implementation of the benchmark experiments is conducted on TensorFlow and Keras[10] platforms. The used computational hardware for the computer experiments was CyberpowerPC with a single GPU: Nvidia Geforce GTX 2080Ti and its CPU: Intel Core i7-8700K @3.7 GHz 6 Core with RAM: 32.0 GB.

6.4.2   Results and Discussion

Table 6.10 lists the best reference values for the investigated models. Since the ultimate goal of this study is to validate the feasibility of adopting DL for automated detection in the context of SHM, the accuracy should be of acceptable value. For the easier Tasks (1–4), Sects. 5.2.2.1–5.2.2.4, the test accuracy is very promising, especially for the classification between steel and non-steel materials. For the harder Tasks (5–8), Sects. 5.2.2.5–5.2.2.8, although the best accuracy results are still acceptable, they are expected to be improved with more data, more sophisticated DL models, better training approaches, and tuned hyperparameters.

---

[10] For more details about Keras platform, refer to https://keras.io/.

Table 6.10 Summary of results of the best test accuracy (%) in all tasks of ϕ-Net

Task	Best result	Network	Model	Accuracy increment above: BL	BL-DA	TL	TL-DA
(1) Scene level	93.4	ResNet-50	TL	+5.7	+5.4	–	+0.4
(2) Damage state	88.9	ResNet-50	TL-DA	+12.2	+13.7	+1.7	–
(3) Spalling condition	83.0	ResNet-50	TL-DA	+7.3	+5.9	**+1.8**	–
(4) Material type	98.5	ResNet-50	TL-DA	+3.5	+5.4	+0.2	–
(5) Collapse mode	78.1	ResNet-50	TL	+13.0	+16.3	–	**+3.4**
(6) Component type	77.0	VGG-19	TL	+11.9	+11.9	–	+0.2
(7) Damage level	74.5	ResNet-50	TL	+19.9	+16.1	–	+1.8
(8) Damage type	72.4	VGG-19	TL-DA	**+28.7**	**+28.7**	+1.2	–

Comparing the DL models, ResNet-50 is more generalized because it achieves the highest test accuracy in most cases (except in Tasks 6 and 8). Thus, to achieve the state-of-the-art performance, it is suggested to perform model selection among multiple DL models. From the perspective of the training approach, TL works well in all eight tasks, with significant improvement in many cases, and could be seen as the necessary training strategy for the current scale of the dataset. It is inferred that the ImageNet [1] pre-trained models already have some sense of features related to the basic and natural objects from the source domain. Thus, transferring source-domain knowledge to the target domain of structural images may provide a better understanding of the deep CNN models toward some general structural vision patterns, e.g., object boundary, material texture, etc. Moreover, fine-tuning from the ImageNet [1] pre-trained models obtains better parameter initialization than directly training models from scratch where parameters are initialized randomly or from certain distributions, e.g., Gaussian. These findings are more evident in the harder tasks with limited amount of data or abstract semantics, e.g., the TL model achieves 10~30% enhancements over the BL model cases for Tasks 5–8. Compared with the TL technique, DA does not significantly improve the accuracy of the test and in some cases performs poorly compared with other approaches. This can be attributed to two major issues in the current experiment: (1) randomness and uncertainty due to only having one run for each case and (2) fixed DA settings, e.g., ranges of rotation angle, translation, etc., may not be suitable for all tasks, which may lead to varying performances.

More information about the performance of a model can be extracted from the training and test history plots. Thus, the best results for VGG-16, VGG-19, and ResNet-50 are plotted in one figure for each task as shown in Fig. 6.13a–h. In all benchmark tasks for all models, the training accuracy is high and close to 100%. However, the gap between the training accuracy and that of the test is evident and it varies from one task to another. This is an indicator of over-fitting and *full training*, i.e., the models contained enough parameters and complexity to learn the task well during training. In ML and DL, usually applying regularization, e.g., dropout with varying dropout rate and BatchNorm (Sect. 4.2.4), refer to Sect. 4.2.4, may help to alleviate the over-fitting problem. These regularization techniques are important to carefully investigate when designing new DL models for new tasks. From the perspective of training stability, DL models are less stable in the early training stage when the learning rate is large. During this stage, the parameters are explored and largely updated with the SGD algorithm, causing the history plots to oscillate. With a scheduled decaying learning rate (especially after 20 epochs), the training plots become flat and smooth, indicating that the model performance is stabilizing.

The computational efficiency is an important metric in DL applications. Table 6.11 lists the average computational times per training epoch for each model under different training approaches, providing a reference for future studies. Without DA, ResNet-50 has the least computational time, VGG-16 has the second least time, and VGG-19 has the most time as it contains more parameters than the other two networks. There are roughly 21 million, 26 million, and 25 million parameters in VGG-16, VGG-19, and ResNet-50, respectively. Even though ResNet-50 is deeper than the VGGNets, using a smaller filter size (1×1 rather than the 3×3 applied in the VGGNets) and the use of the shortcut connections make the gradient flows faster and leads to accelerating the training process for the ResNet-50. With DA, all models have increasing time and similar training time per epoch due to the extra time needed to perform the transformation for the DA on the raw images. As for the TL, it does not affect the average training time because it neither introduces new parameters nor performs image-processing operations. Thus, TL is a more efficient and economical model in the experiments reported herein.

The reported results above provide a reference for relevant future DL applications where improvement of these results can be sought. In order to obtain more accurate, more efficient, and more resilient AI-based SHM models, tuning hyper-parameters (e.g., learning rate, dropout rate, and number of neurons in the FC-layers), developing new models, and use of advanced training approaches can be considered in future studies, refer to Part III of this book for examples of such studies. It is recommended to make full use of the existing ϕ-Net dataset [17] to enlarge the user's own dataset, establish and quantify finer categories, and relabel images for new vision tasks.

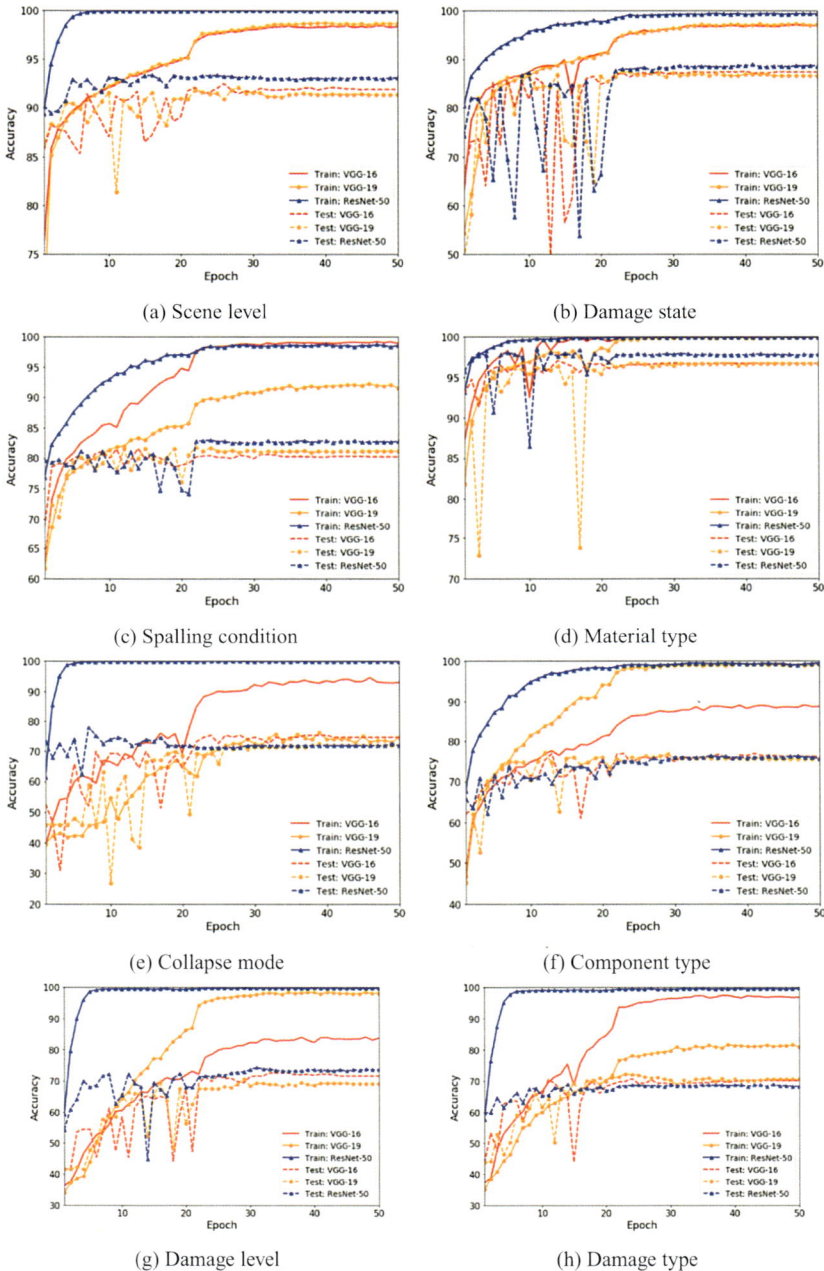

(a) Scene level

(b) Damage state

(c) Spalling condition

(d) Material type

(e) Collapse mode

(f) Component type

(g) Damage level

(h) Damage type

Fig. 6.13 Accuracy histories of the best models for 50 epochs in all tasks of ϕ-Net

Table 6.11 Average computational time in seconds per training epoch in all tasks

Task	Network	Model			
		BL	BL-DA	TL	TL-DA
(1) Scene level	VGG-16	145	178	146	187
	VGG-19	169	185	169	188
	ResNet-50	131	183	123	190
(2) Damage state	VGG-16	71	79	71	81
	VGG-19	82	83	82	83
	ResNet-50	64	83	63	83
(3) Spalling condition	VGG-16	42	46	42	46
	VGG-19	48	49	48	49
	ResNet-50	37	47	37	47
(4) Material type	VGG-16	52	67	52	67
	VGG-19	59	69	59	69
	ResNet-50	46	68	46	68
(5) Collapse mode	VGG-16	8	8	8	8
	VGG-19	9	9	9	9
	ResNet-50	7	8	7	8
(6) Component type	VGG-16	28	30	28	30
	VGG-19	33	33	33	33
	ResNet-50	26	30	26	31
(7) Damage level	VGG-16	25	26	25	26
	VGG-19	29	29	29	29
	ResNet-50	22	27	22	27
(8) Damage type	VGG-16	25	25	25	25
	VGG-19	28	29	28	29
	ResNet-50	22	27	22	27

6.5   Hierarchical Transfer Learning

In the above experiments, each task and its attributes is treated as independent from other tasks, where the relationships between tasks are ignored. However, based on domain knowledge, these tasks have certain relationships with each other, especially considering the hierarchy shown in Figs. 5.1 and 5.13. Such hierarchical relationships have already been utilized in developing the ϕ-Net dataset. In addition, previous experimental results indicate the

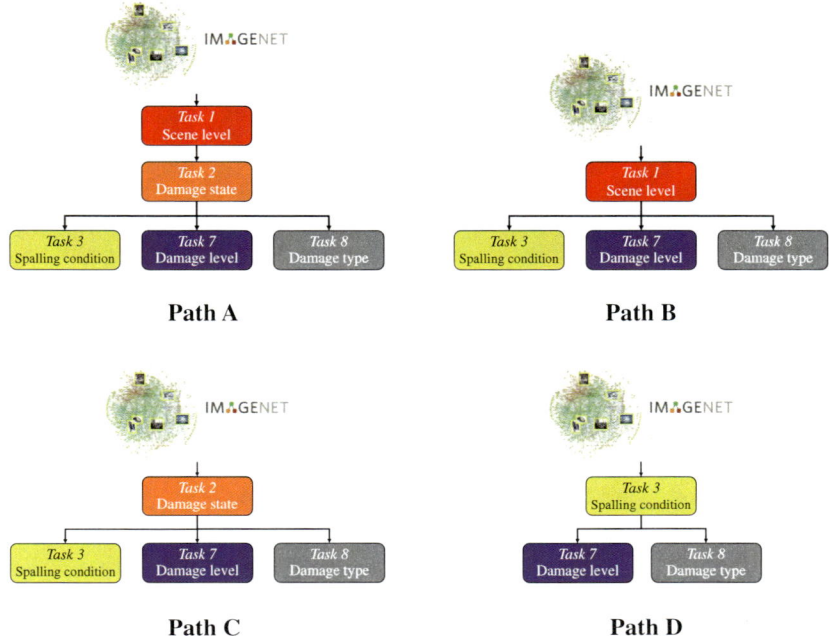

Fig. 6.14 Hierarchical transfer learning paths

feasibility of a large amount of knowledge transfer from the ImageNet [1] to the ϕ-Net tasks [17] and a significant improvement of the baseline (BL) model, refer to Table 6.10, can be obtained.

In this section, the above mentioned hierarchy is used to investigate whether domain expertise can enhance the accuracy of the DL models. Accordingly, one relevant approach, namely, hierarchical transfer learning (HTL), is introduced, which conducts the TL through *transfer paths* determined based on domain knowledge, Fig. 6.14. In other words, in HTL, the relationships between tasks are transferred through the designed hierarchical paths, which aim to improve the model performance in hard Tasks (e.g., 7 and 8) via information transferred from easy Tasks (e.g., 1 and 2). Refer to Sect. 5.2.2 for definitions of the different tasks.

6.5.1   Hierarchical Transfer Paths

Figures 5.1, 5.13, and 5.15 present certain choices of task sequence to reflect the relationships between each task. In practice, there are many other choices. For benchmarking purposes, four transfer paths, Fig. 6.14, including Tasks 1, 2, 3, 7 and 8 (refer to Sect. 5.2.2), are designed based on the hierarchical relationships of the ϕ-NeXt framework, Fig. 5.15. Path A is directly obtained from the ϕ-NeXt framework via eliminating the general Task 4 (material type), which already achieved very accurate results (98.5%, as listed in Table 6.10). Moreover, two other tasks are eliminated, namely, Task 5 (collapse mode) and Task 6 (component type), due to the constraints that Tasks 3, 7 and 8 are *subsequent tasks* if the image is in the object level, ruling out the need for Tasks 5 and 6 in path A. Path B follows the main idea of path A but skips Task 2 (damage state). On the contrary, path C uses Task 2 (damage state) to replace Task 1 (scene level). Finally, Path D does not follow the original hierarchy of the ϕ-NeXt, Fig. 5.15. Instead, it investigates the relationships between several subsequent tasks of the object level images. For example, if one object level image is known as "Spalling" in Task 3, it is likely related to the label "Heavy damage" in Task 7 and "Combined damage" in Task 8. The reasoning for this dependency is that, if the spalling area is large, i.e., a large portion of the cover material is lost from the surface of the structural component, it is likely that the object is heavily damaged. Moreover, it is expected in this case that the cause of this excessive spalling formation is due to combined effects of complex deformation and force boundary conditions, which is manifested in the definition of heavy and combined damage scenarios.

The harder Tasks 7 and 8 are expected to benefit from the knowledge related to the easier Tasks 1, 2 and 3. The intuition behind this is that Task 1 contains the most images compared with other tasks and it can provide more information for subsequent tasks. Moreover, Task 2 is the most important attribute in the vision-based SHM and it has a direct relationship with spalling, damage level, and damage type. For example, if one structural image has no damage patterns, the label "Undamaged" should be strongly and positively correlated to the labels "Non-spalling" in Task 3 and "Undamaged" in Tasks 7 and 8. Thus, knowing labels of the easier tasks of the scene level and the damage state is expected to help the model better understand and learn the harder tasks of damage level and damage type.

For simplicity, only the ResNet-50 model is adopted in this experiment and its settings are the same as the BL model in Sect. 6.3.5. Since TL gains significant improvement over the BL model, the ImageNet [1] pre-trained model is assigned as the first model in the considered path of the HTL. Along with the selected path, the model in a current task is inherited from the previous task and then it is used as the pre-trained model for the next task. In the design of this experiment, each task is repeated three times to select the best model, which is subsequently used as the pre-trained model and fine-tuned in the subsequent task. It is assumed that, along the transfer path, a better pre-trained model with the highest accuracy, among multiple runs, has better performance in fine-tuning.

6.5.2 Results and Discussion

The results for all learning paths are presented in Table 6.12. Compared with the ResNet baseline, in general, Task 8 gains nearly 1–2% enhancement in all paths, but there exists a persistent accuracy down-streaming in Task 7 for all learning paths. Comparing the four

Table 6.12 Hierarchical transfer learning results (test accuracy, %)

Path	Source task sequence	Target task	HTL	Increment above ResNet
A	ImageNet → 1 → 2	3	83.3	+2.1
		7	74.1	−0.4
		8	70.9	+2.0
B	ImageNet → 1	3	82.6	+1.4
		7	72.1	−2.4
		8	70.5	+1.6
C	ImageNet → 2	3	82.6	+1.4
		7	70.3	−1.4
		8	70.3	+1.4
D	ImageNet → 3	7	74.1	−0.4
		8	69.7	+0.8

paths, for the purpose of improving hard tasks, path A is thought to be the best, which has the highest test accuracy in Tasks 7 and 8 and it obtains the largest improvement over the baseline in Tasks 3 (+2.1%) and 8 (+2.0%) and the smallest reduction over the baseline in Task 7 (−0.4%). It is also observed that Task 3, as an intermediate task, benefited from the hierarchical relationships with about 1% to 2% improvements in paths A to C. In paths B and C, with pre-training from Tasks 1 or 2, Task 3 achieved a 1.4% improvement. It should be noted that the performance in Task 8 in all paths using the HTL with the ResNet-50 is still slightly lower than that of the VGG-19 (72.4%), Sect. 6.4.2.

Based on the above results, the HTL using shareable task-dependent characteristics achieves some improvement, e.g., for Tasks 3 and 8, but may slightly increase the risk of rendering worse results in some cases, e.g., Task 7, and thus its performance is highly dependent on both the predefined path and the source domain model. Accordingly, with limited choices of the learning paths and discrete number of models, the HTL can not yet fully capture and exploit the inter-task relationships. As for the computational cost, performing HTL only fine-tunes the model parameters based on the previous tasks instead of those directly obtained from the ImageNet [1]. However, HTL requires this fine-tuning to be performed multiple times, e.g., for target Task 3 in path A, the model needs to be retrained three times: (1) ImageNet to Task 1, (2) Task 1 to Task 2, and (3) Task 2 to Task 3, while the conventional TL only needs to be conducted once, i.e., from ImageNet to Task 3 directly. This can significantly increase the computational cost and the performance-to-cost ratio may be rendered impractical. Therefore, more efficient algorithms, e.g., MTL, can be considered, where the network can train all tasks simultaneously while considering their inter-relationships. An accurate, efficient, and resilient MTL method, utilizing the relationships between inputs and inter-tasks, is introduced in Chap. 12.

6.6   Case Study: Post-disaster Damage Assessment

6.6.1   Structural ImageNet Model

Inspired by pre-training deep CNNs [1], e.g., VGGNet [9], InceptionNet [18], and ResNet [12], on the ImageNet [1] and then adapting them to new target-domain tasks, the concept of *structural ImageNet model* (SIM) is introduced in this section. In the SIM, if a DL model is pre-trained on the ϕ-Net and has acceptable performances for its tasks, it is thought of as a good source model to be transferred to other structural vision tasks by further fine-tuning, or can be directly applied to practical scenarios, e.g., post-disaster damage assessment and reconnaissance efforts.

From the best results in ϕ-Net benchmark experiments (Table 6.10), several best models for the easy Tasks (i.e., 1–4, refer to Sect. 5.2.2) achieved promising accuracy values. Thus, they can be directly used in real scenarios. For example, the best model for Task 1 (scene level) has 93.4% accuracy in identifying the scene level of the structural images and that for Task 4 (material type) has 98.5% accuracy in classifying materials between steel and non-steel. These two models can be directly applied to newly collected images (e.g., during reconnaissance efforts following major earthquakes) as part of a preprocessing practical procedure to mitigate laborious human efforts in organizing different scale and material type data. Meanwhile, the best models in hard Tasks (i.e., 5–8, refer to Sect. 5.2.2) achieved acceptable accuracy values. Therefore, they can also be treated as source models for further fine-tuning to more complex tasks with larger datasets and finer class categories. For example, in Task 6 (component type) due to the lack of variant structural images, the component type is limited to four classes, including the general label "other components (OC)". If the users collect more labeled images of the "OC" label, these images can be split into finer classes, e.g., brace, joint, or non-structural components, to be used for subsequent fine-tuning and updating of the initial SIM.

6.6.2   Application of SIM to the 1999 Chi-Chi Earthquake

In this section, an application of the ϕ-Net with the use of its trained SIM is presented for the rapid assessment of structural images from the 1999 Chi-Chi earthquake in Taiwan, where unlabeled images are collected from Google Images.[11] The detection procedure is straightforward via feeding the collected images into the trained SIM corresponding to the task of interest, then the model outputs the predicted labels and the probabilities of all

---

[11] https://images.google.com.

	# 1	# 2	# 3	# 4	# 5
Real image					
Scene Level	P: 0.0% O: 0.0% *S: 100.0%*	P: 0.0% O: 0.6% *S: 99.4%*	P: 0.0% *O: 100.0%* S: 0.0%	P: 0.6% *O: 99.4%* S: 0.0%	P: 0.0% *O: 99.9%* S: 0.1%
Damage State	*D: 99.3%* UD: 0.7%	*D: 99.8%* UD: 0.2%	*D: 100.0%* UD: 0.0%	*D: 99.7%* UD: 0.3%	*D: 99.5%* UD: 0.5%
Spalling condition	-	-	NSP: 0.9% *SP: 99.1%*	NSP: 0.1% *SP: 99.9%*	NSP: 0.1% *SP: 99.9%*
Collapse Mode	NC: 6.3% PC: 14.3% *GC: 79.4%*	NC: 0.1% PC: 5.6% *GC: 94.3%*	-	-	-
Component Type	-	-	Beam: 0.3% Col: 45.2% *Wall: 49.7%* Else: 4.8%	Beam: 0.3% Col: 12.2% *Wall: 79.9%* Else: 7.6%	Beam: 2.7% *Col: 57.9%* Wall: 14.1% Else: 25.3%

Fig. 6.15 Results predicted by the SIMs for the 1999 Chi-Chi earthquake images

possible label choices of the task. This operation is repeated for multiple tasks and then a comprehensive set of results is obtained for decision-making purposes. Based on the fact that most collected images belong to the object and scene levels, this application is simplified to focus on the following five tasks among the original ϕ-Net eight benchmark classification tasks: (1) scene level, (2) damage state, (3) spalling condition, (4) collapse mode, and (5) component type. If the image corresponds to the object level, it is explored for the tasks of spalling condition and component type only. On the other hand, if the image corresponds to the structural level, it is explored for the collapse mode task only.

Figure 6.15 lists the attributes of five sample images from the 1999 Chi-Chi earthquake, where the SIMs are the best models for the corresponding tasks, identified in Table 6.10. The predicted labels corresponding to the highest probabilities are shown in bold italic font, which are consistent with the visually-determined ground truth. However, the single underlined value represents a wrong prediction. In this case study, the SIMs achieved satisfactory results for these newly seen images indicating good model generalization. Especially in the easier tasks (i.e., scene level, damage state, and spalling condition), the predicted probabilities for the correct labels are almost 100% showing the high confidence of the used SIMs toward such tasks. The one incorrect example corresponds to the SIM mistakenly predicting the component type of image #3 as a wall instead of a column. From the perspective of the visual pattern, this column is a boundary one on the right side of a wall. Thus, it is highly possible that the texture and boundary of the wall have confused the SIM. Moreover, labels "column" and "wall" have very closely predicted probabilities, 45.2% and 49.7%, respectively, reflecting the low confidence of the model under the ambiguity of the input image #3. On the contrary, images #4 and #5 with clear vision patterns lead to more accurate results with higher confidence. It is noted that the average computational time for a single image in each task is within 0.1 s (i.e., this application is considered as rapid "real-time" or at least "near real-time" assessment) including the numerical encoding of the image and performing the classification.

In summary, the SIMs can achieve good performance in rapid assessment even though benchmarking (not necessarily state-of-the-art) models are adopted to develop these SIMs. The multiple attributes extracted by the SIMs can provide information for decision-making purposes and the detection accuracy can be further improved with the use of better-trained DL models. This application illustrates the practical use and benefit of the ϕ-Net and its well-trained SIMs for vision-based SHM and for better streamlining of future reconnaissance efforts.

6.7   Exercises

1. Provide a brief description of popular methods and strategies used to improve the training of AI models, particularly in DL. Include their key characteristics and the general process involved.
2. Explore the influence on model performance when employing a VGG-16 architecture for a specific task and altering the input size. Based on the network configuration in Table 6.2, compute the number of trainable parameters (including the bias terms) in each considered scenario and list them in a table. Discuss how the change in input size affects the model's behavior and outcome.
3. Discuss the consequences of retraining a varying number of Conv blocks while fine-tuning a VGG-16 model. Compute the number of trainable parameters (including the bias terms) in each scenario and list them in a table.
4. How to judge the occurrence of over-fitting? Discuss two solutions of how to prevent or alleviate this issue.
5. Considering the 4 TL paths outlined in Fig. 6.14, design 2 new unique paths and provide a discussion regarding the rationale behind their design.

References

1. J. Deng et al., Imagenet: a large-scale hierarchical image database, in *2009 IEEE conference on computer vision and pattern recognition* (2009), pp. 248–255
2. Y. LeCun et al., Gradient-based learning applied to document recognition. Proce. of the IEEE **86**(11), 2278–2324 (1998)
3. A. Krizhevsky, Learning multiple layers of features from tiny images. Technical Report TR-2009 (2009)
4. S.J. Pan, Q. Yang, A survey on transfer learning. IEEE Trans. Knowl. Data Eng. **22**(10), 1345–1359 (2009)
5. R. Raina et al., Self-taught learning: transfer learning from unlabeled data, in *Proceedings of the 24th International Conference on Machine Learning* (2007), pp. 759–766

6. M.D. Zeiler, R. Fergus, Visualizing and understanding convolutional networks, in *European Conference on Computer Vision* (Springer, 2014), pp. 818–833

7. L. Shao, F. Zhu, X. Li, Transfer learning for visual categorization: a survey. IEEE Trans. Neural Netw. Learn. Syst. **26**(5), 1019–1034 (2014)

8. A. Krizhevsky, I. Sutskever, G.E. Hinton, Imagenet classification with deep convolutional neural networks, in *Advances in Neural Information Processing Systems* (2012), pp. 1097–1105

9. K. Simonyan, A. Zisserman, Very deep convolutional networks for large-scale image recognition (2014). arXiv:1409.1556

10. H.-C. Shin et al., Deep convolutional neural networks for computeraided detection: CNN architectures, dataset characteristics and transfer learning. IEEE Trans. Med. Imag. **35**(5), 1285–1298 (2016)

11. Y. Song, I. McLoughLin, L. Dai, Deep bottleneck feature for image classification, in *Proceedings of the 5th ACM on International Conference on Multimedia Retrieval* (2015), pp. 491–494

12. K. He et al., Deep residual learning for image recognition, in *Proceedings of The IEEE Conference on Computer Vision and Pattern Recognition* (2016), pp. 770–778

13. M. Oquab et al., Learning and transferring mid-level image representations using convolutional neural networks, in *Proceedings of the IEEE Conference on Computer Vision and Pattern Recognition* (2014), pp. 1717–1724

14. D.P. Kingma, J. Ba, Adam: a method for stochastic optimization (2014). arXiv:1412.6980

15. D.E. Rumelhart, G.E. Hinton, R.J. Williams, Learning representations by back-propagating errors. Nature **323**(6088), 533–536 (1986)

16. K. Simonyan, A. Vedaldi, A. Zisserman, Deep inside convolutional networks: visualising image classification models and saliency maps (2013). arXiv:1312.6034

17. Y. Gao, K.M. Mosalam, PEER Hub ImageNet: a large-scale multiattribute benchmark data set of structural images. J. Struct. Eng. **146**(10), 04020198 (2020)

18. C. Szegedy et al., Going deeper with convolutions, in *Proceedings of the IEEE Conference on Computer Vision and Pattern Recognition* (2015), pp. 1–9

19. F. Zhuang et al., A comprehensive survey on transfer learning. Proc. IEEE **109**(1), 43–76 (2020)

Structural Damage Localization 7

In addition to the classical classification tasks, structural damage localization (following the usual detection task) plays a crucial role in vision-based SHM. As introduced in Chap. 2, damage locations are represented by bounding boxes with coordinates and localization algorithms and models aim to conduct a regression task to accurately predict the coordinates, which closely match the ground truth. Based on their working principle, the localization models can be categorized into two types: single-stage and two-stage. Representative models for the single-stage type include the SSD [1] and YOLO series [2–6]. On the other hand, the R-CNN series [7–9], refer to Sect. 2.2, includes typical models for the two-stage type.

This chapter builds upon the knowledge of the classification tasks, discussed in Chap. 6, and expands the scope to include structural damage localization tasks. There exist several differences between damage classification and localization tasks. One major distinction is the labeling format, where the localization task uses multi-dimensional vectors to record the box coordinate information, unlike the discrete numbers used in the classification tasks. Consequently, specific localization evaluation metrics, namely, intersection over union (IoU) and average precision (AP) are introduced. Subsequently, several typical localization models are presented. These include the single-stage type of YOLO series, which comprises vanilla YOLO, YOLO v2, v3, and v4, as well as the two-stage type of R-CNN series, which includes R-CNN, Fast R-CNN, and Faster R-CNN. To further explore the feasibility of using these methods in damage localization tasks, two case studies are presented focusing on the following two tasks: (1) bolt loosening detection and (2) concrete spalling detection.

© The Author(s), under exclusive license to Springer Nature Switzerland AG 2024   181
K. M. Mosalam and Y. Gao, *Artificial Intelligence in Vision-Based Structural Health Monitoring*, Synthesis Lectures on Mechanical Engineering,
https://doi.org/10.1007/978-3-031-52407-3_7

7.1   Localization Evaluation Metrics

7.1.1   Intersection over Union

The IoU, referred to as the Jaccard index or Jaccard similarity coefficient [10], is a classical metric used in localization and segmentation tasks to estimate how well a predicted bounding box or mask matches the ground truth data. It is defined as the ratio of the intersection (overlap) area and the union area of the ground truth and the predicted bounding boxes in localization or masks in segmentation, refer to Fig. 7.1.

The value range of IoU is from 0.0 (no overlap) to 1.0 (perfect fit) indicating the degree of fitness between the prediction and the corresponding ground truth. Empirically, $IoU \geq 0.95$ is an excellent score, $0.95 > IoU \geq 0.7$ is a good score, and $IoU = 0.5$ is commonly treated as a borderline score. For localization problems involving multiple classes, where the number of target object classes exceeds 2, the mean IoU (mIoU) is calculated by averaging the IoU values across all classes.

7.1.2   Average Precision

In several widely recognized CV benchmark datasets for localization and segmentation tasks, such as the MS COCO dataset [11], another standard evaluation metric for both localization and segmentation is the AP. This metric is calculated using varying IoU thresholds, which are denoted by ϵ's. It should be noted that when computing the AP, the calculations of the recall and precision follow Eqs. 3.14 and 3.16, respectively, in the classification task, with the exception of the definition of the model predictions: TP, FP, TN, and FN. For one data sample, redefine TP as the number of bounding boxes or masks with IoU greater than ϵ. On the other hand, FP is redefined as the number of bounding boxes or masks with IoU equal to or less than ϵ, while TN corresponds to the model detecting no bounding boxes or masks when the ground truth indeed has no such bounding boxes or masks and FN represents the number of missed detection cases (i.e., bounding boxes or masks present in the ground truth, but not detected by the model). Consequently, the computed recall and precision enable the generation of the so-called precision-recall (PR) curve.

Fig. 7.1 IoU computation

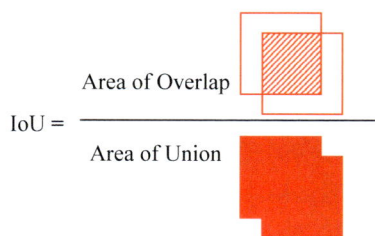

$$IoU = \frac{\text{Area of Overlap}}{\text{Area of Union}}$$

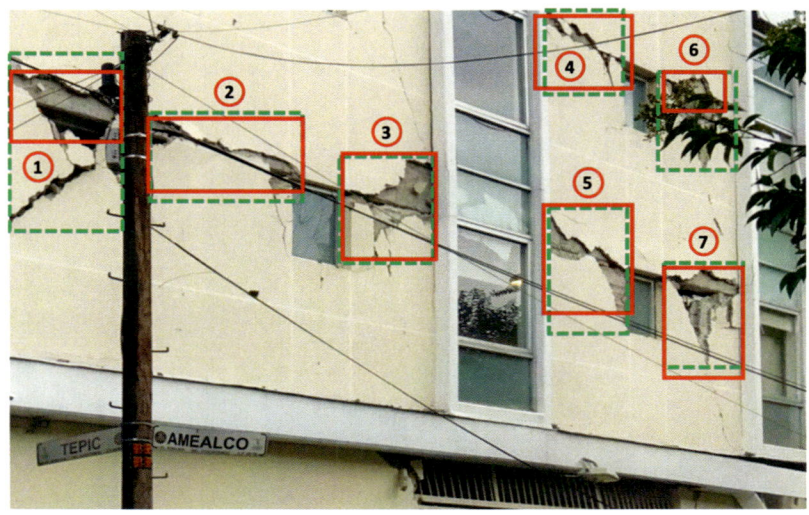

Fig. 7.2 Example of 7 instances of spalling damage with both ground truth and prediction (*Note* green dashed box represents the ground truth and red solid box represents the prediction by the AI model)

Table 7.1 Bounding box predictions for the example in Fig. 7.2

Rank	Confidence	IoU > 0.5	Outcome	Box #
1	0.98	True	TP	3
2	0.89	True	TP	4
3	0.88	True	TP	5
4	0.78	True	TP	7
5	0.66	False	FP	1
6	0.61	False	FP	6
7	0.52	True	TP	2

The most commonly used IoU threshold is $\epsilon = 0.5$, which is often set as the default value, and the resulting AP is referred to as AP^{50}. To illustrate the calculation of AP^{50}, Fig. 7.2 presents an example where there are 7 instances of spalling damage in a structural image and the localization network predicts 7 bounding boxes along with the corresponding confidence scores. As discussed in Sect. 7.4.3, the confidence score reflects the degree of the model's confidence that the box contains an object. The IoU threshold, ϵ, is set as 0.5 and thus the label "True" is assigned for $\epsilon > 0.5$ and "False" otherwise. The results, sorted by the confidence score of each bounding box, are presented in Table 7.1.

Initially, the threshold for the confidence is set to 0.98 from the top-ranked prediction in Table 7.1, where only one correct box prediction with a "True" label is identified. Using

Table 7.2 Results of the precision and recall for the example in Fig. 7.2

Rank	# TP	# FP	Precision (P)	Recall (R)
1	1	0	1.00	0.14
2	2	0	1.00	0.28
3	3	0	1.00	0.42
4	4	0	1.00	0.57
5	4	1	0.80	0.57
6	4	2	0.67	0.57
7	5	2	0.71	0.71

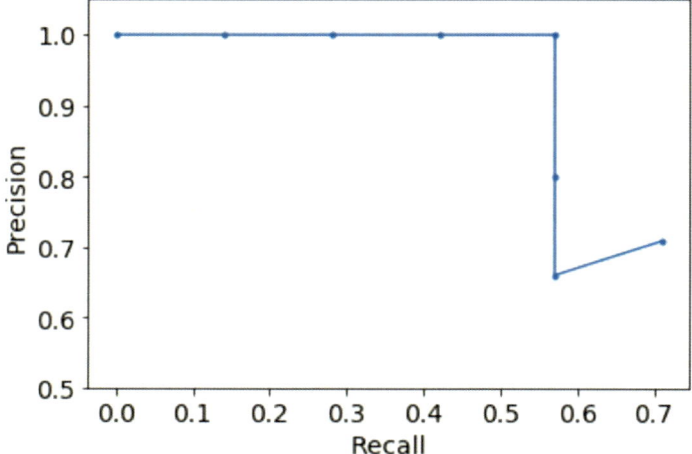

Fig. 7.3 PR curve of the predictions for the example in Fig. 7.2

Eqs. 3.14 and 3.16, a PR pair is calculated as $P = 1/1 = 1.00$ and $R = 1/7 \approx 0.14$. Moving forward, the threshold for confidence is adjusted to 0.89 for the second-ranked prediction in Table 7.1, where in this case, two correct predictions are found and this results in a PR pair of $P = 2/2 = 1.00$ and $R = 2/7 \approx 0.28$. This process is repeated until the seventh rank in Table 7.1, with a confidence threshold of 0.52. The corresponding values of P and R are coincidentally the same, where $P = R = 5/7 \approx 0.71$. Values of P and R for all 7 boxes are listed in Table 7.2. Finally, the PR pairs are treated as coordinates, (R, P), for the PR curve shown in Fig. 7.3.

There are many ways to compute the AP and one of the popular approaches [11] is to compute the AP using the AUC of the PR curve for a specific choice of the IoU threshold value. However, in real applications, the PR curve is often not smooth but jagged and conducting integration to calculate the AUC of such a practical PR curve, having many points, is time-consuming. Thus, a smoothing operation is typically adopted before calculating the

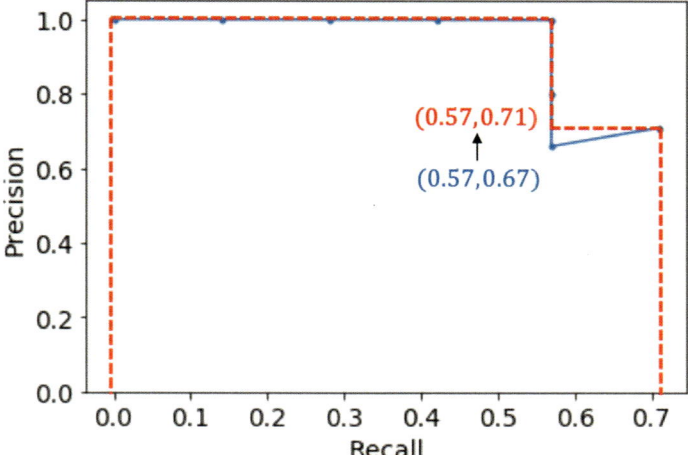

Fig. 7.4 Smoothing the PR curve of the predictions for the example in Fig. 7.2

AUC. Therefore, for each point on the PR curve, the P value is taken as the maximum precision on the right side of that point. For example, as illustrated in Fig. 7.4, the point $(0.57, 0.67)$ is elevated to $(0.57, 0.71)$. Ultimately, given the smoothed PR curve in Fig. 7.4, i.e., the red dashed line, the AP score is computed as the AUC. Thus, in this example, $AP^{50} = 0.57 \times 1.00 + (0.71 - 0.57) \times 0.71 = 0.67$. It is noted that different AP methods yield different AP scores due to different assumptions and used simplifications, e.g., the previously discussed smoothing process.

In several benchmarking studies [12–14], a set of AP variations, e.g., AP^{75}, AP^{S}, AP^{M}, and AP^{L}, were adopted for specific purposes. AP^{75} is similar to AP^{50} but calculated with $\epsilon = 0.75$. It applies a more strict threshold, reducing the chances of FP and FN in the detection results. This stricter evaluation provides a more rigorous assessment, particularly useful when high precision and accurate localization are crucial. On the other hand, AP^{S}, AP^{M}, and AP^{L} are used for evaluating the model's detection ability for *small* objects (ground truth bounding box resolution $< 32 \times 32$), *medium-sized* objects ($32 \times 32 \leq$ ground truth bounding box resolution $\leq 96 \times 96$), and *large* objects (ground truth bounding box resolution $> 96 \times 96$), respectively.

Extending to multi-class (with the number of classes larger than 2) problems, such as multiple damage pattern localization, the computation of the AP differs slightly from the approach mentioned above. In this case, the calculation of the AP considers the predictions and ground truth annotations for each class individually. In other words, for a specific class, only the predicted bounding boxes, the corresponding confidence scores, and the ground truth annotations are related to that class. Assume there are C classes and the AP_c, where $c \in \{1, 2, \ldots, C\}$, is computed for each class using the same process as in the binary cases ($C = 2$). The final AP, which is commonly referred to as the mean average precision (mAP), is then calculated as the mean of all AP_c scores, i.e., $mAP = \sum_{c=1}^{C} AP_c / C$.

7.2   YOLO Series

Traditional object detection or localization methods involve running a classification model on various regions of an image, which, in general, is computationally intense. YOLO is the first single-stage regression-based method proposed in 2016 by Redmon et al. [2]. It takes a different approach by reformulating the localization problem as a regression one and directly predicting the bounding boxes (note that Bbox is used in the sequel as a short for bounding box) and their confidence scores (explained in Sect. 7.2.1) from the entire image in a single pass.

YOLO has been continuously evolving going through multiple versions and improvements, such as YOLO v2, YOLO v3, and YOLO v4. Each version introduced architectural changes and enhancements to improve the accuracy and speed. The YOLO series gained popularity due to its ability to provide real-time object detection on resource-constrained devices, making it suitable for applications like self-driving cars, surveillance systems, and real-time video analysis. In this section, the vanilla YOLO is discussed in detail and illustrated using SHM scenarios. Besides, three classical variants, i.e., YOLO v2, YOLO v3, and YOLO v4, are briefly introduced.

7.2.1   YOLO Framework

The vanilla YOLO builds an end-to-end framework, which uses the entire image as an input to the network, directly regressing the position and category of the Bbox in the output layer. The network reasons globally about the full image and all the objects in the image [2]. Unlike other methods and algorithms that employ sliding windows or region proposals (such as the R-CNN series introduced in Sect. 7.4), YOLO performs the detection in a single pass through the NN. In addition, running the vanilla YOLO requires the use of three key features, namely, (1) a grid-based approach, (2) Bbox and class prediction, and (3) non-maximum suppression[1] (NMS).

The detailed working principle of YOLO is introduced below and illustrated with a structural image for damage localization, as an example in Fig. 7.5. Given an arbitrary structural image input, it is firstly resized to a pre-determined resolution, $H \times W$, where $H = W = 448$ is adopted in the original YOLO study [2]. Subsequently, YOLO divides the input image into multiple grid cells with a fixed size of $S \times S$, e.g., 7×7 is used in [2]. Each grid cell is responsible for predicting objects if the center of an object falls within it.

YOLO predicts N_B Bboxes and their corresponding confidence scores for each grid cell. The Bboxes are defined relative to the cell's location, size, and aspect ratio. Each Bbox consists of 5 components, denoted by (x, y, w, h, CS). Given an image with width W and height H and taking the top-left corner coordinates as the origin, i.e., $(0, 0)$, then

---

[1] Non-maximum suppression is a CV method that selects a single entity out of many overlapping entities, by typically adopting the criterion of discarding entities below a given probability bound.

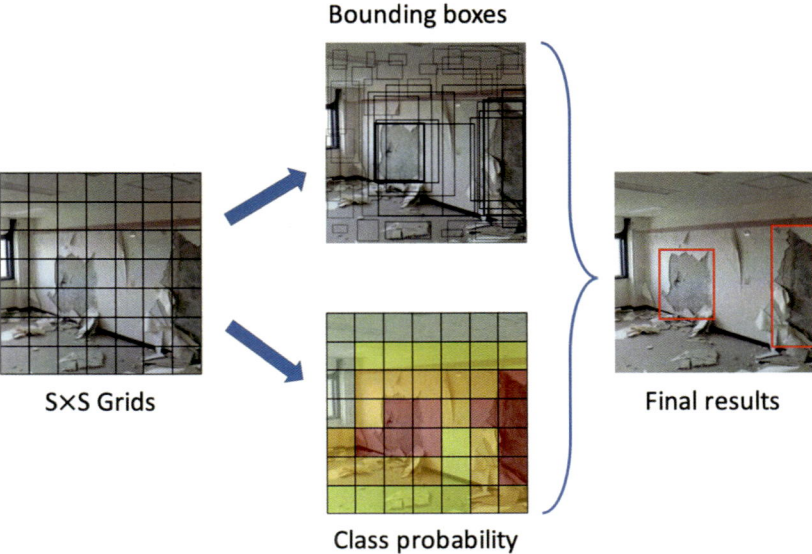

Fig. 7.5 YOLO localization for an image with spalling damage ($S = 7$)

the bottom-right corner coordinates of the image become (W, H). Consider a Bbox with global coordinates of its geometric center (x_B, y_B), width w_B, and height h_B. This Bbox is assumed to be responsible for one particular cell whose top-left corner coordinates are (x_g^{tl}, y_g^{tl}) (considered as the origin of the cell). The (x, y) pair is the coordinates of the center of the Bbox relative to the origin of the grid cell. This pair of coordinates is parameterized to be relative offsets of the Bbox with respect to the cell at hand, i.e., $x = \left(x_B - x_g^{tl}\right)/x_g^{tl}$ and $y = \left(y_B - y_g^{tl}\right)/y_g^{tl}$. It is to be noted that this normalization (unlike, e.g., dividing by the cell size, W/S or H/S) is important to tie the Bbox with the cell, which the Bbox is responsible for predicting its objects. This is inspired by the positional encoding, discussed in Sect. 4.5.3.1. Moreover, the Bbox dimensions are normalized by the image dimensions to obtain $w = w_B/W$ and $h = h_B/H$. The example in Fig. 7.6 illustrates these normalization rules. Assume the input image is 224×224, i.e., $W = H = 224$, and divided into a $S \times S = 4 \times 4$ grid, i.e., the cell size in pixels is $W/S \times H/S = 224/4 \times 224/4 = 56 \times 56$. Suppose one Bbox is predicted with a pair of center coordinates (x_B, y_B) = (122, 95) and dimensions $w_B = 118$ and $h_B = 90$. This Bbox is responsible for a cell at the second row and third column of the grid (counted from the top left cell of the grid), whose pair of coordinates of its top-left corner is (x_g^{tl}, y_g^{tl}) = (112, 56), as shown in Fig. 7.6. Based on the above normalization rules, $x = (122 - 112)/112 \approx 0.09$, $y = (95 - 56)/56 \approx 0.69$, $w = 118/224 \approx 0.53$, and $h = 90/224 \approx 0.40$.

The confidence score (CS) reflects the likelihood (in terms of probability) of the Bbox to contain an object, which is calculated as "$P(\text{obj}) \cdot \text{IoU}_{\text{obj}}$". It is to be noted that $P(\text{obj})$

Fig. 7.6 Example of Bbox
coordinates ($S = 4$)

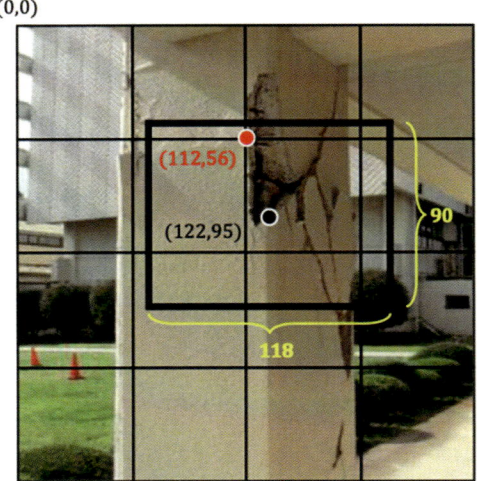

judges the existence of the object in the grid cell. If there is no object in the cell, $P(\text{obj}) = 0.0$, otherwise $P(\text{obj}) = 1.0$. The specific metric IoU_{obj} calculates the IoU value (refer to Fig. 7.1) between the ground truth and the predicted Bboxes. In addition, YOLO predicts C (number of classes) conditional class probabilities $P(\text{class}|\text{obj})$ for each grid cell, indicating the presence of different object categories within that box. These probabilities are computed by a backbone DL model, e.g., CNN, using the *softmax* activation. In other words, only one set of class probabilities is calculated per each grid cell, regardless of the number of Bboxes. Therefore, for each input image, YOLO outputs a tensor with the shape of $(S, S, 5 \times N_B + C)$. For example, suppose $S = 7$, $N_B = 4$, where each Bbox is defined with 5 components, as discussed above, and number of classes is $C = 3$, e.g., Task 5 in Sect. 5.2.2.5, the output has the shape of $(7, 7, 23)$.

From a statistical point of view, the ground truth is only available during training, i.e., $P(\text{obj})$ is known for each grid cell, and the DL model outputs the $P(\text{class}|\text{obj})$. While predicting a new input test image, since the model is well-trained, the DL model can directly output the conditional class probability $P(\text{class}|\text{obj})$ and its CS, which actually encodes the information of $P(\text{obj}) \cdot \text{IoU}_{\text{obj}}$ in training. Therefore, a class-specific CS for class i is obtained for each Bbox as follows:

$$CS_i = P(\text{class}|\text{obj}) \cdot P(\text{obj}) \cdot \text{IoU}_{\text{obj}}. \tag{7.1}$$

After generating the Bbox predictions, there may exist multiple Bboxes with different sizes related to one object. Thus, the NMS technique is applied to eliminate overlapping Bboxes with lower confidence scores, keeping only the most confident ones, as illustrated in Fig. 7.7. There are several different ways for the NMS computation and for more details refer to [15].

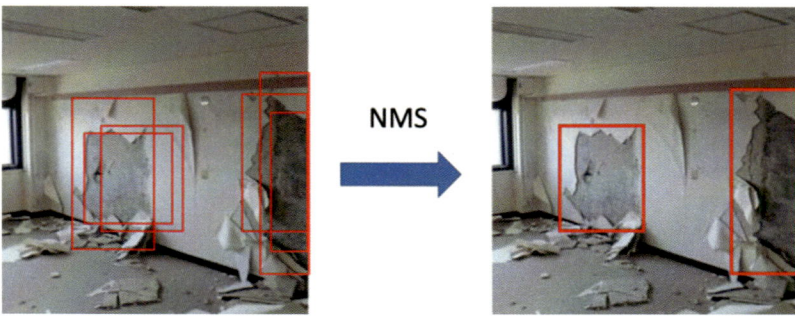

NMS

Fig. 7.7 NMS for overlapped Bboxes

7.2.2 Network Configuration and Loss

Vanilla YOLO adopts InceptionNet-like [16] CNN as the backbone to extract features and is followed by FC-layers to determine the tensor output, discussed in Sect. 7.2.1, including class and Bbox information. YOLO treats the target detection as a regression problem and it uses the MSE as the loss function. As shown in Eq. 7.2, the loss, L, is composed of two parts with respect to localization error and classification error with different weights. The localization error computes the coordinate difference between the ground truth and predicted Bboxes, which has a weight factor λ_{coord}. On the other hand, the classification error measures the confidence predictions for the Bboxes that do not contain objects, which is weighted by λ_{nobj}.

$$
\begin{aligned}
L = {} & \lambda_{\text{coord}} \sum_{i=0}^{S^2} \sum_{j=0}^{N_B} \mathbb{1}_{i,j}(\text{obj}) \cdot \left[\left(x_i - \hat{x}_i \right)^2 + \left(y_i - \hat{y}_i \right)^2 \right] \\
& + \lambda_{\text{coord}} \sum_{i=0}^{S^2} \sum_{j=0}^{N_B} \mathbb{1}_{i,j}(\text{obj}) \cdot \left[\left(\sqrt{w_i} - \sqrt{\hat{w}_i} \right)^2 + \left(\sqrt{h_i} - \sqrt{\hat{h}_i} \right)^2 \right] \\
& + \sum_{i=0}^{S^2} \sum_{j=0}^{N_B} \mathbb{1}_{i,j}(\text{obj}) \cdot \left(CS_i - \widehat{CS}_i \right)^2 \\
& + \lambda_{\text{nobj}} \sum_{i=0}^{S^2} \sum_{j=0}^{N_B} \mathbb{1}_{i,j}(\text{nobj}) \cdot \left(CS_i - \widehat{CS}_i \right)^2 \\
& + \sum_{i=0}^{S^2} \mathbb{1}_i(\text{obj}) \cdot \sum_{c \in \text{classes}} \left(P_i(c) - \hat{P}_i(c) \right)^2,
\end{aligned}
\tag{7.2}
$$

where $P_i(c)$ is the class probability of the grid cell i and "$\hat{\bullet}$" represents the predicted quantity $\bullet$. The indicator function $\mathbb{1}_i(\text{obj})$ is 1 if the object, e.g., spalling, occurs in grid cell i and 0 otherwise. Similarly, $\mathbb{1}_{i,j}(\text{obj})$ is 1 if there exists the object, e.g., spalling, in grid cell i with the jth Bbox predictor in this cell is indeed responsible for that cell and 0 otherwise. Moreover, $\mathbb{1}_{i,j}(\text{nobj})$ is the opposite of $\mathbb{1}_{i,j}(\text{obj})$, i.e., it is 1 when there does not exist the object in cell i considering the Bbox j and 0 otherwise. In Eq. 7.2, the first two terms are related to the coordinate prediction, the third term is related to the CS prediction when the Bbox includes the object, the fourth term is about the CS prediction when the Bbox does not have the object, and the final term is used for penalizing the class prediction mismatch. It is worth mentioning that the weights for the third and fifth terms are set to 1.0, which are dropped from the expression. Moreover, the use of the square roots in the second term, to some extent, reduces the importance of variations on larger bounding boxes compared to smaller ones. For more details about the vanilla YOLO, the reader is referred to [2].

7.2.3   Advantages and Limitations

Vanilla YOLO has the following advantages:

+ Simple and fast: It builds a single-stage framework, i.e., pipeline, with a CNN, whose training and prediction are both end-to-end. Therefore, it is relatively simple and costs less time than two-stage models.
+ Low false positive rate of the background: It performs convolution on the entire image, which leads to a large view, making it less prone to misjudgment of the image background.
+ Strong generalization and robustness: It works for many non-natural objects, e.g., artworks, which provides the potential to detect damage in structural images for vision-based SHM.

On the other hand, vanilla YOLO has several limitations as listed below:

− Strong spatial constraints: Even though each grid predicts N_B Bboxes, only one Bbox with the highest confidence score is ultimately selected as the output, i.e., each grid can only predict at most one object. When the proportion of objects in the image is small, such as certain grids containing multiple small spalling regions, only one of them can be detected.
− Low generalization rate for new or unusual object aspect ratios: YOLO learns the object aspect ratio only from the training data. Thus, for new data with different or unusual aspect ratios, the YOLO model may struggle to generalize well with such lack of information. This means that it may be unable to locate objects with unusual proportions, e.g., long cracks.

– Lower localization accuracy compared to two-stage models: This is due to the loss function used to estimate the detection performance, which treats the errors in the same way for both small and large bounding boxes (refer to the second term in Eq. 7.2 with the square root quantities). As a result, it can lead to incorrect localization results.

7.2.4 YOLO v2/v3/v4

7.2.4.1 YOLO v2

YOLO v2 [3] is an improved version of the vanilla YOLO and it achieves better prediction accuracy, higher speed, and more recognition of the objects. It can support detecting 9,000 different objects and a YOLO9000 model is obtained when it is trained simultaneously on both object detection and image classification datasets.

Compared to vanilla YOLO, YOLO v2 introduces many new designs and techniques for performance enhancement. YOLO v2 added the BatchNorm (Sect. 4.2.4) (BN) to solve the gradient vanishing and explosion problems in the backpropagation process and to reduce the sensitivity to some hyper-parameters (e.g., learning rate, range of network parameters, and selection of the activation function). It introduced the concept of the high-resolution classifier that fine-tunes the classifier on images with a resolution of 448×448 (vanilla YOLO trains a classifier on the resolution of 224×224) and then re-traines the network for localization using 448×448 images, adjusting the backbone to work better on higher resolution inputs. By removing the FC-layer, YOLO v2 adopts the classification network called Darknet[2] as the backbone network and uses the prior box to predict the Bboxes, so as to achieve better detection efficiency [3]. The prior box is also called the *anchor box* in Faster R-CNN [9], refer to Sect. 7.4.3, where a set of boxes with pre-determined sizes and aspect ratios are assigned to each grid cell. For example, in vanilla YOLO using $S = 7$ and $N_B = 2$, the total number of prior boxes to be evaluated is $7 \times 7 \times 2 = 98$. For a more general example, if evaluating 3 different sizes and 2 different aspect ratios of the prior boxes, the number of prior boxes to be evaluated is increased to $7 \times 7 \times 3 \times 2 = 294$. Similar to the vanilla YOLO, the NMS technique is applied to eliminate overlapping boxes to generate the final Bbox for the prediction. To alleviate two concerning issues that occur when adopting the idea of the prior box, namely, (1) manual selection of box dimensions and (2) model instability, especially in the early training stage, *dimension clusters*, and *direct location prediction* techniques are proposed to respectively resolve these two issues, which are also adopted in later versions of YOLO, e.g., YOLO v3. Besides, YOLO v2 includes other new techniques, such as using fine-grained features, multi-scale training, and hierarchical classification. For more details, the reader is referred to [3].

---

[2] Darknet is an open-source framework to train NNs and is the basis for YOLO to setup its architecture. The original repository of Darknet can be found in https://github.com/pjreddie/darknet.

7.2.4.2 YOLO v3 and YOLO v4

The model of YOLO v3 is much more complex than previous models and the running speed and prediction accuracy can be balanced by changing the size of the model structure. Three major improvements are made: (1) consider multi-scale prediction, (2) adopt a better backbone classification network, and (3) avoid using *softmax* in the classifier and instead use the binary cross-entropy as the loss function, Eq. 3.7.

To consider a multi-scale prediction, a FPN [17] is introduced in YOLO v3. As illustrated in Fig. 4.26, FPN builds a pyramid of multi-scale feature maps from an input image. It combines high-resolution features with low-resolution and semantically strong features to create a feature pyramid that preserves both fine-grained details and high-level semantic information. For more details of FPN, refer to [9].

The whole YOLO v3 framework is depicted in the example shown in Fig. 7.8. For improved performance in feature extraction, the YOLO v3 model adopts Darknet-53, a deep CNN having 53 Conv layers without FC-layers, as the main backbone network. Darknet-53 is similar to ResNet in using the residual unit (refer to Sect. 4.3.3), denoted as "Resx" in Fig. 7.8 and its recognition accuracy is close to ResNet-101 and ResNet-152 but is faster [4]. Note that "x" in "Resx" stands for # of residual units in each Resblock as shown in the bottom left insert of Fig. 7.8. Besides the residual units, Darknet introduces a new module, namely, CBL,[3] which compresses the model and accelerates its convergence. In the neck layer, the FPN is used to up-sample ("US" in Figs. 7.8 and 7.9) and fuse the extracted features from the backbone. Finally, the head layer uses CNNs for feature decoding to achieve the desired object detection.

On the basis of YOLO v3, YOLO v4 is proposed by Bochkovskiy et al. [5]. The network structure of YOLO v4 is very similar to that of YOLO v3. However, YOLO v4 improves several aspects, e.g., backbone, feature fusion module, and network training techniques.

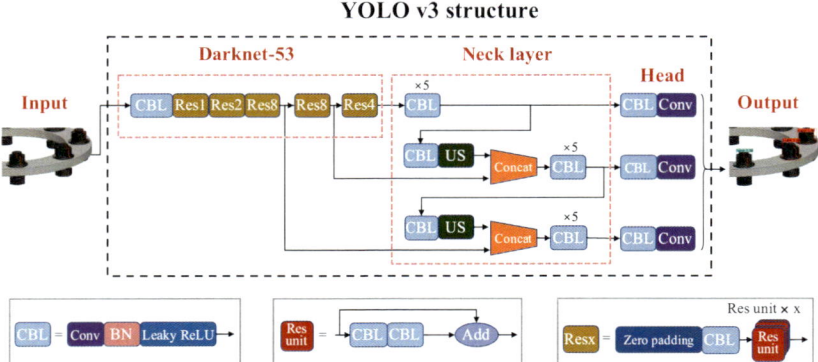

Fig. 7.8 YOLO v3 network

---

[3] C, B, and L stand for Conv, BatchNorm, and *leaky ReLU* (*LReLU*) in Eq. 4.2, respectively.

YOLO v4 Structure

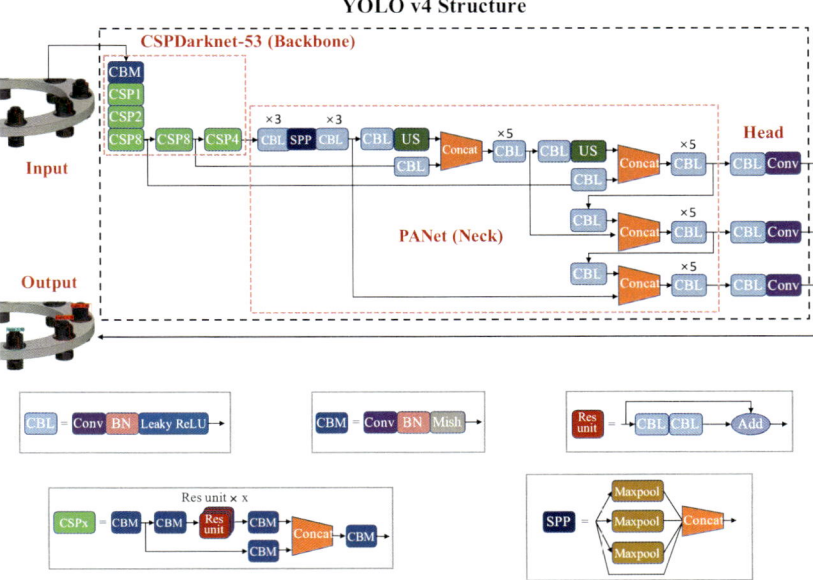

Fig. 7.9 YOLO v4 network

YOLO v4 is characterized by its integration of many advanced techniques and modules. As shown in Fig. 7.9, the YOLO v4 model adopts CSPDarknet-53[4] as the backbone network, supplemented with the Mish activation function[5] to extract the image features. Note that in Fig. 7.9, Mish is part of the CBM (Conv, BN, Mish) module, see related insert in Fig. 7.9. Similar to Fig. 7.8, the "x" in CSPx stands for the # of residual units in each Resblock as shown in the bottom right insert of Fig. 7.9. The path aggregation network (PANet) [20] is combined with the spatial pyramid pooling (SPP) network (SPPNet) [21] as an additional module to form a neck network to achieve a better fusion of the extracted features. These improvements enhance the learning ability of the network, achieve data augmentation, and improve the regression speed.

In the ever-changing landscape of AI and CV domains, the YOLO series models have been proliferating at an accelerating pace and their evolution shows no signs of slowing down. Due to the rapid advancements in this field, this section does not provide an exhaustive coverage of all the existing YOLO models. Nevertheless, the above coverage represents sufficient background for avid readers to pursue further readings of the evolving literature and remain updated on the latest YOLO models.

---

[4] CSPDarknet-53 employs a "cross stage partial (CSP) network" (CSPNet) [18] within the DarkNet-53 [4].

[5] Mish is a self-regularized non-monotonic function [19], which is mathematically expressed as $f(x) = x \cdot \tanh(softplus(x))$, where $softplus(x) = \ln(1 + e^x)$.

7.3   Case Study: Bolt Loosening Localization

Wind is a renewable source of energy. The wind energy system harnesses the power of wind to generate electricity. One of the key components of a wind energy system is the wind turbine tower (WTT), which plays a vital role in capturing this energy. Due to the rapid expansion in the field of wind energy, there has been a significant increase in the number of accidents involving the collapses of WTTs and many of them are due to the loosening of mechanical bolt connections [22]. Therefore, these accidents warn that bolt loosening is a critical factor that triggers the collapse of WTTs and highlight the importance of effective monitoring of the loosening status of the bolts.

In most of the current WTTs inspection projects, hand-held equipment is used (e.g., ultrasonic flaw detector and digital torque wrench) to assist in manual inspection detection to obtain information on the loosening status of the bolts. However, many bolts are concealed, difficult to reach, and detect their conditions by such equipment. Moreover, in some wind farm sites, the manual inspection approach is not only expensive and time-consuming but also hazardous (and hence may violate health and safety regulations) during situations of extreme weather conditions. In addition, due to the influence of the external environment affecting the inspector and/or the hand-held equipment, there may be misjudgment for bolts with small angles of looseness. Therefore, in this case study, a YOLO-based method is introduced to realize a fast and accurate bolt-loosening detection technique for WTTs.

7.3.1   Two-Step YOLO-Based Localization Framework

As mentioned in Chap. 1, applying AI to real civil and structural applications is not straightforward. Two issues are raised for this case study, namely, (1) *labeled data deficiency*: the dataset collection process of WTTs' bolt loosening detection is often time-consuming and sometimes real images with specific features are difficult to obtain and (2) *small angel sensitivity*: existing methods lack accurate and straightforward detection of bolts with small angles of looseness. Therefore, a two-step YOLO-based framework for WTTs' bolt loosening detection is developed, as depicted in Fig. 7.10.

The core idea of the developed framework is that, in step I, high-quality synthetic images are generated using a 3D graphical modeling software to mimic the real images for training and validation, where an assumption is made that these synthetic images share similar properties with the real ones. Subsequently, in step II, the trained model is evaluated using the images collected from different real scenarios. Accordingly, the underlying principle of this framework involves training on a synthetic dataset and subsequently making predictions on a real-world dataset. By iteratively repeating these two steps and comparing the detection results of various versions of the YOLO model, one identifies the most appropriate YOLO-based model for future predictions.

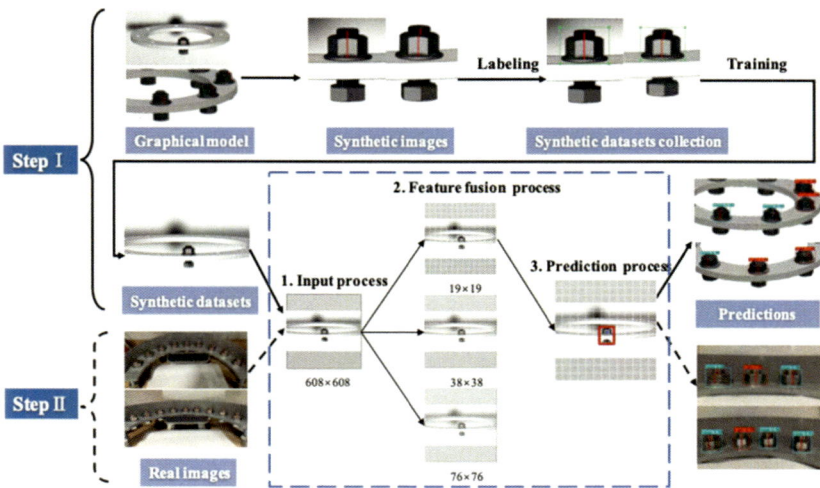

Fig. 7.10 YOLO-based bolt loosening detection framework for WTTs

In step I, the features contained in the datasets can be adjusted according to the detection results of different versions of the used YOLO models to achieve better detection results. The workflow for step I includes the following steps:

1. Building the graphical model: Create the graphical model of a generic bolted connection using the 3D modeling software, namely, SolidWorks [23].
2. Forming the dataset: Varying the size, loosening angle, and perspective of the graphical model, images containing a single bolt each are collected to form the training dataset and images containing multiple bolts are collected to form the validation dataset. Subsequently, the labeled datasets are constructed using the *LabelImg* software.[6]
3. Training the YOLO models: The YOLO-based models, i.e., YOLO v3, YOLO, or YOLO v4-Tiny (to be discussed in Sect. 7.3.3), are selected as candidates. They are trained by the synthetic training dataset and validated on the synthetic validation dataset to obtain the validation results of each YOLO-based model.
4. Selecting the best candidate models: Based on the validation results of the different versions of the YOLO models, the hyper-parameters are adjusted and/or the size of the synthetic training dataset is varied. Finally, the training and validation are repeated to determine a set of candidate models with the best performance.

In step II, a new dataset containing images of WTTs' bolts taken from real environments is collected as a test dataset to comprehensively evaluate the candidate models trained and

---

[6] *LabelImg* is an open-source graphical annotation tool used for labeling object bounding boxes in images, refer to https://github.com/tzutalin/labelImg. It is part of the Label Studio community, https://huggingface.co/LabelStudio.

validated using the synthetic images in step I for real-world applications. The test dataset is developed using real photographs or videos of bolt connections taken via smartphones, cameras, or UAVs. The workflow for step II includes the following steps:

1. Setting the key environmental variables: In order to mimic the detection performance in real environments, various environmental variables are set when collecting these bolt images, including different capture distances, perspective angles, resolutions, lighting conditions, and indoor and outdoor environments.
2. Evaluating the trained models: The collected real images under different environmental conditions are fed into the trained YOLO-based models for a series of comprehensive evaluations.

7.3.2   Bolt Loosening Datasets

In this case study, two types of datasets are developed, i.e., synthetic dataset and real dataset, where the former is generated from screenshots of a graphical model and the latter is built from photographs and video frames taken from laboratory (indoor) and field (outdoor) environments. These two datasets are discussed in detail in the following two sections.

7.3.2.1 Synthetic Image Dataset

The 3D modeling software SolidWorks [23] is utilized to develop a graphical model of the bolted connections, where a flange of the WTTs' bolted connections is selected as the target object, Fig. 7.11. This model consists of a 10 mm thick flange ring with inner and outer diameters of 220 mm and 320 mm, respectively. It incorporates eight sets of M20 nuts[7] and bolts with visible bolt loosening marks (red lines). To train and validate the DL models, synthetic images are generated using SolidWorks, forming the basis for constructing the datasets. The graphical model can be conveniently and effectively reconfigured to generate new images of various types of bolts for loosening detection. Furthermore, SolidWorks offers the capability to simulate real scenes by incorporating different materials and colors and adjusting the environmental conditions, such as shadows and lighting. Although this case study does not explore these features extensively, it demonstrates the potential of using such graphical models in data collection to achieve diverse attributes, e.g., in the damage localization tasks of real-life SHM applications.

A large number of images (a total of 490 images) is collected based on different distances and views as listed in Table 7.3. The relative movement of the marks is used as a criterion for determining the looseness of the bolt and two classification labels are defined: (1) tight (intact) and (2) loose (damaged). These high-quality images are then manually labeled using

---

[7] M20 Hexagon Nuts (Chinese standards [24]) are Mild steel (Grade 8), 18 mm nut thickness, 30 mm nut socket size, 2.5 mm thread pitch, and 20 mm thread size (nominal diameter).

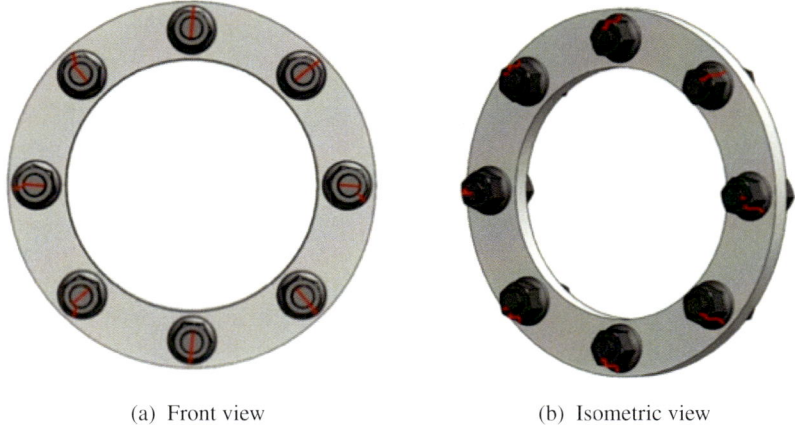

(a) Front view   (b) Isometric view

Fig. 7.11 A sample of the graphical model of the bolted connection flange

Table 7.3 Number of images in the synthetic dataset of the WTT bolted connections

View angle (°)	Loosening angle (°)			Total #
	0	0–5	5–20	
0	80	50	40	170
15	70	50	40	160
30	70	50	40	160

LabelImg to obtain the ground truth bounding boxes (representing the nuts and their bolts). Figure 7.12 shows two typical training images with a resolution of 1494 × 828 pixels.

7.3.2.2 Real Image Dataset

To further evaluate the robustness of different trained DL models and assess their applicability under varied environmental conditions beyond those considered in the generated synthetic images, a laboratory specimen replicating a typical flange of the WTT bolted connection was designed. The specimen consisted of a flange ring and nine sets of standard bolts and their nuts (M20) featuring visible bolt loosening marks as shown in Fig. 7.13. The dimensions of the flange, bolts, and nuts can be found in Table 7.4. Images were taken using a smartphone camera (iPhone X model) equipped with a 12-Megapixel sensor, $f/1.8$

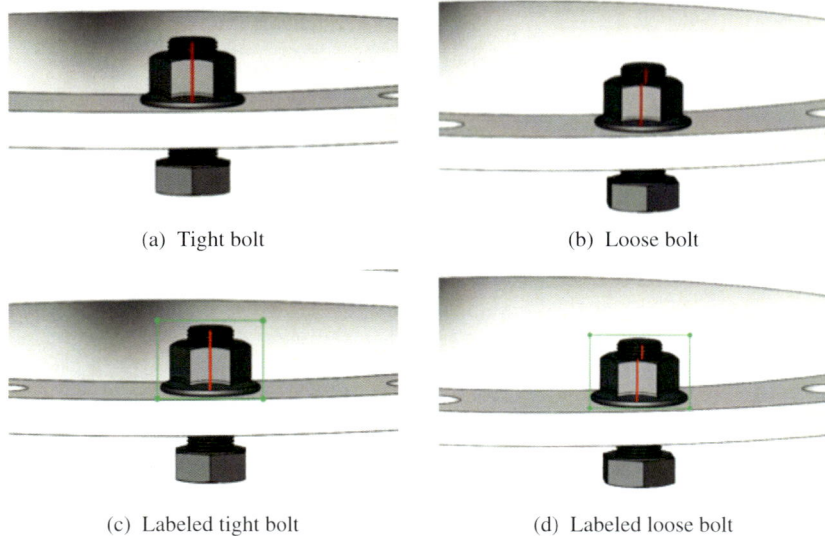

(a) Tight bolt   (b) Loose bolt

(c) Labeled tight bolt   (d) Labeled loose bolt

Fig. 7.12 Typical training images of the WTT bolted connections

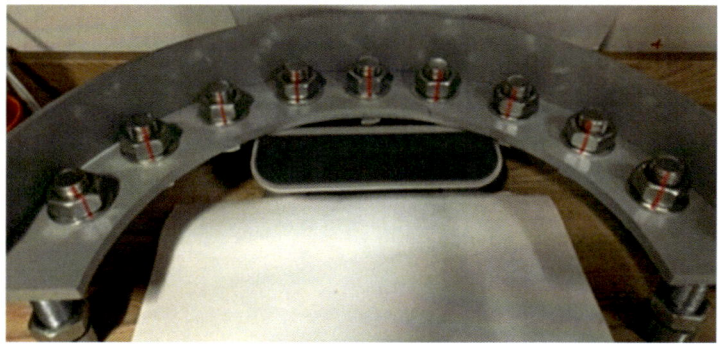

Fig. 7.13 Laboratory specimen of a WTT connection with bolts loosening marks

aperture size,[8] and focal length, $f = 24$ mm. The images have a resolution of 1440×1080 pixels.

To evaluate the robustness of the trained model under different environmental conditions, the bolt images were captured under varied conditions, encompassing different states of looseness, shooting angles, light intensity, and both indoor and outdoor settings. As shown in Fig. 7.14, the vertical and horizontal angles of the cell phone shooting view are defined

---

[8] The aperture is the opening of the lens that light goes through to land on the image sensor of the camera to make a picture. Aperture sizes are measured by f-stops. A high f-stop, e.g., $f/22$ is for a very small aperture, and a low f-stop, e.g., $f/1.8$ is for a wide aperture, where much more light is allowed into the camera than if the aperture is closed down for $f/22$.

Table 7.4 Dimensions of the laboratory specimen of the WTT connection

Dimension	Value (mm)
Flange inner diameter	320
Flange outer diameter	430
Flange thickness	10
Drum thickness	5
Drum height	60
Bolt length	50
Bolt diameter (nominal thread diameter)	20
Nut width (socket size)	30
Nut thickness	18

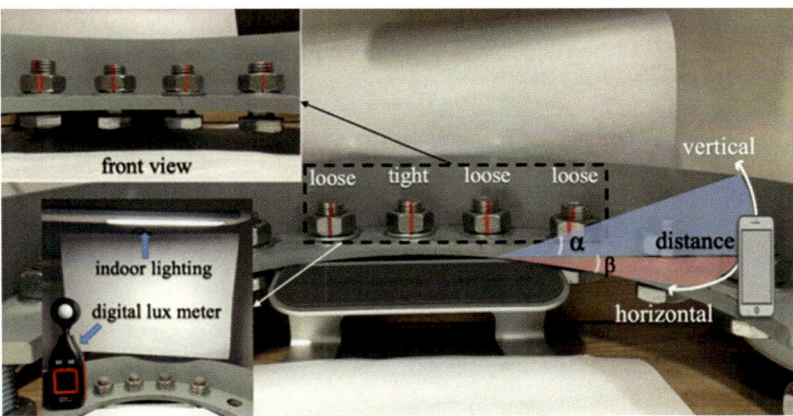

Fig. 7.14 Laboratory setup of a WTT connection and cell phone positioning

by α and β, respectively, and the indoor light intensity was measured by a digital lux meter, placed near the specimen.

7.3.3   YOLO Model Configurations and Settings

Three YOLO models are adopted in this case study. These are YOLO v3, YOLO v4, and YOLO v4-Tiny.[9] Considering the small size of the bolt, the network input size was set to 608×608 pixels. Moreover, TL was adopted where all models were pre-trained on the

---

[9] YOLO v4-tiny is the compressed version of YOLO v4. It reduces the number of parameters and is designed to be performed on machines that have low computing power, e.g., mobile and embedded devices. For more details, refer to [25].

Table 7.5 Training parameters of YOLO-based models for the WTT bolt loosening

Parameter	YOLO v3	YOLO v4	YOLO v4-Tiny
Input size	608×608		
# epochs	200		
$\beta_1 = \beta_2$	0.92		
Learning rate	Initial: 10^{-3}; after 100 epochs: 10^{-4}		
Decay rate	5×10^{-4}		
Batch size	$8 \rightarrow 4$	$4 \rightarrow 2$	$32 \rightarrow 16$

PASCAL VOC 2007 dataset [33]. Based on [22], the number of training epochs was 200 and batch sizes of 8, 4, and 32 were assigned for YOLO v3, v4, and v4-Tiny, respectively. Adam optimizer [26] was adopted, refer to Sect. 6.3.2.2, where its hyper-parameters β_1 and β_2 were taken as 0.92. The initial learning rate and weight decay rate were set to be 10^{-3} and 5×10^{-4}, respectively. After 100 epochs, the learning rate was adjusted to 10^{-4} and the batch sizes of the three models were accordingly reduced by a factor of 4, 2, and 16 for YOLO v3, v4, and v4-Tiny, respectively. The training parameters for the three YOLO models are listed in Table 7.5. All models were trained on 435 (89% of the total 490 images) synthetic images and validated on the remaining 55 (11% of the total 490 images) synthetic images from the dataset discussed in Sect. 7.3.2.1, refer to Table 7.3. These models were evaluated using multiple criteria: (1) accuracy (Sect. 3.4.1) (2) precision (Sect. 7.1.2), (3) recall (Sect. 3.4.2), (4) F_1 score (Sect. 3.4.4), and (5) mAP (Sect. 7.1.2). In addition, all numerical experiments were carried out on the Pytorch platform [27] and a desktop with a single GPU (Intel i7-10700, with 16 GB RAM and NVIDIA RTX 3070 GPU).

7.3.4   Localization Results Under Different Conditions

The validation results for step I are listed in Table 7.6. In terms of model size, YOLO v4-Tiny is much smaller (less than 10%) compared to YOLO v3. However, the former exhibits a mAP value that is $95.71\% - 64.15\% = 31.56\%$ lower than the latter. Additionally, despite YOLO v4 being a newer version than YOLO v3, the results indicate lower accuracy for YOLO v4 in this specific case study. This can be attributed to the fact that YOLO v4 focuses on enhancing the detection accuracy of complex small target objects, which may not contribute significantly to the simple object recognition in this case study. It is noted that similar observations to those from the mAP criterion can be inferred from the other criteria, namely, precision, recall, and F_1 score, as listed in Table 7.6. Furthermore, the YOLO-based models are generally sensitive to the characteristics of the data, which was mostly not the case herein, except for the effect of the image shooting angle and brightness

Table 7.6 The validation results of YOLO-based models for the WTT bolt loosening

Model	Model size	Precision (%)		Recall (%)		F_1 Score		mAP
	(MB)	Loose	Tight	Loose	Tight	Loose	Tight	(%)
YOLO v3	234.71	98.10	77.27	92.79	96.23	0.95	0.86	95.71
YOLO v4	244.29	85.05	71.32	81.98	91.51	0.83	0.80	85.98
YOLO v4-Tiny	22.58	61.94	65.98	86.49	60.38	0.72	0.63	64.15

(as discussed in detail in the following four sections). Moreover, for this particular bolt-loosening problem without complex backgrounds of the images, YOLO v3 is shown to be more suitable. This case study highlights the importance of the *candidate mechanism*, i.e., evaluating several candidate models to select the best one, as newer and more complex models may not necessarily yield better results in all cases.

7.3.4.1 Angles of Looseness

As shown in Fig. 7.15, the bolt with a looseness of 2° or more is correctly identified as "loose", whereas the image of the bolt with a looseness of 1° cannot be accurately classified. It is noted that the number next to the predicted label (loose or tight) indicates the probability $(0.0 \rightarrow 1.0)$ of the label. This indicates that the current model can detect bolt looseness with a minimum detectable angle of 2°. From Table 7.3, the synthetic training set consists of images of bolts with only a range of $(0°–20°)$ looseness. However, the test results encouragingly show that the YOLO v3 model was able to identify bolts with angles of looseness of small values, e.g., 2°, and large values, e.g., 30°. It is noted that when a significant rotation angle occurs, the mark on the nut may be positioned beyond the viewing angle. To assess the model's ability to detect bolt looseness when one of the two marks disappears, further tests were conducted with a bolt rotation of 40°, 60°, 90°, 120°b and 150°, refer to Fig. 7.16. These tests indicate that the model's recognition capability is not limited solely to the features contained within the training and validation datasets but also exhibits strong generalization abilities.

7.3.4.2 Perspective Angles

In this section, the effectiveness of the YOLO v3 model is assessed across various perspective angles (vertical, "α" and horizontal, "β"). Three broad categories of perspective angles were defined: (a) low $(0°–30°)$, (b) medium $(30°–60°)$, and (c) high $(60°–90°)$. As shown in

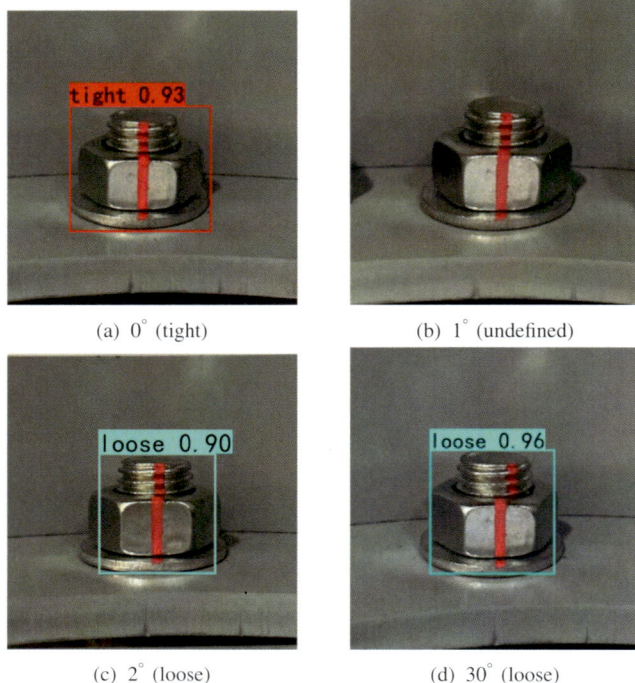

(a) $0°$ (tight)   (b) $1°$ (undefined)

(c) $2°$ (loose)   (d) $30°$ (loose)

Fig. 7.15 Localization results of YOLO v3 for bolts with different loosening angles

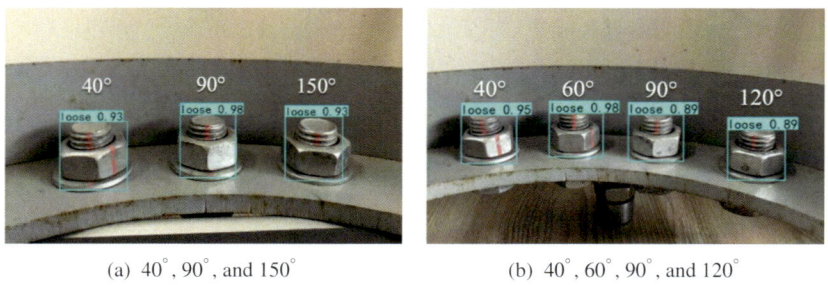

(a) $40°$, $90°$, and $150°$ (b) $40°$, $60°$, $90°$, and $120°$

Fig. 7.16 Localization results of YOLO v3 for bolts with large loosening angles

Fig. 7.17, the model can successfully classify bolts (ground truth, from left to right, is loose, tight and loose) at different α angles. Due to the characteristics of the marks, the relative position of the straight lines can be distorted under different horizontal perspective angles, β, which can lead to incorrect detection. Therefore, the detection of tight and loose bolts at different β angles ($0°$, $15°$, $45°$ and $60°$) was investigated. As shown in Fig. 7.18, the method can achieve accurate detection of bolt loosening (ground truth, from left to right, is tight, loose and loose) at small-to-medium horizontal perspective angles ($0°$ to $45°$) and

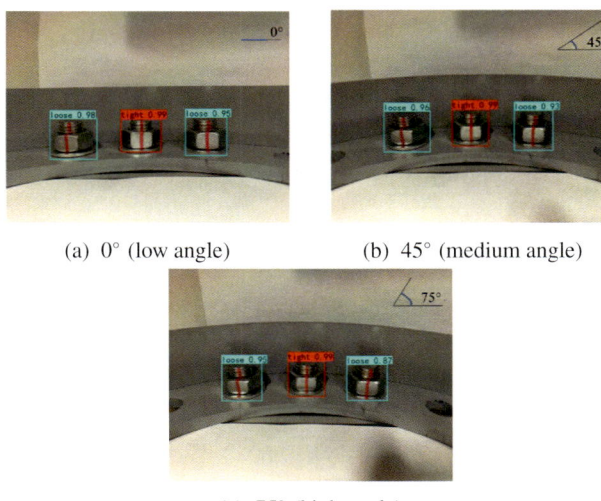

(a) 0° (low angle)   (b) 45° (medium angle)

(c) 75° (high angle)

Fig. 7.17 Localization results of YOLO v3 (different vertical perspective angles)

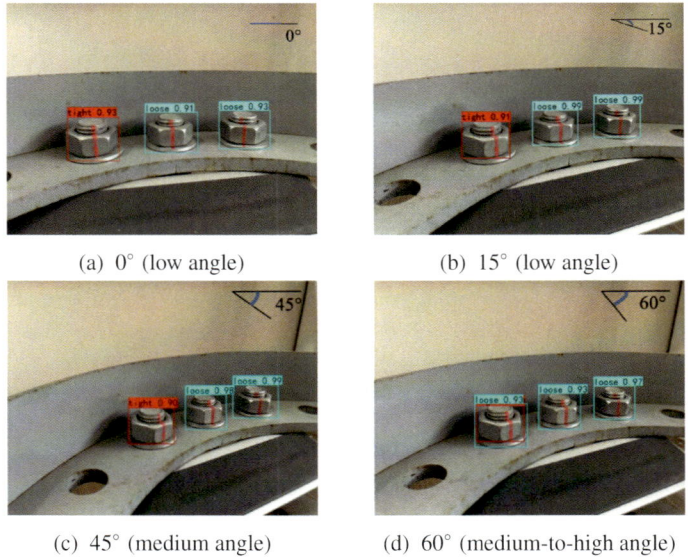

(a) 0° (low angle)   (b) 15° (low angle)

(c) 45° (medium angle)   (d) 60° (medium-to-high angle)

Fig. 7.18 Localization results of YOLO v3 (different horizontal perspective angles)

false detection occurred at medium-to-high angles (>45°). Therefore, acquiring images at large horizontal perspective angles should be avoided. In fact, in practical applications, the camera's shooting angle can be fixed or automatically adjusted based on demands to avoid this adverse effect.

7.3.4.3 Lighting Conditions

In this section, a set of 24 images depicting bolts in various states of loosening (ground truth, from left to right, is loose, tight, loose and loose) were captured under different lighting conditions, namely, dark (0–100 lux), normal (300–800 lux), and bright (1,000–2,000 lux). A selection of the detection results is given in Fig. 7.19 and specific detection metrics are listed in Table 7.7. It is evident that under dark conditions, although the model managed to detect some bolts, there were instances of missed detections, resulting in an accuracy rate of only 68.7%. In contrast, under normal lighting, the YOLO v3 model achieved perfect detection results with 100% accuracy, precision, and recall. However, in bright lighting

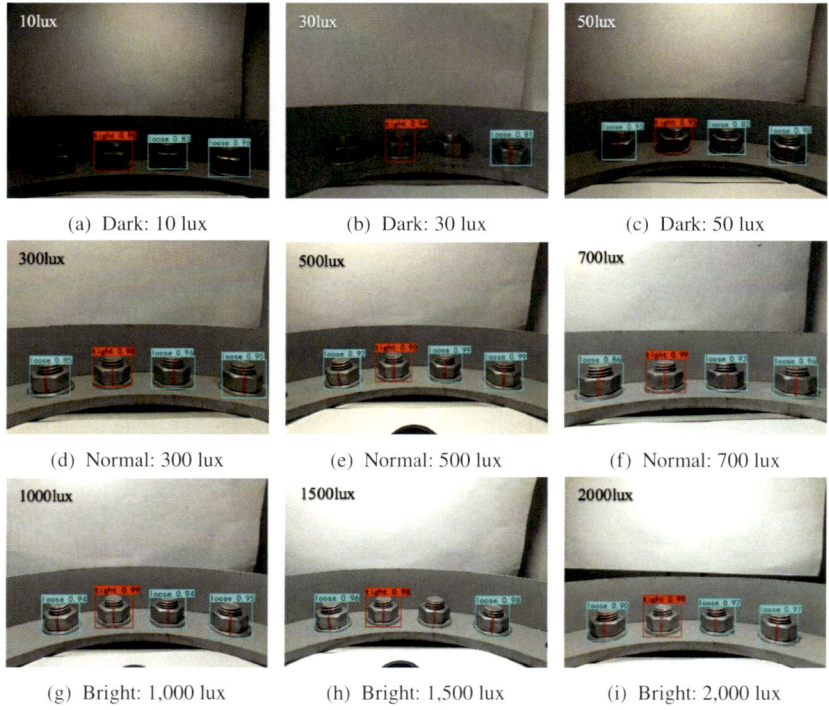

(a) Dark: 10 lux	(b) Dark: 30 lux	(c) Dark: 50 lux
(d) Normal: 300 lux	(e) Normal: 500 lux	(f) Normal: 700 lux
(g) Bright: 1,000 lux	(h) Bright: 1,500 lux	(i) Bright: 2,000 lux

Fig. 7.19 Localization results of YOLO v3 (different lighting conditions)

Table 7.7 Localization results of YOLO v3 under different lighting conditions of 24 test images for the WTT bolt loosening (%)

Condition	Accuracy	Precision	Recall
Dark	68.7	100.0	83.3
Normal	100.0	100.0	100.0
Bright	87.5	100.0	95.4

conditions, overexposure of the images led to a slight increase in missed detections, resulting in an accuracy drop to 87.5%. These results demonstrate that illumination conditions indeed impact the effectiveness of the model recognition.

7.3.4.4 Outdoor Environment

To evaluate the robustness of the trained model with synthetic data for its application in real-world scenarios, a substantial number of images were captured in diverse outdoor backgrounds and the corresponding weather conditions (specifically, all images were captured on days with clear and sunny skies) were collected. Several samples are presented in Fig. 7.20, revealing that the YOLO-based models, trained with a synthetic dataset or based on a single background in a laboratory setting, exhibited remarkable accuracy in locating and identifying instances of bolt loosening across various backgrounds. This outcome underscores the high generalization capability of the trained YOLO-based models in this case study. It establishes a solid basis for the practical implementation of bolt-loosening detection using CV and DL techniques.

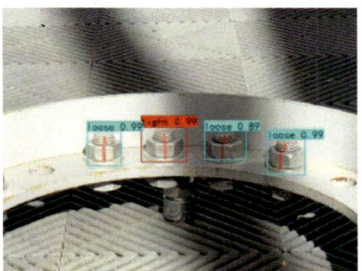

(a) Sample #1 (tiled floor with shadow)

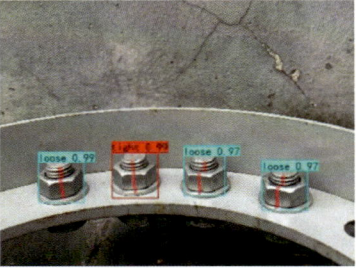

(b) Sample #2 (cracked concrete floor)

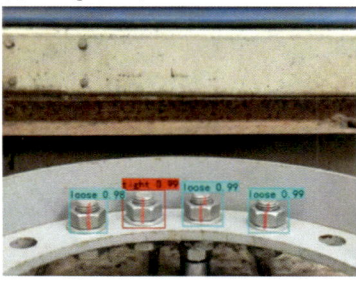

(c) Sample #3 (vertical metallic frame with rusted members)

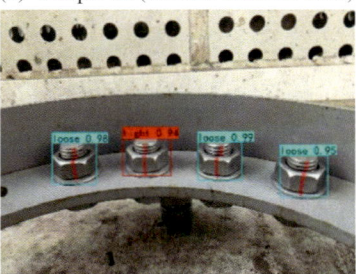

(d) Sample #4 (tiled wall and stained concrete floor)

Fig. 7.20 Localization results of YOLO v3 for different outdoor environments (backgrounds)

7.3.5   Case Study Summary

This case study introduced an automated framework with YOLO-based DL models to achieve rapid localization of bolts with small angles of looseness in WTTs. The framework allows for swift and precise localization and detection of bolt looseness in a laboratory setting. It also demonstrates fast detection and high recognition accuracy, even with only a 2° rotational looseness. The main conclusions of this case study can be summarized as follows:

- Through the candidate mechanism, the YOLO v3 model, trained using a synthetic training dataset, achieved the best mAP of 95.71% on the validation dataset, with a remarkable detection speed of 24 ms per single image. This method stands out by significantly reducing the time and cost involved in collecting real-world images, in contrast to other approaches that necessitate collecting extensive real-environment datasets. Moreover, the utilization of synthetic training data enhances the generalization capability of the DL model.
- The trained model was tested using images of specimens containing bolts with different angles of looseness obtained under various environmental conditions, i.e., different states of looseness, shooting (vertical and horizontal perspective) angles, light intensity, and both indoor and outdoor settings. The results demonstrated that the detection method achieved the desired accuracy level for engineering application scenarios and exhibited remarkable robustness.

7.4   R-CNN Series

7.4.1   R-CNN

R-CNN, which stands for "Region-based CNN", is a popular two-stage approach for object detection in CV, which was the first model in the R-CNN series introduced in 2014 [7]. As illustrated in Fig. 7.21, R-CNN consists of five steps: (1) input the image, (2) generate the region proposal, (3) extract the CNN features, (4) classify the regions, and (5) adjust the Bboxes via regression, where steps 2–5 are described in detail as follows:

- Region proposal generation: A set of potential region proposals is generated using a *selective search algorithm* [28]. This algorithm starts by over-segmenting the image into multiple small regions based on pixel intensity and then employs merging strategies to combine these regions, resulting in hierarchical regional proposals that may contain objects. In the original R-CNN, 2,000 candidate regions were selected with the possibility of having varying sizes.

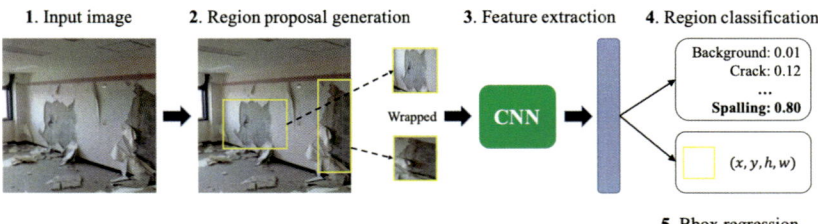

1. Input image   **2.** Region proposal generation   **3.** Feature extraction   **4.** Region classification

Wrapped   **CNN**

Background: 0.01
Crack: 0.12
...
Spalling: 0.80

(x, y, h, w)

5. Bbox regression

Fig. 7.21 Overview of R-CNN for plaster spalling localization

- <u>CNN features extraction</u>: For each candidate region, deep CNNs are used to extract features. The CNN is typically pre-trained on a large image classification dataset such as ImageNet [29], where the original R-CNN adopted AlexNet [30]. Specifically, this involves scaling the 2,000 candidate regions (corresponding to the wrapped regions in Fig. 7.21) to 227×227 pixels, then inputting these candidate regions into the trained network (e.g., AlexNet [30]) to obtain 4,096-dimensional features.[10] resulting in a feature matrix of dimensions 2000×4096.

- <u>Region classification</u>: The features in the previous step are fed into K SVM classifiers, where $K + 1$ is the number of K classes of the objects plus the background and the weight matrix of the SVM classifiers is accordingly $4096 \times (K + 1)$. The dot product of the feature matrix with the weight matrix results in the probability output matrix of $2000 \times (K + 1)$ for the 2,000 regions and their $K + 1$ classes, including the background of each region.

- <u>Bbox regression</u>: Bbox regression is performed to refine the proposed candidate region's coordinates, i.e., (x_1, y_1), (x_2, y_2) and (x_0, y_0) shown in Fig. 2.4, into Bboxes of proper sizes, i.e., height, h_B, and width, w_B, to align them more accurately with the objects, refer to the example in Fig. 7.6 and its related discussion in Sect. 7.2.1 and Fig. 7.21, where for simplicity of the illustration: $x = x_i$, $y = y_i$, $h = h_B$ & $w = w_B$. In this process, the NMS operation is applied to remove redundant candidate regions, where the candidate regions with IoU lower than a selected threshold of a certain class are rejected.

R-CNN outperformed conventional CV detection methods in many benchmark datasets [7] demonstrating several key characteristics. Leveraging the selective search algorithm and employing CNNs for feature extraction significantly enhance the efficiency of the framework, particularly in terms of the speed of performing the candidate region selection. The R-CNN backbone (AlexNet [30]) is pre-trained on ImageNet [29], mitigating issues related to limited size of the available labeled dataset during the localization training. Nonetheless, training the network remains computationally intensive due to the inclusion of 2,000 region proposals per image and the possible existence of many overlapped candidate regions, result-

---

[10] The visual knowledge of AlexNet can be probed by considering the feature activations induced by an image at the last 4,096-dimensional hidden layer [30].

ing in many duplicate calculations. Accordingly, the computational speed is slow and the original R-CNN model can not be applied for real-time predictions and decision-making purposes. Furthermore, the selective search algorithm lacks a learning process, which can occasionally lead to the generation of inadequate candidate region proposals. Consequently, more advanced versions of R-CNN have been developed to address these limitations, as discussed in the following two sections.

7.4.2   Fast R-CNN

Fast R-CNN is a faster and more efficient alternative to the original R-CNN to address the issues of its slow speed and the need for repeated convolutional calculations due to a large number of overlapping region proposals. As illustrated in Fig. 7.22, instead of performing separate convolutional calculations for each region proposal, Fast R-CNN inputs the entire image into the CNN backbone to obtain the corresponding feature map. Then, the candidate region proposals are generated by the selective search algorithm and projected onto these feature maps to obtain the corresponding feature matrix. Thus, the convolutional calculations are effectively shared. These projected region proposals on the feature maps are called regions of interest (RoIs). Subsequently, each RoI is followed by a special pooling layer operation, namely, RoI pooling, to obtain a fixed-dimension feature representation. Finally, FC-layers are added for predicting the Bboxes and their corresponding classes.

Fast R-CNN no longer extracts features separately for each candidate region but rather maps each candidate region onto the feature map after extracting the features from the entire image. This mapping is mainly based on the characteristic that the relative position of the image remains unchanged after multi-layer convolution and pooling, i.e., an invariant property, as discussed in Sect. 4.2.3. For the example illustrated in Fig. 7.23, after the convolution of an image, a corresponding feature map (in some feature space) is obtained and each pixel (x', y') in the feature map can be mapped back (using an inverse mapping) to the original image space (x, y). Therefore, for any RoI on the feature map space, as long as the top-left

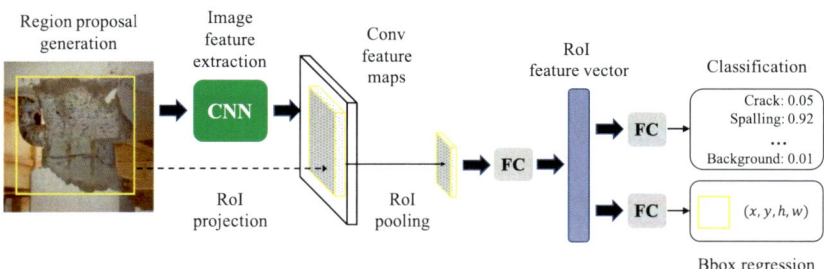

Fig. 7.22 Overview of Fast R-CNN for concrete cover spalling localization

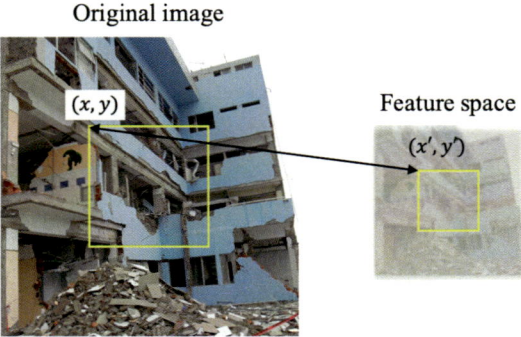

Fig. 7.23 Example of a regional proposal mapping between the original image and the feature maps

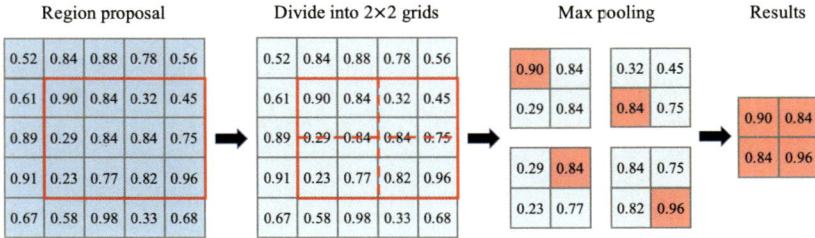

Fig. 7.24 Example of RoI pooling with 2×2 grids

and bottom-right positions of the Bbox are known, it can be mapped back to the original image space.

The RoI pooling plays the role of mapping any size of the region proposal to a fixed-size feature map while preserving the image features. The example illustrating the working principle of the RoI pooling is shown in Fig. 7.24. Given the feature maps of the entire image and the region proposals, these regions are divided into a fixed number of grids, e.g., a 2×2 grid is applied to a 4×3 region proposal in Fig. 7.24 (note that 7×7 grids are commonly used in real applications). In this figure, only one such grid is shown, dashed red lines, for the identified, red solid box, single RoI. It is noted that if the grid boundary coordinates are not integers, they are rounded up to the closest integers. Subsequently, pooling operations, e.g., max pooling or average pooling, are performed on each grid cell, as in the application of max pooling on the 2×2 grid cells of Fig. 7.24. In the original Faster R-CNN and several common application cases when a 7×7 grid layout is adopted, a fixed-size 7×7 output from each region proposal is obtained, which is also denoted as RoI for short.

The Fast R-CNN has two output layers. The first layer outputs a discrete probability distribution for each RoI, represented by a $(K + 1)$-dimensional probability vector $P = (P_0, P_1, \ldots, P_k, \ldots, P_K)$ across K classes plus the background ($k = 0$). This $(K + 1)$-dimensional vector P is obtained by the *softmax* operation toward outputs of the last FC-layers (top FC branch as illustrated in Fig. 7.22). The second layer predicts the Bbox regression offsets, denoted by $O = (O_x, O_y, O_w, O_h)$. According to [7], these offsets are parameterized for the translation (O_x^k, O_y^k) and size (O_w^k, O_h^k) adjustments for the Bbox associated with the respective kth class. Specifically, for the kth class out of the K classes, the corresponding offset components are denoted by $O^k = (O_x^k, O_y^k, O_w^k, O_h^k)$. For training purposes, each RoI is labeled in advance with a ground-truth class, $u \in \{0, 1, \ldots, K\}$, and a ground-truth Bbox regression offsets $V = (V_x, V_y, V_w, V_h)$, as expressed below:

$$
\begin{aligned}
V_x &= (G_x - R_x)/R_w, \\
V_y &= (G_y - R_y)/R_h, \\
V_w &= \ln(G_w/R_w), \\
V_h &= \ln(G_h/R_h),
\end{aligned}
\tag{7.3}
$$

where the two tetrads, (R_x, R_y, R_w, R_h) and (G_x, G_y, G_w, G_h), stand for the geometric center coordinates (x, y), width, w, and height, h, of the selected RoI and its ground truth Bbox, respectively.

A multi-task loss, L, Eq. 7.4, in Fast R-CNN is established based on the previously discussed two outputs. It consists of the classification loss, L_{cls}, and the localization loss, L_{loc}. Denote N_{RoI} as the number of RoIs and both L_{cls} and L_{loc} are computed for each RoI. L_{cls} refers to Eq. 3.6 using q as the probability distribution of the model prediction, $\hat{u}$, and the corresponding ground truth class label, u, having its own probability distribution, $p(u)$. This loss term is denoted by L_{cls} and is expressed in Eq. 7.5.

$$
L = \frac{1}{N_{RoI}} \sum_{N_{RoI}} \left\{ L_{cls} + \lambda \cdot [u \geq 1] \cdot L_{loc} \right\}.
\tag{7.4}
$$

$$
L_{cls} = - \sum_{\text{all classes}} p(u) \ln\left(q(\hat{u})\right).
\tag{7.5}
$$

On the other hand, L_{loc} is computed in Eq. 7.6 by a smooth $L1$ loss [8], denoted as $SL1$, expressed in Eq. 7.7 with $x = O_i^u - V_i$ for ground truth class u and $i \in \{x, y, w, h\}$. It is noted that $[u \geq 1]$ uses Iverson bracket indicator [31], i.e., it equals 1 when $u \geq 1$ and 0 otherwise, to rule out the background RoIs, corresponding to $u = 0$, because there is no ground truth Bbox for the background. Finally, λ is a hyper-parameter that controls the balance between classification loss and the localization loss, which has a default value of 1.0. By using the multi-task loss, the model training is transformed from multi-stage (training for classification and localization separately) to single-stage training. The concept and formulation of the multi-task loss introduced in Fast R-CNN have become a cornerstone in the R-CNN series of models, which has been widely adopted and inherited by subsequent R-CNN models, such as Faster R-CNN.

$$L_{loc}(O^u, V) = \sum_{i \in \{x,y,w,h\}} SL1\left(O_i^k - V_i\right). \tag{7.6}$$

$$SL1(x) = \begin{cases} 0.5x^2 & |x| < 1, \\ |x| - 0.5 & \text{Otherwise.} \end{cases} \tag{7.7}$$

In summary, compared with the original R-CNN, Fast R-CNN achieves higher accuracy and faster speed with less computational cost. This is achieved by conducting the expensive convolution operation only once, reducing the computational cost. Additionally, during training, all network layers can be updated simultaneously, eliminating the need for incremental parameter updates. By utilizing a CNN to extract the features from the entire image and employing RoI pooling, Fast R-CNN allows the RoI feature vectors to capture more informative context than those extracted in R-CNN, resulting in enhanced localization accuracy. The integration of the multi-task loss, which considers both classification and localization, further enhances the model's performance. However, it is worth noting that the selective search algorithm still incurs significant computational overhead. Additionally, rounding up non-integer boundaries during the RoI pooling introduces a positional deviation, refer to Fig. 7.24, which can occasionally lead to inaccuracies in the results. These challenges are addressed in subsequent versions of the R-CNN models. In conclusion, the main three differences between Fast R-CNN and R-CNN are summarized as: (1) Feature-based region proposal generation, (2) RoI pooling, and (3) multi-task loss.

7.4.3   Faster R-CNN

Faster R-CNN stands out as the pinnacle model in the R-CNN series and it can be considered the cornerstone of the two-stage method in DL. A key highlight is its ability to effectively address the inefficiency of the selective search algorithm by introducing the region proposal network (RPN). The RPN leverages NNs, particularly CNN, to autonomously learn and generate candidate region proposals, streamlining the localization process in an end-to-end manner. In essence, Faster R-CNN seamlessly integrates the strengths of both Fast R-CNN and RPN. The entire model comprises three main parts: (1) backbone, (2) RPN, and (3) RoI pooling. Notably, VGGNet and ResNet serve as the commonly used backbones in Faster R-CNN and RoI pooling is inherited from Fast R-CNN.

Figure 7.25 illustrates that the RPN usually works on the feature maps extracted by the CNN backbone and it adopts a sliding window to generate k different rectangles for each point of the feature map, known as the *anchor point*, where the series of rectangles, called *anchor boxes*, are generated using these anchor points as their centers. Each anchor box is represented by its coordinates, where two common coordinate systems are adopted in practice: (1) coordinates of the top-left and bottom-right corners and (2) center point coordinates with specified box width and height, similar to the respective Fig. 2.3a and c in Sect. 2.2.

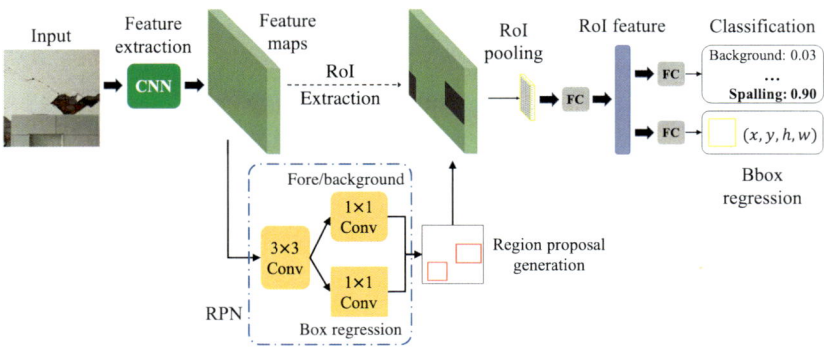

Fig. 7.25 Overview of Faster R-CNN for masonry wall plaster spalling localization

In the original work [9], a base anchor box size was defined as a resolution of 16×16 (measured in the input image's pixel space), which can be adjusted for different localization tasks. The number of anchor boxes, k, is controlled by two parameters: (i) scale and (ii) aspect ratio. Scale parameters are responsible for adjusting the width and height of the anchor box, while the aspect ratio parameters determine the specific shapes of the boxes. It is worth noting that the changes in aspect ratio should keep the area of the anchor boxes constant.[11] For example, consider the default setting with three scale parameters (8, 16, and 32) and three aspect ratio parameters (0.5 for 1:2, 1.0 for 1:1, and 2.0 for 2:1). Consequently, this set of parameters generates 9 anchor boxes, i.e., $k = 3 \times 3 = 9$, for each point on the feature map. If the feature map has a size of 60×40 in the pixel space, the total number of anchor boxes generated is $60 \times 40 \times 9 = 21,600$.

Each anchor box is passed through the RPN, typically comprising a $n \times n$ Conv layer, followed by two branches with a 1×1 Conv layer each, referred to as *cls* for the *classification* layer and *reg* for the *regression* layer. It is noted that $n = 3$ is commonly adopted as shown in Fig. 7.25. The first branch judges whether the anchor box contains an object, which is equivalent to performing a binary classification between foreground and background. In other words, for each point on the feature map, the *cls* layer yields $2k$ confidence scores, computed by *softmax* after the convolution operations. Simultaneously, the second branch handles the task of correcting the position of the anchor boxes, which is a regression problem. Four parameters, namely, d_x, d_y, d_w, and d_h, are computed through regression for each anchor box. The two parameters d_x and d_y represent the offset of the anchor box on the x-axis and y-axis, respectively, while the two parameters d_w and d_h represent the scaling adjustments of the anchor box in width, w_a, and height, h_a, respectively. Consequently, for each point on

---

[11] Two aspect ratios: $R_1 = W_1/H_1 \Rightarrow W_1 = R_1 \times H_1$ and $R_2 = W_2/H_2 \Rightarrow W_2 = R_2 \times H_2$. Constant area: $A = W_1 \times H_1 = R_1 \times H_1^2 = W_2 \times H_2 = R_2 \times H_2^2 \Rightarrow H_2 = H_1 \times \sqrt{R_1/R_2}$. Example: $R_1 = 1.00$ for 1:1 and $R_2 = 0.25$ for 1:4, $H_2 = H_1 \times \sqrt{1.00/0.25} = 2H_1$, $W_1 = H_1$ & $W_2 = 0.25H_2$. For $H_1 = 5.0 \Rightarrow H_2 = 10.0$, $W_1 = 5.0$ & $W_2 = 2.5$, giving $A = 5.0 \times 5.0 = 2.5 \times 10.0 = 25.0$.

the feature map, a total of $4k$ coordinate-related outputs are generated. After obtaining the correction parameters for each anchor box, refined anchor boxes are derived. Subsequently, these boxes are sorted according to the object's confidence score assigned by the *cls* layer, arranging them in a descending order (i.e., from highest to lowest). Following this step, a NMS operation is applied to obtain a series of candidate region proposals.

The subsequent steps of the Faster R-CNN are similar to those of the Fast R-CNN. With the candidate region proposals and feature maps extracted from the backbone, RoI pooling is applied to these regions to efficiently transform the variable-sized regions into fixed-size feature maps. Subsequently, the feature maps are further passed through a series of FC-layers and then split into two branches for object classification and Bbox regression. Faster R-CNN inherits the concept of multi-task loss and the total loss, L, is expressed in Eq. 7.8. It consists of the following four parts: (1) classification loss of the RPN, L_{cls}^{RPN}, (2) regression loss of the RPN, L_{reg}^{RPN}, (3) classification loss of the predicted Bbox, L_{cls}, identical to that of Fast R-CNN, i.e., using Eq. 7.4, which is based on Eq. 3.6, and (4) regression loss of the Bbox, L_{loc}, identical to that of Fast R-CNN, i.e., using Eq. 7.6.

$$L = L_{cls}^{RPN} + \lambda \cdot L_{reg}^{RPN} + L_{cls} + L_{loc}. \tag{7.8}$$

The classification loss of the RPN, L_{cls}^{RPN}, is a binary-entropy loss, Eq. 3.7. Suppose there are N data points (images) for the localization problem, the number of anchor boxes generated in the proposal of the ith data point is denoted by N_{anc}^i, where the total number of anchor boxes is N_{anc}. While in labeling, two types of anchor boxes can be assigned with *positive* labels: (i) the anchor box which has the highest IoU with a ground truth Bbox[12] and (ii) the anchor box which has an IoU overlap higher than 0.7 (a threshold that can be treated as a hyper-parameter) with any ground truth Bbox. If the jth anchor box in the ith data point is positive, it is denoted by $y_{i,j} = 1.0$; otherwise, it is 0.0. In addition, $P(\hat{y})$ denotes the predicted probability of the anchor box by the *cls* layers. Consequently, the L_{cls}^{RPN} is computed as follows:

$$L_{cls}^{RPN} = -\frac{1}{N_{anc}} \sum_{i=1}^{N} \sum_{j=1}^{N_{anc}^i} \left\{ y_{i,j} \ln \left[P\left(\hat{y}_{i,j}\right) \right] + (1 - y_{i,j}) \ln \left[1 - P\left(\hat{y}_{i,j}\right) \right] \right\}. \tag{7.9}$$

To compute the RPN regression loss, L_{reg}^{RPN}, the Bbox coordinates of the prediction and ground truth are parameterized via the anchor box coordinates, i.e., b_k and b_k^*, respectively, as expressed in Eq. 7.10, where the subscript $k \in \{x, y, w, h\}$ represents the box center x-axis and y-axis coordinates, width, and height, respectively. In Eq. 7.10, the subscript a indicates the anchor box and the superscript $*$ represents the ground truth. It is important to note

---

[12] For an example of type (i), suppose there exist three spalling areas in the image and thus three ground truth Bboxes are labeled. For each Bbox, the anchor box with the highest IoU corresponding to that Bbox is treated as the closest prediction and has a high probability of containing the spalling damage. Thus, this anchor box is assigned to the *positive* class.

that, in the computation of L_{reg}^{RPN}, only positive anchor boxes are taken into consideration and their total number is denoted by N_{reg}. The RPN regression box loss is computed in Eq. 7.11 using $SL1$ from Eq. 7.7 with $b_k - b_k^*$ for all coordinate components in all anchor boxes corresponding to a particular, e.g., ith, data point, i.e., $j \in \{1, 2, \ldots, N_{anc}^i\}$. Finally, the hyper-parameter, λ, in Eq. 7.8 is used as a balancing weight between classification and regression, ensuring that $1/N_{anc} \approx \lambda/N_{reg}$. This balance aims to optimize the RPN effectively during training. More details, e.g., related to the training process of Faster R-CNN, can be found in [9].

$$
\begin{aligned}
b_x &= (x - x_a)/w_a, \\
b_y &= (y - y_a)/h_a, \\
b_w &= \ln(w/w_a), \\
b_h &= \ln(h/h_a), \\
b_x^* &= (x^* - x_a)/w_a, \\
b_y^* &= (y^* - y_a)/h_a, \\
b_w^* &= \ln(w^*/w_a), \\
b_h^* &= \ln(h^*/h_a).
\end{aligned}
\tag{7.10}
$$

$$
L_{reg}^{RPN} = \frac{1}{N_{reg}} \sum_{i=1}^{N} \sum_{j=1}^{N_{anc}^i} \sum_{k \in \{x,y,w,h\}} SL1\left(b_k - b_k^*\right).
\tag{7.11}
$$

7.5   Case Study: Concrete Spalling Detection

As discussed in Sects. 2.2 and 5.6, one of the crucial applications in vision-based SHM is the detection and localization of structural damage. For this case study, concrete cover spalling localization is investigated, which is treated as an extension of Task 3 (spalling condition classification) in ϕ-Net, Sect. 5.2.2.3, and can be viewed as one special scenario of Task 9 (damage localization) in ϕ-NeXt, Sect. 5.6.

Concrete spalling is a common problem in structures, particularly in regions with extreme weather conditions or where the concrete is exposed to corrosive environmental elements. It involves the deterioration or chipping of the concrete surface, often caused by factors such as moisture infiltration, freeze-thaw cycles, chemical exposure, or steel reinforcement corrosion. Detecting concrete spalling at an early stage is of utmost importance to prevent further damage and ensure the overall structural integrity of buildings and bridges. In this case study, the Faster R-CNN model is considered to achieve a fast and accurate localization of spalled areas of the RC cover, which can be further leveraged in subsequent damage quantification and decision-making processes.

7.5.1   Spalling Localization Dataset

A total of 1,350 images labeled as "SP" (spalling) from Task 3 are re-labeled using the "*Labelme*" [32] labeling tool. For each spalling region, a rectangular ground truth Bbox is defined and the x and y coordinates of the top-left corner, as well as the width and height of all the Bboxes, are recorded, following the definition (2) in Fig. 2.4, Sect. 2.2. Three sample images are illustrated in Fig. 7.26. Each image has a resolution of 224×224 and the origin of the coordinate system is located at the top-left corner. For evaluation purposes, 1,200 images are used for training and the remaining 150 images are reserved for testing.

7.5.2   Spalling Localization Model Settings

To evaluate the performance of Faster R-CNN in spalling detection, various configurations are adopted. Firstly, a baseline model is built using VGG-16, Sect. 4.3.1. It is pre-trained on ImageNet [29] as the backbone and its parameters are close to the default setting mentioned in the original Fast R-CNN [9]. Secondly, to investigate the optimal backbone for spalling feature extraction, another classical backbone, namely, ResNet-50, Sect. 4.3.3, is employed. Furthermore, TL techniques are utilized, where the backbone is either pre-trained on ImageNet [29] or on MS COCO [11] dataset to investigate the impact of parameter initialization.

All models are trained and evaluated under similar settings. The used scale parameters are 2, 4, 8, 16 and 32 and the aspect ratios for the VGG-16 backbone are set to 0.5, 1.0 and 2.0, while the aspect ratios for the ResNet-50 backbone are set to 0.7, 1.0 and 1.5 based on a preliminary study for better and stable performance. During the training, SGD is employed as the optimizer. The initial learning rate is set to 0.01 with a momentum of 0.9 and the weight decay rate is set to 5×10^{-4}. Furthermore, a batch size of 5 is adopted and the models are trained for a total of 20 epochs. The evaluation is performed using the

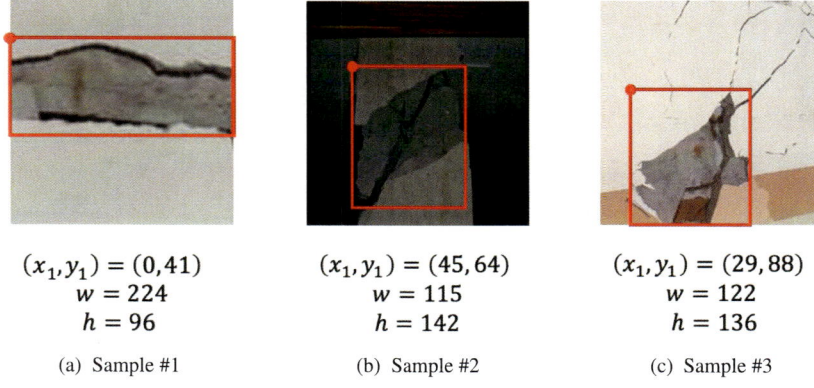

$(x_1, y_1) = (0, 41)$ $(x_1, y_1) = (45, 64)$ $(x_1, y_1) = (29, 88)$
$w = 224$ $w = 115$ $w = 122$
$h = 96$ $h = 142$ $h = 136$

(a) Sample #1   (b) Sample #2   (c) Sample #3

Fig. 7.26 Samples of labeled concrete spalling images

AP^{50} metric introduced in Sect. 7.1.2. The implementation of these experiments is based on PyTorch [27] and performed on a NUC13RNGi7 PC with a single GPU (CPU: 13th Gen Intel(R) Core(TM) i7-13700K 3.40 GHz, RAM: 32 GB and GPU: Nvidia Geforce RTX 4080, GPU memory: 16 GB).

7.5.3 Baseline Spalling Localization

The localization results are compared based on the coordinates and shape of the rectangular region from both the ground truth and the predicted Bboxes. In terms of the quantitative evaluation, the baseline model achieves a numerical AP^{50} metric of 0.559 on the test dataset, a value deemed acceptable given its surpassing 0.5 based on previous observations in CV benchmark experiments [8, 9, 14]. However, its practical utility remains limited, considering that recent studies resulting in commendable visual results of AP^{50} values nearing 0.7 and 0.8 [14]. This assertion is further corroborated by the visual evidence, wherein approximately 40% of the test results exhibit issues such as redundant Bboxes, incomplete coverage, or erroneous detection. The visual representations of image samples are shown in Figs. 7.27 and 7.28, where the former illustrates samples of successful localization results, while the latter shows the unsuccessful samples.

Examining Fig. 7.27 reveals that the baseline model has the capability to capture various sizes and shapes of spalling. For example, Fig. 7.27a highlights the baseline model's successful detection of the classical shear-type spalling pattern stemming from the diagonal cracking. Similarly, the model is able to detect both horizontal spalling with an aspect ratio exceeding 1:1 (Fig. 7.27b) as well as vertical spalling with an aspect ratio less than 1:1 (Fig. 7.27c). Figures 7.27d–f demonstrate the ability of the baseline model to detect varying spalling sizes, from large to medium to small, respectively. Furthermore, the results indicate that the baseline model is capable of localizing multiple spalling areas within one image, as evident from Fig. 7.27g–i. It is important to note that while some spalling samples are detected according to the IoU criterion by making use of the default threshold value of $\epsilon = 0.5$ (Sects. 7.1.1 and 7.1.2), the predicted Bboxes still necessitate refinement, e.g., Fig. 7.27a does not exactly cover all the spalled areas. Therefore, enhancing the model's precision demands a more intricate fine-tuning or leveraging more potent models, while ensuring practical efficacy.

Continuing the discussion on the imperfect predictions, Fig. 7.28 plots a collection of unsuccessful spalling localization samples. There exist three typical failure cases: (1) redundant Bboxes, (2) inaccurate coverage, and (3) erroneous predictions. In Fig. 7.28a and b, the largest Bbox successfully detects the target, yet several smaller Bboxes redundantly accompany it. Alongside the use of other backbones, one straightforward way for improvement involves setting a higher threshold in conjunction with applying the NMS algorithm for eliminating the Bboxes with low-confidence. Figure 7.28c and d are two samples that illustrate the inaccurate coverage. While some partial Bboxes adeptly localize the spalling

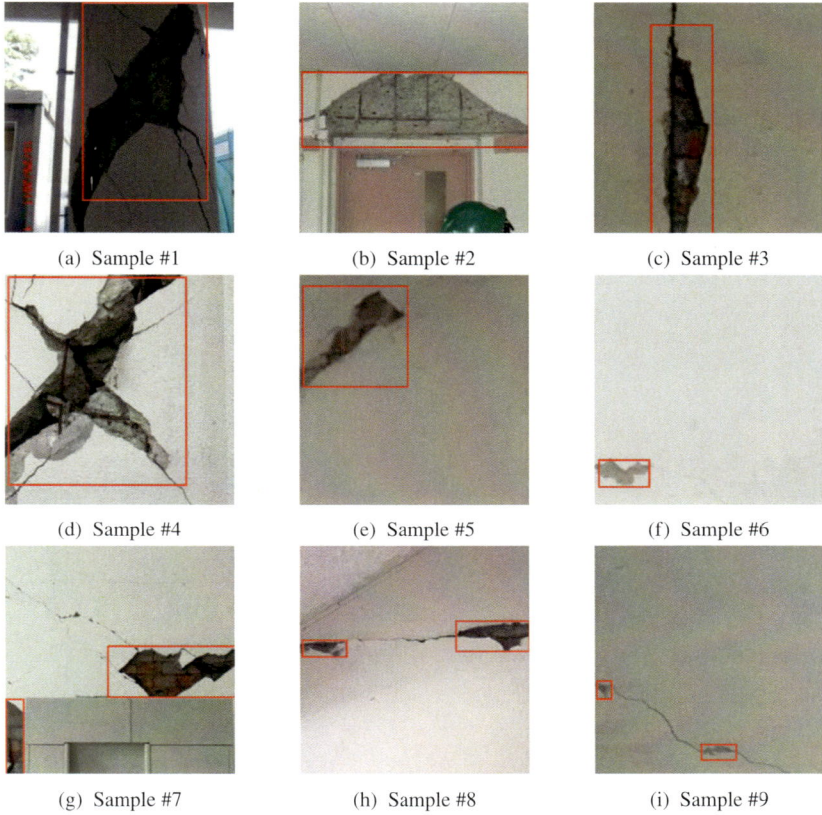

(a) Sample #1   (b) Sample #2   (c) Sample #3

(d) Sample #4   (e) Sample #5   (f) Sample #6

(g) Sample #7   (h) Sample #8   (i) Sample #9

Fig. 7.27 Successful spalling localization samples of the baseline model

regions, other Bboxes fall short to accurately capture the complete spalling areas. Moreover, Fig. 7.28e and f depict two cases with erroneous predictions. Notably, Fig. 7.28e portrays an occurrence where a Bbox incorrectly identifies a window on a door as a spalling area due to its resemblance, with a dark color, to actual spalling. Both inaccurate coverage and wrong prediction can be mainly attributed to the limited feature representation and extraction capability of the VGG-16 backbone, particularly under insufficient labeled training data, which can be improved by utilizing alternate backbones such as ResNet-50 or training with better strategies.

In summary, the baseline model demonstrates commendable performance on the test dataset by effectively detecting typical spalling patterns, accommodating various sizes, and localizing multiple spalled areas within an image. However, its suitability for practical applications remains constrained. This limitation mainly stems from the inadequacy of the used labeled training dataset, which subsequently constrains the backbone's capacity for robust feature representation and extraction. Thus, a pivotal approach for enhancing the

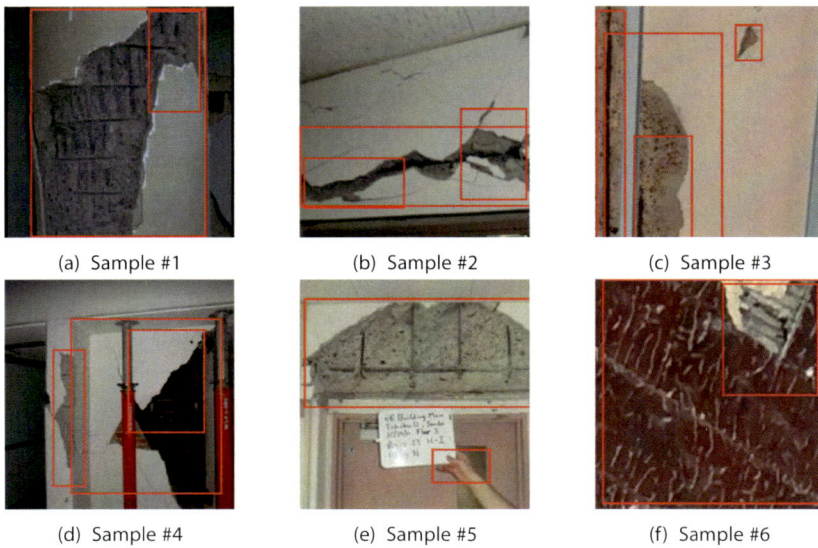

(a) Sample #1   (b) Sample #2   (c) Sample #3

(d) Sample #4   (e) Sample #5   (f) Sample #6

Fig. 7.28 Unsuccessful spalling localization samples of the baseline model

performance entails the integration of a more robust and potent backbone architectures and training techniques.

7.5.4   Influence of the ResNet Backbone

Regarding the performance enhancement, a comparative model of Faster R-CNN, adopting ResNet-50 (pre-trained on ImageNet [29]) as the backbone, is used. It is referred to in the sequel as the B̲aseline M̲odel, or BM-Res50 for short. Evaluated using the same metric, BM-Res50 achieves an AP^{50} value of 0.739, showcasing an improvement of $(0.739 - 0.559)/0.559 = 32.2\%$ over the baseline model using VGG-16.

The significant performance enhancement of BM-Res50 can be directly reflected from the visual results in Fig. 7.29. For example, BM-Res50 can accurately detect all Bboxes in Fig. 7.29i, countering the baseline model's issue of redundant Bboxes in Fig. 7.28c. Another example is illustrated by Fig. 7.29c where the BM-Res50 model provides more superior Bbox coverage. This is in contrast to the results of the baseline model in Fig. 7.27a where the predicted Bboxes fail to adequately cover the spalling area. Moreover, a closer inspection of the samples in Fig. 7.29, BM-Res50 inherits the fundamental traits of the baseline model, where it can detect spalling across various scales and even multiple locations. As illustrated in Fig. 7.29g–i, BM-Res50 can accurately locate 3 or more spalling regions, while the baseline model can partially handle locating 1 or 2 spalling regions and its performance deteriorates with the increase of the number of the spalling regions and their pattern intricacies.

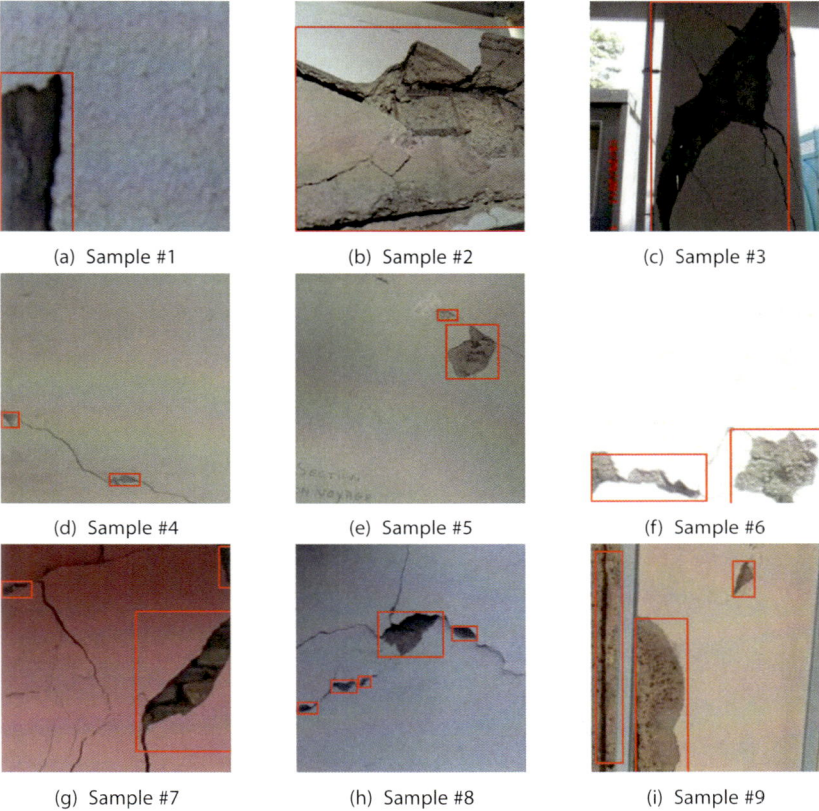

(a) Sample #1   (b) Sample #2   (c) Sample #3

(d) Sample #4   (e) Sample #5   (f) Sample #6

(g) Sample #7   (h) Sample #8   (i) Sample #9

Fig. 7.29 Successful spalling localization samples of BM-Res50

Despite a significant improvement over the baseline model in terms of reducing both missing and erroneous localizations, it is noteworthy that BM-Res50 still encounters similar failure cases, e.g., redundant Bboxes in Fig. 7.30b, c and e, incomplete coverage in Fig. 7.30d and f, and erroneous predictions in Fig. 7.30. Therefore, this observation implies that these three types of issues are recurrent challenges in spalling damage localization. Nonetheless, it is reasonable to deduce that a more sophisticated backbone model might mitigate these issues to some extent, thereby reducing the occurrence of such failures. The pursuit of additional methodologies to alleviate these concerns warrants further investigation.

7.5.5   Impact of TL on Localization

To explore the impact of TL in Faster R-CNN on localization tasks, a comparative study is conducted between two pre-trained ResNet-50 backbones, i.e., pre-trained on either (i)

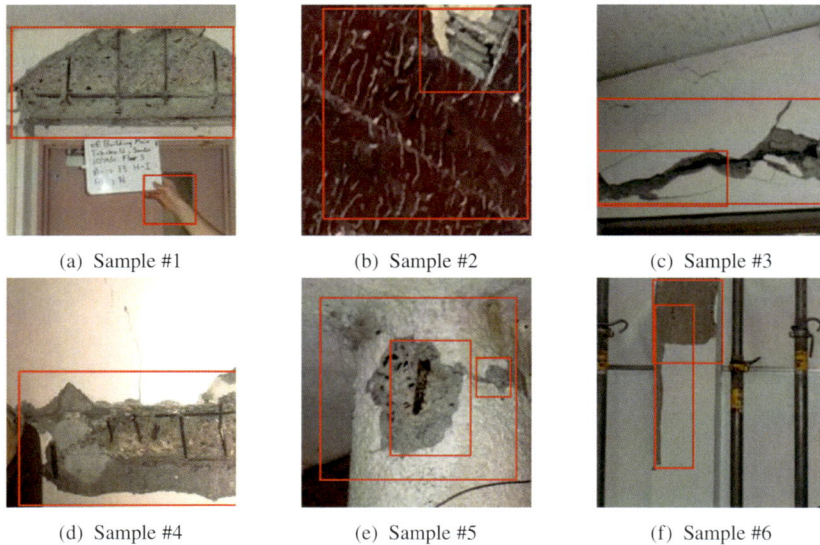

(a) Sample #1   (b) Sample #2   (c) Sample #3

(d) Sample #4   (e) Sample #5   (f) Sample #6

Fig. 7.30 Unsuccessful spalling localization samples of BM-Res50

ImageNet [29] (results detailed in Sect. 7.5.4) or (ii) the MS COCO dataset [11], where the model in this later case is referred to as BM-Res50-CO for brevity. This model achieves an AP^{50} value of 0.754, exhibiting a slight superiority over the BM-Res50 by $(0.754 - 0.739)/0.739 = 2.0\%$.

As illustrated in Fig. 7.31, BM-Res50-CO shares a similar performance with BM-Res50. In particular, it achieves more accurate localization Bboxes for large spalling areas (Fig. 7.31a–d). Actually, its slight improvement over BM-Res50 stems from mitigating issues related to incomplete Bbox coverage. The most recurrent failure cases are redundant Bboxes and erroneous predictions, as presented in Fig. 7.32. However, compared to BM-Res50, the severity of the failure cases of BM-Res50-CO is diminished, with a higher number of Bboxes displaying higher accuracy. For example, in Fig. 7.32b, two Bboxes precisely locate the spalling regions, whereas a single Bbox erroneously detects an inspector. Interestingly, a failure sample like the one depicted in Fig. 7.32d is a common occurrence, consistent with both the baseline and BM-Res50 models. Therefore, the source dataset used for pre-training does exert a discernible impact on the localization performance, where the prior knowledge and domain-specific features in the source domain can be inherited and fine-tuned to the target domain. In other words, certain features within the MS COCO dataset demonstrate greater proficiency in localizing spalling when contrasted with those present in ImageNet [29]. The optimal selection of the pre-trained source dataset can consequently furnish Faster R-CNN with an improved initialization, leading to an elevated performance level.

An additional complementary investigation is conducted utilizing the ResNet-50 backbone without any pre-training. To elaborate, both the backbone and the remaining layers

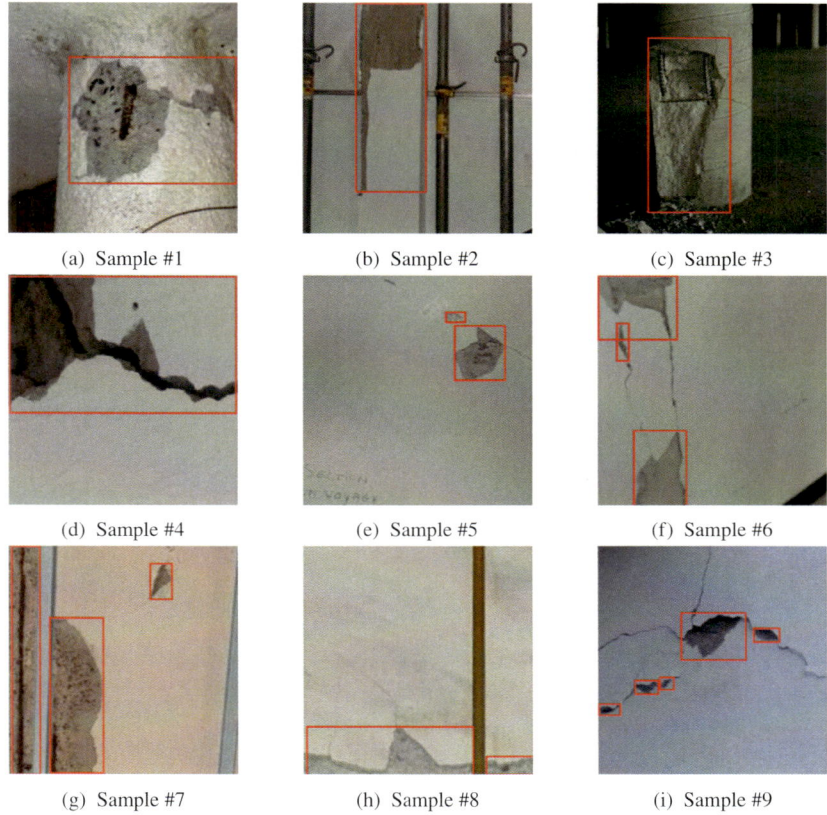

(a) Sample #1   (b) Sample #2   (c) Sample #3

(d) Sample #4   (e) Sample #5   (f) Sample #6

(g) Sample #7   (h) Sample #8   (i) Sample #9

Fig. 7.31 Successful spalling localization samples of BM-Res50-CO

within the entire Faster R-CNN are trained from scratch, i.e., starting with random initialization. This model has a low AP^{50} value of 0.284, which is significantly lower than the performance of BM-Res50 and BM-Res50-CO. This investigation effectively demonstrates the necessity of TL for tasks with limited training dataset and highlights the pivotal role of well-informed parameter initialization.

To further relieve the persistent challenges of redundant Bboxes and erroneous predictions, it becomes imperative to explore advanced alternatives for backbone architectures. For instance, the integration of cutting-edge backbones like Transformers, as elaborated in Chap. 12, presents a promising alternative. Furthermore, augmenting the training process with a larger volume of labeled data emerges as a straightforward, yet effective, strategy for better performance.

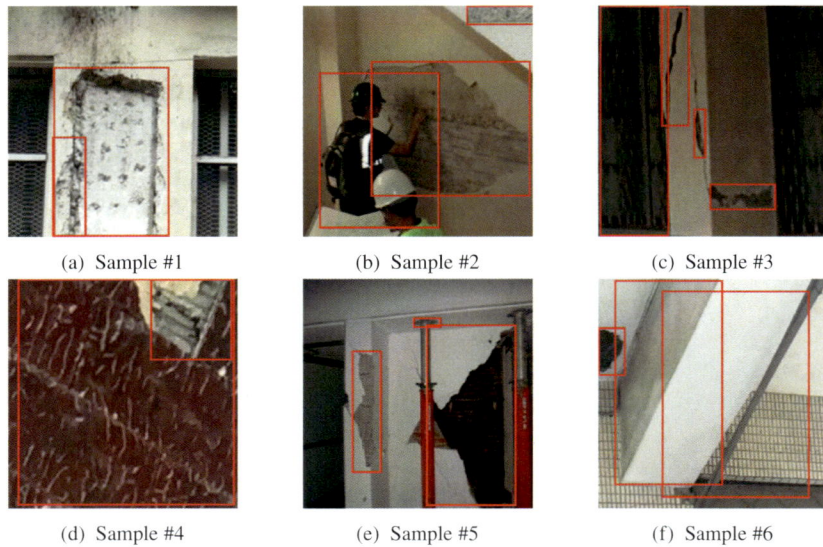

(a) Sample #1   (b) Sample #2   (c) Sample #3

(d) Sample #4   (e) Sample #5   (f) Sample #6

Fig. 7.32 Unsuccessful spalling localization samples of BM-Res50-CO

7.5.6 Summary

This case study introduced a classic two-stage localization framework known as Faster R-CNN for concrete spalling damage localization. To validate its efficacy, the Faster R-CNN model was evaluated across various configurations, including backbone selection and parameter initialization using different pre-trained weights through TL. The results obtained from 150 test images are listed in Table 7.8, where BM-Res50-No refers to the model with a backbone trained from scratch without any pre-training. Additionally, Fig. 7.33 showcases selected samples from different models to facilitate the comparative analysis. The main conclusions of this case study are summarized as follows:

- The baseline model, utilizing a VGG-16 backbone pre-trained on the ImageNet dataset [29], successfully detects typical spalling patterns, various sizes, and even multiple spalled areas. These results validate the feasibility of using Faster R-CNN for spalling localization. However, the achieved $AP^{50} = 0.559$ falls short, limiting its practical utility.

Table 7.8 Summary of Faster R-CNN results for spalling localization

Model	Baseline	BM-Res50	BM-Res50-CO	BM-Res50-No
AP^{50}	0.559	0.739	0.754	0.284

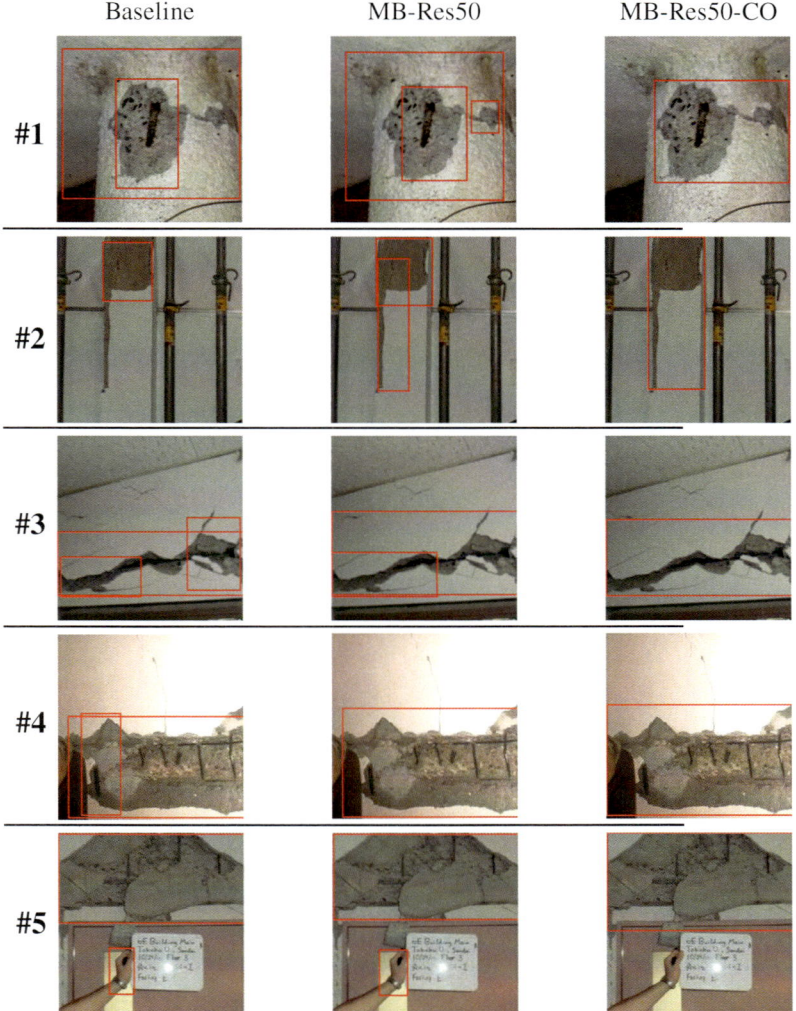

Fig. 7.33 Selected comparative samples between baseline model, BM-Res50, and BM-Res50-CO for concrete spalling localization from 5 images

- Substituting the VGGNet backbone with a ResNet-50 pre-trained on ImageNet [29] significantly enhances the localization performance. Moreover, through TL with pre-trained parameters from different source domains, the model gains a slight enhancement. Herein, the model pre-trained on the MS COCO dataset outperforms the one pre-trained on ImageNet [29], underscoring the critical role of parameter initialization from an appropriate source domain.

- Importantly, if the ResNet-50 backbone is not pre-trained and is instead initialized randomly, the AP^{50} performance drops considerably, which demonstrates the necessity of TL for tasks constrained by a limited training dataset.
- The visual results of Bboxes mirror the quantitative performance via AP^{50} and the investigated models exhibit 3 recurrent failure cases: (i) redundant Bboxes, (ii) inaccurate coverage, and (iii) erroneous predictions. Compared to the baseline model and BM-Res50, the use of a suitable backbone and pre-trained weights in BM-Res50-CO reduces the failure cases, particularly mitigating the issue of redundant Bboxes.

To further enhance the spalling localization performance, exploring advanced alternatives for the backbone or investigating new localization frameworks are promising approaches. Additionally, augmenting the training dataset by incorporating diverse and high-variety data stands out as a viable solution, warranting dedicated and additional future research efforts.

7.6   Exercises

1. Based on three types of definitions for the Bbox in Sect. 2.2, compute the other two types of Bbox coordinates (definitions (1) and (3)) for the samples in Fig. 7.26.
2. Suppose the input feature map size is 5×5, which is the same as the one in Fig. 7.24, compute the following two scenarios of RoI pooling (max pooling) as illustrated in Fig. 7.34: (a) the extracted region is 3×4 using a 2×2 grid and (b) the extracted region is 4×4 using a 2×2 grid.
3. Conduct a literature review and create a timeline plot to depict the evolving history of the YOLO and R-CNN series.
4. Provide a concise summary of the characteristics of the YOLO and R-CNN series, followed by a discussion of their respective advantages and disadvantages.

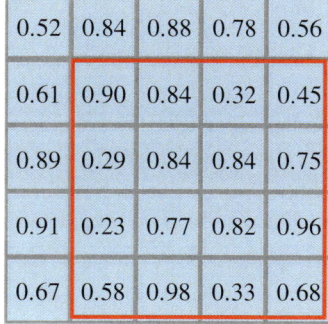

(a) Scenario #1   (b) Scenario #2

Fig. 7.34 Two scenarios of RoI pooling

References

1. W. Liu et al., Ssd: single shot multibox detector, in *Computer Vision- ECCV 2016: 14th European Conference, Amsterdam, The Netherlands, October 11–14, 2016, Proceedings, Part I* 14 (Springer, 2016), pp. 21–37
2. J. Redmon et al., You only look once: unified, real-time object detection, in *Proceedings of the IEEE Conference on Computer Vision and Pattern Recognition* (2016), pp. 779–788
3. J. Redmon, A. Farhadi, YOLO9000: better, faster, stronger, in *Proceedings of the IEEE Conference on Computer Vision and Pattern Recognition* (2017), pp. 7263–7271
4. J. Redmon, A. Farhadi, Yolov3: an incremental improvement (2018). arXiv:1804.02767
5. A. Bochkovskiy, C.-Y. Wang, H.-Y.M. Liao, Yolov4: optimal speed and accuracy of object detection (2020). arXiv:2004.10934
6. G. Jocher, A. Chaurasia, J. Qiu, YOLO by Ultralytics. Version 8.0.0. (2023). https://github.com/ultralytics/ultralytics
7. R. Girshick et al., Rich feature hierarchies for accurate object detection and semantic segmentation, in *Proceedings of the IEEE Conference on Computer Vision and Pattern Recognition* (2014), pp. 580–587
8. R. Girshick, Fast r-cnn, in *Proceedings of the IEEE International Conference on Computer Vision* (2015), pp. 1440–1448
9. S. Ren et al., Faster r-cnn: towards real-time object detection with region proposal networks, in *Advances in Neural Information Processing Systems* 28 (2015)
10. P. Jaccard, The distribution of the flora in the alpine zone. New Phytologist. **11**(2), 37–50 (1912). https://doi.org/10.1111/j.1469-8137.1912.tb05611.x
11. T.-Y. Lin et al., Microsoft coco: common objects in context, in *Computer Vision-ECCV 2014: 13th European Conference, Zurich, Switzerland, September 6–12, 2014, Proceedings, Part V* 13 (Springer, 2014), pp. 740–755
12. K. He et al., Mask r-cnn, in *Proceedings of the IEEE International Conference on Computer Vision* (2017), pp. 2961–2969
13. Z. Liu et al., Swin transformer: hierarchical vision transformer using shifted windows, in *Proceedings of the IEEE/CVF International Conference on Computer Vision* (2021), pp. 10012–10022
14. Y. Gao et al., Multi-attribute multi-task transformer framework for vision-based structural health monitoring. Comput.-Aided Civil Infrast. Eng. (2023)
15. J. Hosang, R. Benenson, B. Schiele, Learning non-maximum suppression, in *Proceedings of the IEEE Conference on Computer Vision and Pattern Recognition* (2017), pp. 4507–4515
16. C. Szegedy et al., Going deeper with convolutions, in *Proceedings of the IEEE Conference on Computer Vision and Pattern Recognition* (2015), pp. 1–9
17. T.-Y. Lin et al., Feature pyramid networks for object detection, in *Proceedings of the IEEE Conference on Computer Vision and Pattern Recognition* (2017), pp. 2117–2125
18. C.-Y. Wang et al., CSPNet: a new backbone that can enhance learning capability of CNN, in *Proceedings of the IEEE/CVF Conference on Computer Vision and Pattern Recognition Workshops* (2020), pp. 390–391
19. D. Misra, Mish: a self regularized non-monotonic activation function (2019). arXiv:1908.08681
20. S. Liu et al., Path aggregation network for instance segmentation, in *Proceedings of the IEEE Conference on Computer Vision and Pattern Recognition* (2018), pp. 8759–8768
21. K. He et al., Spatial pyramid pooling in deep convolutional networks for visual recognition. IEEE Trans. Pattern Anal. Mach. Intell. **37**(9), 1904–1916 (2015)
22. X. Yang et al., Deep learning-based bolt loosening detection for wind turbine towers. Struct. Control Health Monit. **29**(6), e2943 (2022)
23. J.E. Akin, *Finite Element Analysis Concepts: via SolidWorks* (World Scientific, 2010)

24. GB6170-86, *Hexagon Nuts, Style 1-Product Grades A and B* (China Standard Press, Beijing, 1986)
25. C.-Y. Wang, A. Bochkovskiy, H.-Y.M. Liao, Scaledyolov4: scaling cross stage partial network, in *Proceedings of the IEEE/cvf Conference on Computer Vision and Pattern Recognition* (2021), pp. 13029–13038
26. D.P. Kingma, J. Ba, Adam: a method for stochastic optimization (2014). arXiv:1412.6980
27. A. Paszke et al., Pytorch: an imperative style, high-performance deep learning library, in *Advances in Neural Information Processing Systems* 32 (2019)
28. Jasper RR. Uijlings et al., Selective search for object recognition. Int. J. Comput. Vis. **104**, 154–171 (2013)
29. J. Deng et al., Imagenet: a large-scale hierarchical image database, in *2009 IEEE Conference on Computer Vision and Pattern Recognition* (2009), pp. 248–255
30. A. Krizhevsky, I. Sutskever, G.E. Hinton, Imagenet classification with deep convolutional neural networks, in *Advances in Neural Information Processing Systems* (2012), pp. 1097–1105
31. Ian Goodfellow, Yoshua Bengio, Aaron Courville, *Deep Learning* (MIT Press, Cambridge, 2016)
32. K. Wada, labelme: Image Polygonal Annotation with Python (2018). https://github.com/wkentaro/labelme
33. M. Everingham et al., The PASCAL Visual Object Classes Challenge 2007 (VOC2007) Results. http://www.pascal-network.org/

Structural Damage Segmentation

8

Structural damage segmentation is another key task in vision-based SHM. As introduced in Chap. 2, images are labeled pixel by pixel and segmentation algorithms and models aim to recognize all pixels, group a region of pixels with the same label, and assign a class label for each region to match the ground truth. Complementary to classification (Chap. 6) and localization (Chap. 7) tasks, segmentation further provides additional information for damage quantification, which is useful for the post-disaster decision-making process.

This chapter presents a brief overview of how AI technologies can be used to tackle the structural damage segmentation task. Firstly, the common segmentation performance evaluation metrics are introduced, i.e., mIoU and AP. Subsequently, two classical segmentation models are presented, namely, FCN (refer to Chap. 2) and U-Net. In order to delve deeper into the practicality of employing these approaches for the damage segmentation purposes, one case study of recognizing concrete spalling areas, complementing the case study in Sect. 7.5, is investigated.

8.1 Segmentation Evaluation Metrics

There are many metrics suitable for evaluating the segmentation performance. Two commonly used ones, namely, mIoU and AP (specifically for masks and is referred to as "Mask AP"), are introduced in the subsequent sections. In practice, the choice of the evaluation metric depends on the specific goals and characteristics of the segmentation task.

8.1.1 mIoU

As introduced in Sect. 7.1.1, the IoU metric can be used in both object localization and image semantic segmentation tasks, but the way it is applied differs slightly between the

© The Author(s), under exclusive license to Springer Nature Switzerland AG 2024
K. M. Mosalam and Y. Gao, *Artificial Intelligence in Vision-Based Structural Health Monitoring*, Synthesis Lectures on Mechanical Engineering,
https://doi.org/10.1007/978-3-031-52407-3_8

two. Unlike using Bboxes in localization, the main outcome of conducting segmentation is to generate a mask image, the evaluation metric is usually based on the comparison between the mask image and the ground truth pixel level labeling. For example, the pixel level label may only contain binary information, i.e., the occurrence of the crack is 1, otherwise, it is 0. Denoting the recognized image as I, where (x, y) represents an arbitrary pixel, P, the mask is generated by the segmentation model as M_s and the ground truth pixel level is labeled as M_g. Define $\mathbb{1}\{\cdot\}$ as the indicator function, i.e., if the inside condition is true, it is 1, otherwise it is 0. Thus, IoU is computed as follows:

$$\text{Intersection} = \sum_{(x,y) \in P} \mathbb{1}\{M_s(x, y) + M_g(x, y) = 2\} \tag{8.1}$$

$$\text{Union} = \sum_{(x,y) \in P} \left[\mathbb{1}\{M_s(x, y) = 1\} + \mathbb{1}\{M_g(x, y) = 1\} \right] - \text{Intersection} \tag{8.2}$$

$$\text{IoU} = \text{Intersection}/\text{Union} \tag{8.3}$$

Furthermore, while conducting segmentation for objects such as damage with multiple classes (i.e., number of classes $K \geq 2$), a mean IoU, denoted as mIoU, is obtained by averaging the IoU values for each class including the background. Let $P_{i,j}$ be the number of pixels that the ground truth is class i and the prediction is class j. Moreover, T_i is the total number of pixels of class i, i.e., $T_i = \sum_j P_{i,j}$, mIoU is calculated as follows:

$$\text{mIoU} = \frac{1}{K+1} \cdot \sum_i \frac{P_{i,i}}{T_i + \sum_j P_{j,i} - P_{i,i}} \tag{8.4}$$

An example is illustrated in Fig. 8.1, where "0" represents the background and "1" & "2" are two classes, i.e., $K = 2$. Note that the red pixels in Fig. 8.1c are the mismatches between the ground truth classes (Fig. 8.1a) and predicted classes (Fig. 8.1b). Accordingly, the pixels' CM is computed in Fig. 8.2. According to Eq. 8.4, mIoU is computed by averaging the IoU values for each class. Starting from class 0, from CM, the number of pixels correctly predicting class 0 is 32, i.e., $P_{0,0} = 32$. The number of pixels, whose ground truth label is class 0, is 35, i.e., $T_0 = P_{0,0} + P_{0,1} + P_{0,2} = 32 + 2 + 1 = 35$. The sum of pixels, whose prediction is class 0, is $\sum_j P_{j,0} = P_{0,0} + P_{1,0} + P_{2,0} = 32 + 2 + 0 = 34$. Therefore, IoU for class 0 is $\text{IoU}_0 = 32/(35 + 34 - 32) = 0.865$. Similarly, $\text{IoU}_1 = 6/(8 + 9 - 6) = 0.545$ and $\text{IoU}_2 = 5/(6 + 6 - 5) = 0.714$. Therefore, mIoU $= (0.865 + 0.545 + 0.714)/3 = 0.708$.

In summary, through the above expression, a higher mIoU indicates better segmentation performance, as it represents a better overlap between the predicted and the ground truth masks. In addition, mIoU has the advantages of non-negativity, symmetry, and scale invariance, which support its wide usage in practical engineering scenarios.

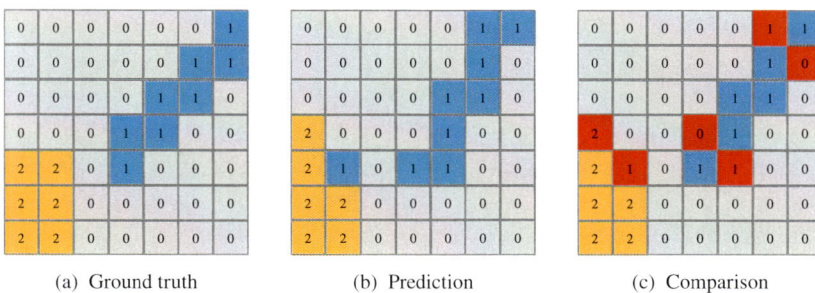

(a) Ground truth   (b) Prediction   (c) Comparison

Fig. 8.1 Example of mIoU computation

Fig. 8.2 CM of the prediction

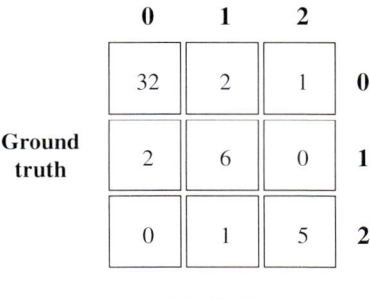

8.1.2 Mask AP

The mIoU provides an overall measure of how well the model segments different classes, while AP focuses on the precision-recall (PR) trade-off, refer to Sect. 7.1.2, which is particularly relevant when evaluating how well objects are segmented across a range of confidence levels. The computation of AP is the same as that in localization in Sect. 7.1.2 except that the target is mask instead of Bbox.

For an arbitrary structural image, by setting a threshold value, ϵ, define TP as the number of predicted masks with IoU (overlap with corresponding ground truth) larger than ϵ, FP as the number of masks equal to or less than ϵ, and FN as the number of missed detection (i.e., there is a mask in ground truth but the model does not detect it). For completeness of the definitions, the TN in the image segmentation case is related to the background region of the image when correctly not detected by the model. Consequently, the recall, precision, and PR curve are computed. The remaining steps are the same as that in the localization task, as introduced in Sect. 7.1.2. Furthermore, for multiple classes, a mean AP is calculated by averaging the AP values of the masks for each class, which is also known as mAP. Similar to the localization task, to investigate the model performance on different sizes of objects or confidence levels, more variant AP can be considered, e.g., AP^{75}, AP^{S}, AP^{M}, and AP^{L}, as discussed in Sect. 7.1.2.

8.2 Transposed Convolution

In the field of CV, because of continuous Conv operations, the input size of each layer decreases leading to extracted informative feature maps. For several vision tasks, such as segmentation, these feature maps need to be restored to their original size for further pixel-wise prediction. This operation uses *up-sampling* for expanding the image size to achieve the mapping of images from small resolution to large resolution. There are a few standard methods for up-sampling, such as nearest neighbor interpolation [1], bilinear interpolation [2], and transposed convolution, where transposed convolution plays a crucial role in image segmentation.

Transposed convolution, TransConv for short, is also known as "deconvolution" (Deconv) or "fractional-strided convolution" [3]. The concept of TranConv for DL vision task was proposed in [4] and was further developed in [5, 6]. With the successful applications of TransConv in NN visualization, it has been gradually adopted for other tasks such as segmentation [3] and for generative models [7].

TransConv has a similar process to the Conv operation and it has a solid mathematical formulation, which is demonstrated using an example below. Suppose an input image X has a size of 4×4 and its matrix form is expressed as follows:

$$X = \begin{bmatrix} x_1 & x_2 & x_3 & x_4 \\ x_5 & x_6 & x_7 & x_8 \\ x_9 & x_{10} & x_{11} & x_{12} \\ x_{13} & x_{14} & x_{15} & x_{16} \end{bmatrix}.$$

A 3×3 Conv filter (kernel), W, is performed on X and its matrix form is as follows:

$$W = \begin{bmatrix} w_{0,0} & w_{0,1} & w_{0,2} \\ w_{1,0} & w_{1,1} & w_{1,2} \\ w_{2,0} & w_{2,1} & w_{2,2} \end{bmatrix}.$$

Based on the Conv computation procedure in Sect. 4.2.1, take an example of using stride, $S = 1$ and no padding, after convolving X with W, the output feature map, Y, has a shape of 2×2 as follows:

$$Y = \begin{bmatrix} y_1 & y_2 \\ y_3 & y_4 \end{bmatrix}.$$

Consider expanding X and Y into vector forms as follows:

$$\vec{X} = [x_1, x_2, x_3, x_4, x_5, x_6, x_7, x_8, x_9, x_{10}, x_{11}, x_{12}, x_{13}, x_{14}, x_{15}, x_{16}]^T,$$

$$\vec{Y} = [y_1, y_2, y_3, y_4]^T.$$

Then, the whole Conv process can be expressed in the vector form, $\vec{Y} = C\vec{X}$, where

$$C = \begin{bmatrix} w_{0,0} & w_{0,1} & w_{0,2} & 0 & w_{1,0} & w_{1,1} & w_{1,2} & 0 & w_{2,0} & w_{2,1} & w_{2,2} & 0 & 0 & 0 & 0 & 0 \\ 0 & w_{0,0} & w_{0,1} & w_{0,2} & 0 & w_{1,0} & w_{1,1} & w_{1,2} & 0 & w_{2,0} & w_{2,1} & w_{2,2} & 0 & 0 & 0 & 0 \\ 0 & 0 & 0 & 0 & w_{0,0} & w_{0,1} & w_{0,2} & 0 & w_{1,0} & w_{1,1} & w_{1,2} & 0 & w_{2,0} & w_{2,1} & w_{2,2} & 0 \\ 0 & 0 & 0 & 0 & 0 & w_{0,0} & w_{0,1} & w_{0,2} & 0 & w_{1,0} & w_{1,1} & w_{1,2} & 0 & w_{2,0} & w_{2,1} & w_{2,2} \end{bmatrix}.$$

While conducting TransConv, firstly C is transposed to C^T and the input, $\vec{X}$, can be approximated as follows:

$$\widehat{\vec{X}} = C^T \vec{Y}.$$

Specifically, $C^T \vec{Y}$ matrix-vector product, with $C^T_{16 \times 4}$ and $\vec{Y}_{4 \times 1}$, results in a vector, $\widehat{\vec{X}}_{16 \times 1}$, which is subsequently unflattened to a 4×4 matrix by every 4 elements from their vector form. It is noted that the TransConv operation does not perfectly recover the values of elements in the vector $\vec{X}$ (same for X) through C^T and $\vec{Y}$ (same as Y) because C^T is not necessarily equal to C^{-1} (if its pseudo-inverse[1] form exists) and the main purpose of TransConv is to recover the shape of $\vec{X}$ and thus X. Furthermore, the matrix multiplication of $C^T \vec{Y}$ can be conducted through the Conv operations with padding tricks on Y. Therefore, these characteristics explain the name of TransConv. More details and examples can be found in [3–6]. Accordingly, TransConv performs an up-sampling from the 2×2 output, Y, back to the original 4×4 shape of the input, X, and its parameters (e.g., $w_{i,j}$, $i, j \in \{0, 1, 2\}$ of the filter) in the matrix C can be learned through network training by the backpropagation (Sect. 4.1.2) and the gradient descent method (Sect. 4.1.3). Therefore, no predefined interpolation methods are required.

Related to the CV implementation, TransConv is performed by conducting Conv operation on the feature map (e.g., Y in the example above) with certain zero padding tricks, where the zeros can be padded within the feature map and on its boundary as well, refer to Fig. 8.3. Suppose the input feature map size is $H_{in} \times W_{in}$ and the target output size is $H_{out} \times W_{out}$. The parameters of the TransConv operation include the size of the Conv filter, $F \times F$, moving stride size, S, and padding size on each side of the map, U. These are the same notations used in Eq. 4.18, with the exception of the input and output sizes. Based on these definitions, the TransConv computation is performed as follows:
1. Pad $(S − 1)$ rows and columns of "0s" within the input feature map.
2. Pad $(F − U − 1)$ rows and columns of "0s" outside of the input feature map.
3. Flip the parameters of the Conv filter upside down, as well as left to right.
4. Conduct Conv operations on the padded input feature map using the flipped $F \times F$ Conv filter with a stride of 1 and no further zero padding operations.

The above steps result in the input-output relationship expressed in Eqs. 8.5. It is to be noted that for the more general case of F, S, and U differing between the height, H, and the width, W, directions of the feature map, their respective values need to be introduced in Eqs. 8.5.

---

[1] A rectangular matrix $[A]_{m \times n}$ with $m \neq n$ has a pseudo-inverse as a generalization of the notion of inverse of a square, invertible matrix where $m = n$. It is computed using a "best fit" in the sense of the *least squares solution* to a system of linear equations that lacks a unique solution.

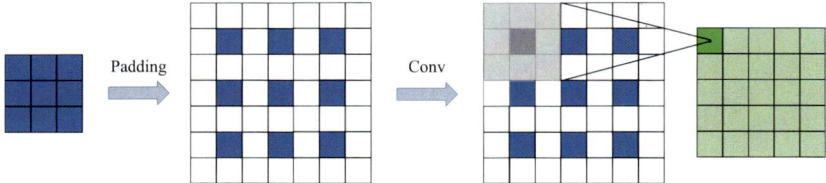

Fig. 8.3 Example of transposed convolution

$$H_{out} = (H_{in} - 1) \times S - 2 \times U + F$$
$$W_{out} = (W_{in} - 1) \times S - 2 \times U + F \qquad (8.5)$$

To illustrate the above TransConv computation, the example illustrated in Fig. 8.3 shows a feature map of size 3×3, i.e., $H_{in} = W_{in} = 3$, and the TransConv up-sampling operation is conducted using a 3×3 filter, i.e., $F = 3$, with a stride, $S = 2$, and padding size, $U = 1$. Based on the above procedure, the feature map is expanded firstly, where $S - 1 = 2 - 1 = 1$ row and column of "0s" are added within the original feature map and $F - U - 1 = 3 - 1 - 1 = 1$ row and column of "0s" are appended to the outside of the original feature map. Therefore, a 7×7 padded input to the TransConv operation is obtained. Subsequently, the Conv operation using the 3×3 filter with a stride of 1 is applied to this padded feature map without any additional padding, which, according to Eqs. 4.16, yields the final up-sampled result with a size of 5×5, i.e., $H_{out} = W_{out} = 5$. This is also easily obtained through the application of Eqs. 8.5, where $H_{out} = W_{out} = (3 - 1) \times 2 - 2 \times 1 + 3 = 5$. Moreover, if the sizes of the input and output are known based on the needs of the model, the user can define the TransConv parameters to satisfy these needs. Note that in applying Eqs. 4.16 in this example, $P' = W_{out}$, $Q' = H_{out}$, $P = 7$, and $Q = 7$.

8.3   FCN

FCN was proposed in 2015 [8], which is the first study in the semantic segmentation problem in the modern CV field. An example of FCN is illustrated in Fig. 8.4. As its name shows FCN discards the FC-layers and replaces them with pure Conv layers. Subsequently, the output after these Conv operations is further up-sampled by the TransConv operation to the same size as the input to realize a pixel-wise prediction. To improve the prediction and up-sampling performance, a *skip architecture* was developed. Therefore, fully convolution, TransConv, and skip architecture are three key novel contributions of this specific network, as discussed in detail below:

1. Fully convolution: In several classical CNNs (e.g., VGGNet), Fig. 4.7, several Conv blocks are followed by FC-layers (Dense layers), which are used for generating a fixed-length feature vector for classification. As a result, the features extracted by the Conv

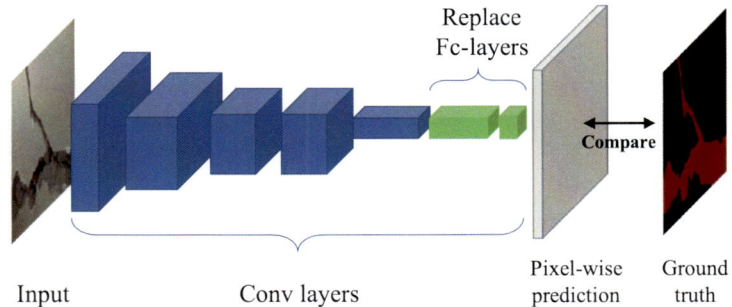

Fig. 8.4 Illustration of FCN

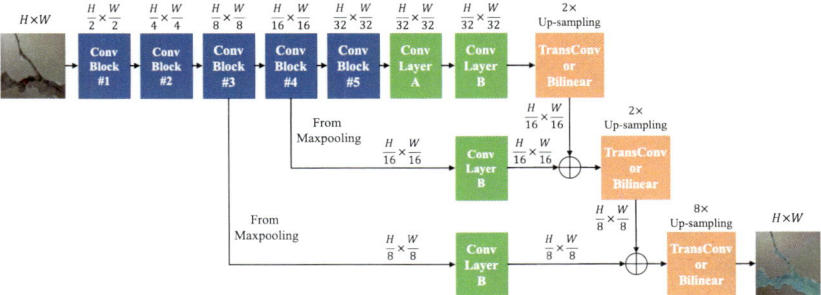

Fig. 8.5 FCN with skip architecture using VGG-type backbone in Table 6.2

blocks are squeezed into a 1D vector, which is not suitable for 2D segmentation problems. FCN replaces these FC-layers with Conv layers, such as 1×1 Conv, which maintains the 2D shape but increases the third channel, i.e., feature dimension, by the number of filters applied. Therefore, the output after these Conv layers keeps the spatial information and can be further used for pixel-wise segmentation predictions.

2. Up-sampling: Because of successive pooling operations after each Conv block, the spatial dimension of the feature maps decreases significantly. To obtain a pixel-wise prediction to compare with the ground truth label shown in Fig. 8.4, FCN adopts the TransConv layer or the bilinear interpolation [2] to up-sample and restore the Conv output to the input size.

3. Skip architecture: To improve the up-sampling performance, a skip architecture (similar to the short-cut connection in ResNet in Sect. 4.3.3) is developed, which is used for fusing feature information from both shallow and deep layers. For example, as illustrated in Fig. 8.5, the VGG-type backbone listed in Table 6.2 is adopted. It contains five Conv blocks and the output size of each block after max pooling is shown above the Conv block box in Fig. 8.5, where the size of the last Conv block is $H/32 \times W/32$. Based on FCN settings, two FC-layers in the original VGGNet are replaced by two Conv layers (Layers A & B in Fig. 8.5) and followed by a TransConv layer. If no skip architecture

is applied, the TransConv layer directly conducts an excessive 32-time up-sampling to restore the extracted features to the input size of $H \times W$. If skip architecture is applied to link Conv blocks # 3 & 4, as illustrated in Fig. 8.5, the output of the last Conv layer (i.e., layer B) is firstly up-sampled to a size of $H/16 \times W/16$. Subsequently, the feature map with a size of $H/16 \times W/16$ from Conv block # 4 (after max pooling) is convolved once and merged with the above TransConv results and the output is further up-sampled to $H/8 \times W/8$. Similarly, it is further merged with output from Conv block # 3 and 8-time up-sampled (much less excessive and more accurate than the 32-time up-sampling when no skip architecture is used) to the input size of $H \times W$, which is the final pixel-wise prediction.

As the first modern deep CNN model in segmentation, the FCN realizes the pixel-wise segmentation in an acceptable manner and the proposed skip architecture further helps improve its performance. However, the segmentation results are not refined enough and some up-sampling results are blurred and missing the details. In addition, because the FCN is modified from CNN, even though it uses Conv operations to replace FC-layers, it still conducts an independent pixel classification without fully considering the relationship between pixels. Therefore, more advanced models are expected to be developed for the purpose of semantic segmentation.

8.4 U-Net

The original proposal of U-Net was made to solve problems in biomedical images [9], where image data shortage is a common issue. However, due to its excellent performance, it has been widely applied in various fields for semantic segmentation, such as satellite image segmentation and industrial defect detection.

As illustrated in Fig. 8.6, the architecture design of the U-Net adopts the concept of EN-DE as mentioned in Sect. 2.3. The left half part, i.e., the EN, is used for feature extraction, while the right half, i.e., the DE, is reserved for up-sampling for pixel-wise prediction. Similar to FCN, U-Net also introduces the *skip connection* between each level with the same spatial size, where the feature maps in the EN are copied and concatenated with the up-sampled output in the DE at the same level. These skip connections help preserve fine-grained details during the segmentation process. Take the example in Fig. 8.6, the EN adopts a series of 3×3 Conv filters and then performs a 2×2 max pooling to shrink the feature map by half. Repeat this process until reaching the bottleneck layer (at the bottom) which has features with dimension of 1024. These features are up-sampled by TransConv or bilinear interpolation [2] with a filter size of 2×2 to double the spatial size. Subsequently, the conventional Conv operation is conducted to process these feature maps with adjusted feature dimensions. Repeat this process until the final output has the same size as the input. It is noted that in Fig. 8.6, each box is a multi-channel feature map where the number

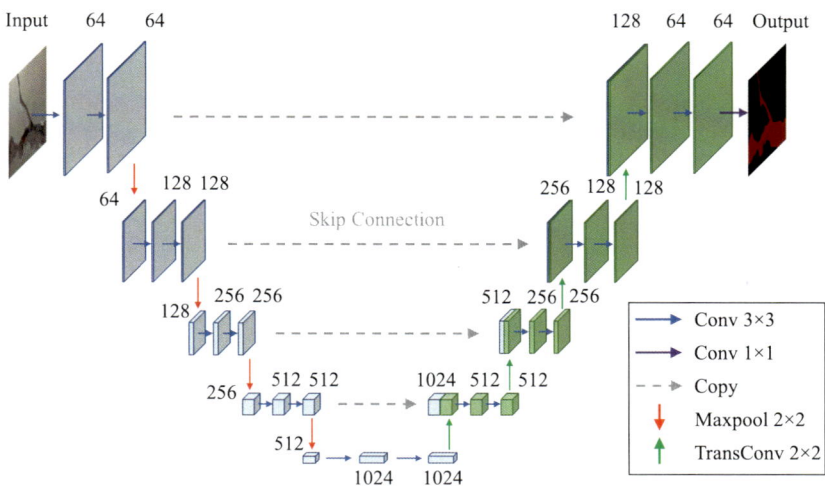

Fig. 8.6 U-Net

of channels is listed on top of the box. Moreover, similar to classical CNN designs, e.g., VGGNet [10], the feature dimension in the shallow layers (e.g., close to the top in Fig. 8.6) is smaller; on the contrary, the feature dimension expanded in the deep layers (e.g., close to the bottom in Fig. 8.6).

In summary, the U-Net architecture resembles a U-shape, with a contracting path (encoder, EN) on one side and an expanding path (decoder, DE) on the other side. This design is characterized by a contracting path, which captures the context and information from the input image and an expansive path, which enables precise localization and segmentation.

From the literature review, U-Net has proven to be adaptable to various domains beyond medical imaging. Its versatility has led to applications in semantic segmentation, image-to-image translation, and other CV tasks. A large number of variants of U-Net-like models were developed and founded on the basis of the original U-Net [9], e.g., V-Net [11], U-Net++ [12, 13], and Attention U-Net [14]. However, there exist several drawbacks to the U-Net, as it possesses a large number of parameters due to the skip connections and the additional layers in the expanding path. Consequently, the model becomes more prone to over-fitting, particularly when operating with small datasets. Moreover, U-Net requires additional computational resources due to the presence of skip connections, making it more computationally demanding compared to alternative architectures. Finally, U-Net is also sensitive to the initialization of the model parameters, which introduces challenges in its training.

8.5   Case Study: Concrete Spalling Segmentation

In Sect. 7.5, concrete spalling damage has been detected and located with spatial information, i.e., coordinates of the Bboxes. In this case study, the spalled areas are further identified pixel by pixel, which helps to quantify the damaged areas. Two classical segmentation models, FCN and U-Net, introduced above, are applied herein.

8.5.1   Spalling Segmentation Dataset

Similar to the localization task discussed in Sect. 7.5, the same 1,350 raw spalling images with a resolution of 224×224 are collected and re-labeled using the "*Labelme*" [15] tool. Unlike the process of drawing Bboxes in the localization task, for segmentation, a distinct labeling format is employed, which involves drawing multiple masks to cover the target, i.e., the spalled area in this case. However, the spalled area typically exhibits irregular shapes, necessitating a unique approach to creating the masks. Therefore, the mask of the spalled area can be determined by drawing an outline consisting of multiple polylines[2] and all pixels in the inner area have the "SP" label, refer to Sect. 5.2.2.3. On the other hand, the remaining out-of-mask pixels are labeled as "background". Three sample images are illustrated in Fig. 8.7. Similar to the localization task, the same 1,200 images are used for training and the remaining 150 images are reserved for testing.

(a) Sample #1   (b) Sample #2   (c) Sample #3

Fig. 8.7   Samples of labeling masks for concrete spalling

---

[2] A polyline is a set of points with line segments in between. It is used in computer-aided design (CAD) software to represent irregular boundaries, e.g., roads or rivers in a map and in geographic information systems (GIS).

8.5.2   Spalling Segmentation Settings of the Models

For FCN, one baseline model without skip architecture is adopted, whose backbone (i.e., its 5 Conv blocks) is inherited from the ImageNet [16] pre-trained VGG-16 model. These are followed by two Conv layers to generate a heatmap-like output. Subsequently, the heatmap output is up-sampled through one TransConv layer to map back to the original input-sized prediction. From the network configuration in Table 8.1, the spatial size of the heatmap output is 7×7 and the original input image size is 224×224, resulting in $224/7 = 32$ times up-sampling. Based on Eqs. 8.5, the parameters: stride, $S = 32$, padding, $U = 0$, and filter size, $F = 32$, are selected. Since the baseline model up-samples the output 32 times, it is denoted as FCN32s. To explore the effectiveness of skip architecture, refer to Fig. 8.5,

Table 8.1 FCN details without skip architecture

Block	Layer type	Filter size (#)/Pooling size	Output size
Input	Input/Image	–	(224, 224, 3)
Conv Block 1	Convolutional	3×3 (64)	(224, 224, 64)
	Convolutional	3×3 (64)	(224, 224, 64)
	Max Pooling	2×2, $S^a = 2$	(112, 112, 64)
Conv Block 2	Convolutional	3×3 (128)	(112, 112, 128)
	Convolutional	3×3 (128)	(112, 112, 128)
	Max Pooling	2×2, $S = 2$	(56, 56, 128)
Conv Block 3	Convolutional	3×3 (256)	(56, 56, 256)
	Convolutional	3×3 (256)	(56, 56, 256)
	Convolutional	3×3 (256)	(56, 56, 256)
	Max Pooling	2×2, $S = 2$	(28, 28, 256)
Conv Block 4	Convolutional	3×3 (512)	(28, 28, 512)
	Convolutional	3×3 (512)	(28, 28, 512)
	Convolutional	3×3 (512)	(28, 28, 512)
	Max Pooling	2×2, $S = 2$	(14, 14, 512)
Conv Block 5	Convolutional	3×3 (512)	(14, 14, 512)
	Convolutional	3×3 (512)	(14, 14, 512)
	Convolutional	3×3 (512)	(14, 14, 512)
	Max Pooling	2×2, $S = 2$	(7, 7, 512)
Conv layer A	Convolutional	7×7 (4096)	(7, 7, 4096)
Conv layer B	Convolutional	1×1 (4096)	(7, 7, 4096)
	Convolutional	1×1 (2)	(7, 7, 2)
Up-sampling layer	Transposed Conv	32×32 (2), $S = 32$	(224, 224, 2)

^{a}Stride, $S = 1$, unless otherwise noted.

two comparison models are used. The first, denoted as FCN16s, connects the features from Conv block # 4 with the two-time up-sampled heatmap output of size 14×14 and then it is further up-sampled $224/14 = 16$ times. The second, denoted as FCN8s, is similar to FCN16s, but it further connects the features from Conv block # 3 with the output from the merged results of Conv block # 4 & the up-sampled heatmap output and the new merged results are up-sampled 8 times to restore to the size of the input image.

For U-Net, the architecture design is the same as Fig. 8.6, which contains 4-time down-sampling and up-sampling. It is noted that padding is used among the Conv operation at the same level to maintain the size of the feature map. Since 2×2 max pooling is used between each level, the feature map size shrinks by half. As a result, with an input size of 224×224, after 4-time down-sampling, the bottleneck feature map at the bottom has a shape of $14 \times 14 \times 1024$ because $224/2^4 = 14$. In the DE part, the TransConv layer is added for a two-time up-sampling between layers, where its parameters are selected as $S = 2$, $U = 0$, and $F = 2$ satisfying Eqs. 8.5.

During training, all models are trained under a similar setting. The batch size is 10 and a total of 70 training epochs is set. The SGD is employed as the optimizer. There are slight differences in the optimizer setting. For FCN models (FCN32s, FCN16s & FCN8s), the initial learning rate is 5×10^{-11} with a momentum of 0.99 and the weight decay rate is 5×10^{-4}. For the U-Net model, the initial learning rate is 5×10^{-3} with a momentum of 0.9 and the weight decay rate is 5×10^{-4}.

The evaluation metric adopted herein is mIoU, Sect. 8.1.1. The implementation of these experiments is based on PyTorch [17] and performed on NUC13RNGi7 PC with a single GPU (CPU: 13th Gen Intel(R) Core(TM) i7–13700K 3.40 GHz, RAM: 32 GB and GPU: Nvidia Geforce RTX 4080, GPU memory: 16 GB).

8.5.3 FCN Segmentation Results

The performance of FCN models is discussed in this section. The mIoU values on the test dataset for FCN32s, FCN16s & FCN8s are 74.22%, 80.16% & 80.18%, respectively. These results demonstrate the effectiveness of the skip architecture in FCN-type models. Specifically, FCN16s and FCN8s demonstrate a notable improvement of about 6% from the baseline FCN32s without skip architecture, showcasing the value of skip architecture. However, this enhancement is not significant when connecting more shallow layers, i.e., FCN8s only marginally outperforms FCN16s. Therefore, in practical applications, considering the computational efficiency, the skip architecture may just be applied to the deeper Conv blocks.

The visual segmentation samples are presented in Figs. 8.8 and 8.9. Figure 8.8 shows five typical segmentation samples using the three models, where almost all models successfully capture the spalling areas. It is observed that both FCN8s and FCN16s achieve accurate results, effectively covering the spalled area, and they exhibit similar performance. On the contrary, the predicted masks generated by FCN32s are less accurate and incomplete. For instance, in samples #2, #3 & #4, FCN32s fails to cover certain corners of the spalling and in sample #5, it misses the spalling on the bottom left corner and overestimates it on the right. This discrepancy explains the previously mentioned 6% gap in numerical results between the FCN with and without the skip architecture. Additionally, Fig. 8.9 displays a few unsuccessful predictions, characterized by incomplete coverage, missed detection, and erroneous prediction. For example, in samples #1 & #2, both FCN8s and FCN16s capture the primary spalled area but miss several small areas at the sharp corners of the spalled area. In samples #3 & #4, while they detect the major spalled area, they also present erroneous pixel-wise predictions unrelated to spalling. In line with the numerical results, FCN32s performs even worse, lacking an accurate mask boundary for spalling and displaying the propensity to miss small objects, such as corner spalling in samples #1 to 3 in Fig. 8.9.

In summary, FCN using the VGG-16 backbone performs well in spalling segmentation tasks, particularly with the incorporation of skip architecture. As depicted in Fig. 8.8, FCN8s and FCN16s effectively cover the spalled areas. However, there is still room for improvement with issues like incompleteness, missed detection, and erroneous detection. Therefore, there is a need to explore more advanced methods to address these limitations.

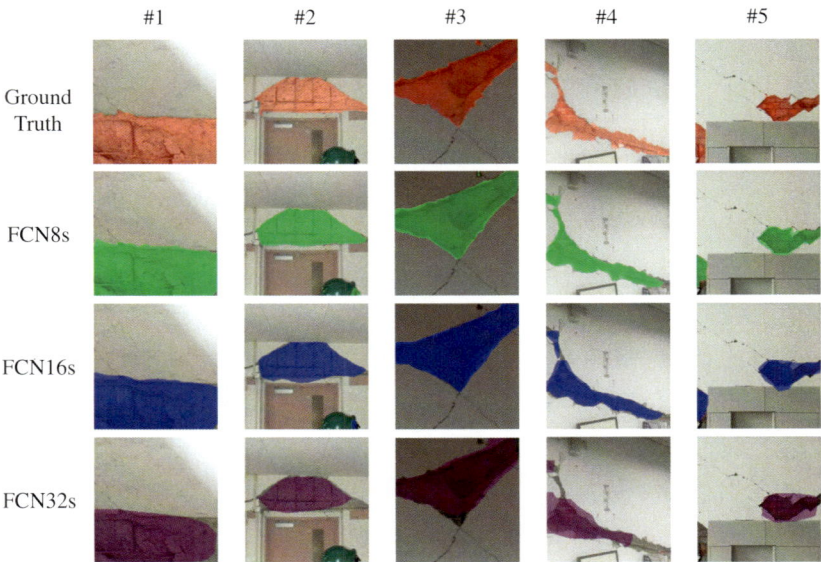

Fig. 8.8 Successful segmentation samples of FCN

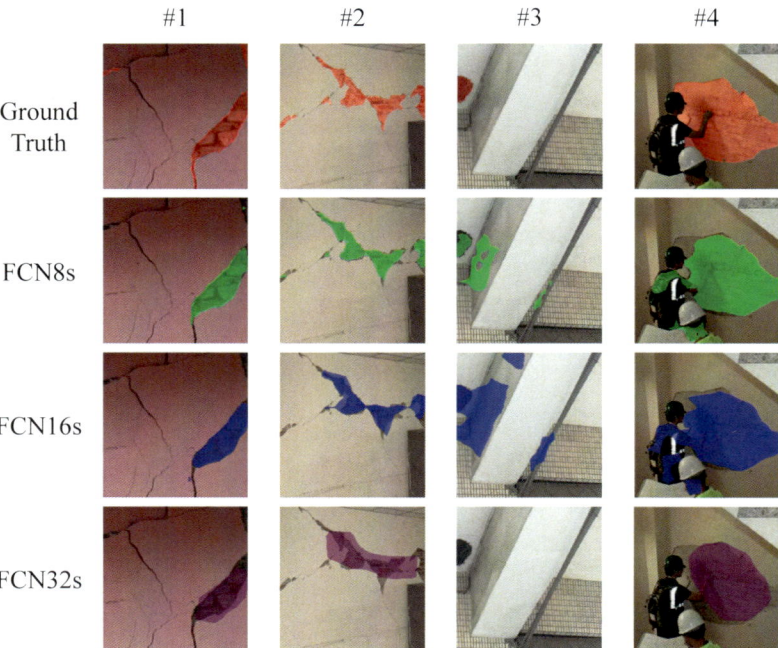

Fig. 8.9 Unsuccessful segmentation samples of FCN

8.5.4   U-Net Segmentation Results

The mIoU achieved by U-Net on the test dataset is 79.42%, demonstrating a level of performance comparable to previous models like FCN8s and FCN16s. Visual segmentation samples showcasing U-Net's results can be seen in Figs. 8.10 and 8.11. In Fig. 8.10, five successful samples of U-Net are presented and the spalling areas with varying sizes are well-covered. It is noted that in sample #3, U-Net even identifies a small spalling on the top, a detail missed by both FCN8s and FCN16s. Therefore, these examples demonstrate the effectiveness of U-Net as well as FCN in certain spalling scenarios. On the other hand, Fig. 8.11 presents several unsuccessful samples, where the U-Net can capture the major spalling area but exhibits incompleteness in the boundary, as seen in samples #1 to #3. Similar to the FCN models, the U-Net also has the same issue of erroneous prediction in sample #4.

In conclusion, U-Net can be viewed as a viable alternative to FCN in certain scenarios. In addition, it is advisable to consider employing U-Net in conjunction with other segmentation models within a candidate mechanism, refer to Sect. 7.3.4, for a more thorough analysis. To further enhance the performance, researchers have been developing more advanced U-Net variants, as mentioned in Sect. 8.4, such as U-Net++ [12, 13] and Attention U-Net [14].

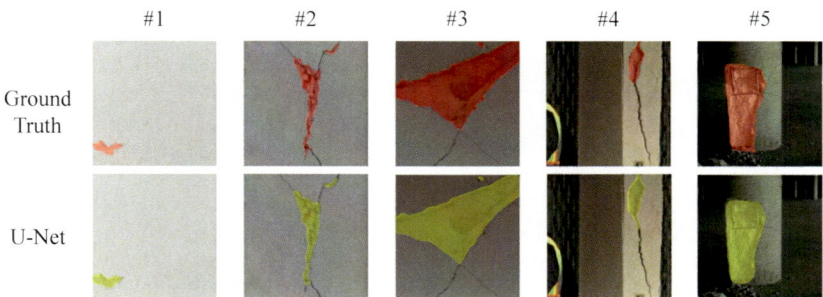

Fig. 8.10 Successful segmentation samples of U-Net

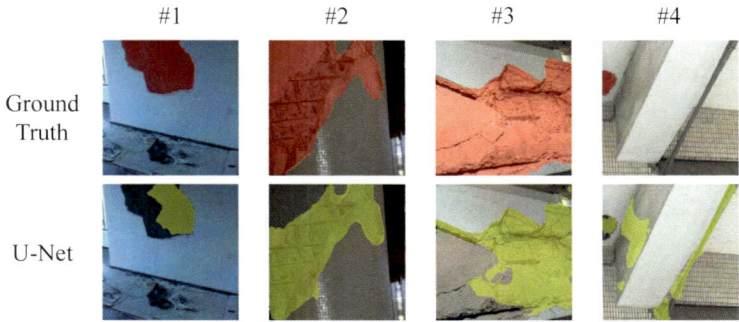

Fig. 8.11 Unsuccessful segmentation samples of U-Net

8.5.5 Summary

In summary, this case study explored the application of two classical segmentation models, namely, FCN and U-Net, for concrete spalling segmentation. The evaluation of these models is based on the mIoU, which quantifies the overlapping between predicted masks and ground truth pixel-wise labels. The experimental results demonstrate that both models, particularly FCN with skip architecture achieve comparable performance, yielding a mIoU value of approximately 80%. Additionally, visual examples illustrate that the predicted masks effectively cover the spalling areas. However, an analysis of unsuccessful predictions reveals certain issues, including incompleteness, missed detection, and erroneous detection. These challenges highlight opportunities for practical improvement. To address these concerns, potential strategies include advancing FCN and U-Net with enhanced versions and exploring alternative NN architectures such as Mask R-CNN [18] and Transformers, Sect. 4.5. Furthermore, mIoU evaluates the results pixel by pixel, focusing more on the *local* performance. On the other hand, AP or mAP of the mask treats the mask of a damaged area as one single object and evaluates the segmentation quality on a *global* scale, analyzing area by area. This approach could provide valuable insights and merits further investigation.

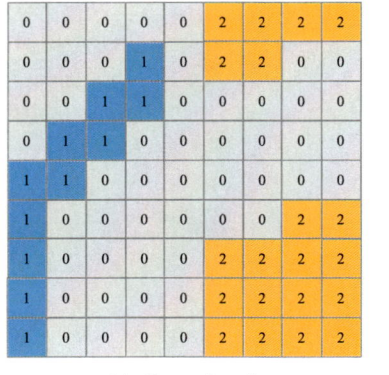   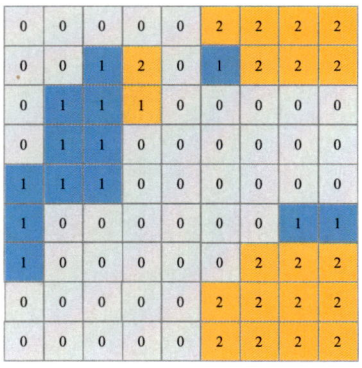

(a) Ground truth   (b) Prediction

Fig. 8.12 A 9×9 resolution of one segmentation example

8.6   Exercises

1. Suppose one image has a resolution of 9×9, both ground truth and prediction by a DL segmentation model are shown in Fig. 8.12, compute the corresponding mIoU value.
2. While conducting TransConv for up-sampling, suppose the input image size is 7×7 and the TransConv stride, $S = 16$, padding size, $U = 0$, and filter size, $F = 16$, compute the output size.
3. While conducting TransConv for up-sampling, suppose the input image size is 14×14 and the target output size is 224×224, provide 3 valid combinations of TransConv stride, S, padding size, U, and filter size, F, to satisfy the relationship between the given input and output sizes.
4. Based on the definitions of FCN16s and FCN8s, following the illustration in Fig. 8.5, draw the architecture illustration of FCN16s and FCN8s. Suppose the input size is 224×224, annotate the output shape of each layer.
5. Suppose the input image has a shape of $256 \times 256 \times 3$ and a U-Net model is adopted with a configuration illustrated in Fig. 8.6. List its detailed configurations for both the contracting (EN) and the expanding (DE) paths, including the output size of each layer in a table similar to Table 8.1.

References

1. A.C. Bovik, *The Essential Guide to Image Processing* (Academic Press, 2009)
2. E.J. Kirkland, E.J. Kirkland, Bilinear interpolation, in *Advanced Computing in Electron Microscopy* (2010), pp. 261–263

3. V. Dumoulin, F. Visin (2016) *A Guide to Convolution Arithmetic for Deep Learning* (Mar 2016). eprint: 1603.07285

4. M.D. Zeiler et al., Deconvolutional networks, in *2010 IEEE Computer Society Conference on Computer Vision and Pattern Recognition* (IEEE, 2010), pp. 2528–2535

5. M.D. Zeiler, G.W. Taylor, R. Fergus, Adaptive deconvolutional networks for mid and high level feature learning, in *International Conference on Computer Vision*, vol. 2011 (IEEE, 2011), pp. 2018–2025

6. M.D. Zeiler, R. Fergus, Visualizing and understanding convolutional networks, in *European Conference on Computer Vision* (Springer, 2014), pp. 818–833

7. Y. Gao, B. Kong, K.M. Mosalam, Deep leaf-bootstrapping generative adversarial network for structural image data augmentation. Comput. Aided Civil Infrastruct. Eng. **34**(9), 755–773 (2019)

8. J. Long, E. Shelhamer, T. Darrell, Fully convolutional networks for semantic segmentation, in *Proceedings of the IEEE Conference on Computer Vision and Pattern Recognition* (2015), pp. 3431–3440

9. O. Ronneberger, P. Fischer, T. Brox, U-net: Convolutional networks for biomedical image segmentation, in *Medical Image Computing and Computer-Assisted Intervention–MICCAI 2015: 18th International Conference, Munich, Germany, October 5-9, 2015, Proceedings, Part III 18* (Springer, 2015), pp. 234–241

10. K. Simonyan, A. Zisserman, *Very Deep Convolutional Networks for Large-scale Image Recognition* (2014). arXiv:1409.1556

11. F. Milletari, N. Navab, S.-A. Ahmadi, V-net: Fully convolutional neural networks for volumetric medical image segmentation, in *Fourth International Conference on 3D Vision (3DV)*, vol. 2016 (IEEE, 2016), pp. 565–571

12. Z. Zhou et al., Unet++: A nested u-net architecture for medical image segmentation, in *Deep Learning in Medical Image Analysis and Multimodal Learning for Clinical Decision Support: 4th International Workshop, DLMIA 2018, and 8th International Workshop, ML-CDS 2018, Held in Conjunction with MICCAI 2018, Granada, Spain, September 20, 2018, Proceedings 4* (Springer, 2018), pp. 3–11

13. Z. Zhou et al., Unet++: Redesigning skip connections to exploit multiscale features in image segmentation. IEEE Trans. Med. Imaging **39**(6), 1856–1867 (2019)

14. O. Oktay et al., *Attention U-Net: Learning Where to Look for the Pancreas* (2018). arXiv:1804.03999

15. K. Wada, *labelme: Image Polygonal Annotation with Python* (2018). https://github.com/wkentaro/labelme

16. J. Deng et al., Imagenet: A large-scale hierarchical image database, in *2009 IEEE Conference on Computer Vision and Pattern Recognition* (2009), pp. 248–255

17. A. Paszke et al., Pytorch: An imperative style, high-performance deep learning library, in *Advances in Neural Information Processing Systems*, vol. 32 (2019)

18. K. He et al., Mask r-cnn, in *Proceedings of the IEEE International Conference on Computer Vision* (2017), pp. 2961–2969

Part III
Advanced Topics of AI in Vision-Based SHM

In Part III, an in-depth consideration is given to complex scenarios, encompassing limited data availability, unprocessed data, potential highly imbalanced class labels, and discrepancies in laboratory data compared to real-world environments. This part introduces a range of advanced AI technologies within the vision-based SHM, namely, generative adversarial network (GAN), semi-supervised learning, and active learning, to enhance the adaptability of AI models to the above mentioned complex scenarios.

Chapter 9 covers the fundamental knowledge of GAN and assesses its performance in generating synthetic structural images alongside the corresponding evaluation metrics. Subsequently, the potential usage of the GAN-based image augmentation method in the classification problems is explored, particularly in scenarios with limited available real data and insufficient computing resources.

Chapter 10 introduces a new approach, namely, semi-supervised learning, which effectively utilizes both labeled and unlabeled data for training. Furthermore, besides the above mentioned limited data and computational constraints, this chapter tackles the imbalanced class/label issues. A novel semi-supervised variant GAN model, named Balanced Semi-supervised GAN (BSS-GAN), is developed by utilizing a balanced batch sampling technique. Through a case study, this new model can thoroughly exploit the features of the unlabeled data and simultaneously increase the model's synthetic data generation and classification capabilities.

Chapter 11 provides insight into another innovative approach, namely, active learning, which aims to reduce manual labeling efforts. Starting from self-training, the essential concepts, training techniques, and procedures are covered. To enhance the model's utilization of unlabeled data, the active learning framework incorporates a stacked convolutional autoencoder (SCAE). These advancements culminate in a new training paradigm, namely, SCAE-based active learning, which is validated through a case study for crack identification of a RC bridge deck.

Generative Adversarial Network for Structural Image Data Augmentation

9

As mentioned in Sects. 1.3.2 and 2.4, the practical usage of AI in SHM encounters challenges related to the scarcity of labeled data or even difficulties in data collection due to complex environments. Moreover, imbalanced class labels can introduce bias in the recognition of the DL models. To mitigate these issues, Chap. 6 discussed TL technologies. Besides TL, the widely adopted solution, namely, DA, is discussed in this chapter. From Sect. 6.1, conventional DA involves several processing operations, e.g., rotation, zoom, and flip, as illustrated in Fig. 6.1. By combining these numerical operations, the number of training data can be increased. However, this method can only generate highly correlated datasets, which fail to enhance data variety and may even lead to inferior performance. Consequently, to introduce more variety in the data, generative model-based methods, e.g., GAN [1], enter the sight of SHM researchers.

Unlike conventional DL models, GAN consists of two networks, namely, the *generator* and the *discriminator*, where the generator creates synthetic data and the discriminator classifies an input sample as "real" or as "synthetic". GAN uses *adversarial training*, where each of the two networks aims to minimize the gain of the opposite network while maximizing its own. Ideally, both the generator and the discriminator converge to the *Nash equilibrium* [2], where the discriminator gives equal predictive probabilities to the real and the synthesized samples. In other words, such equilibrium indicates that the quality of the synthetic images produced by the generator is as good as that of the real images. Thus, the discriminator can no longer distinguish whether the data are real or synthetic.

In summary, this chapter aims to cover the fundamental knowledge of the GAN. It investigates GAN's performance in synthetic structural image generation and explores the potential usage of the GAN-based image augmentation method in classification problems, especially under low-data regimes, typically encountered in past applications related to structural extreme events reconnaissance.

© The Author(s), under exclusive license to Springer Nature Switzerland AG 2024   247
K. M. Mosalam and Y. Gao, *Artificial Intelligence in Vision-Based Structural Health Monitoring*, Synthesis Lectures on Mechanical Engineering,
https://doi.org/10.1007/978-3-031-52407-3_9

9.1   GAN Mechanism

As mentioned above, GAN consists of a parametrized discriminator and a generator, which can be any designed function or NN. The training of GAN is in some sense adversarial by playing the game between the generator and the discriminator where they look forward to iteratively maximizing their own outcomes and minimizing the outcome of the opponent turn by turn. Figure 9.1 illustrates the procedure of how GAN works. The generator aims to produce synthetic data to fool the discriminator. Conversely, the discriminator aims to distinguish between real data and synthetic data produced by the generator. Once the generator fools the discriminator successfully, i.e., G1 fools D1, the discriminator updates itself until it can recognize the synthetic data correctly, i.e., D1 updates itself to D2. Similarly, once the synthetic data are identified by the updated discriminator, the generator updates itself until it can fool the discriminator again, i.e., G1 updates to G2. Finally, the game ends when the discriminator can no longer update itself to identify the synthetic data, i.e., G3 vs. D3. Under such a situation, the probability of classification of the synthetic vs. the real data of the discriminator becomes about 0.5, i.e., a random guess and the synthetic data are thought to be sufficiently close to the real data.

Mathematically, the training process of the GAN can be viewed as a game optimization problem and its optimal solution is the *Nash equilibrium*. However, computing Nash equilibrium has its difficulties, e.g., due to possible non-convex loss functions (refer to Sect. 4.1.2) and high dimensional space of parameters. Thus, in practice, using gradient-based algorithms to minimize the loss of the discriminator and the generator simultaneously is preferred.

Denote x as the real data point, z as a high dimensional random variable (also known as the *code*), p_d, p_g, and p_z as respective distributions of the "true" data, the generative

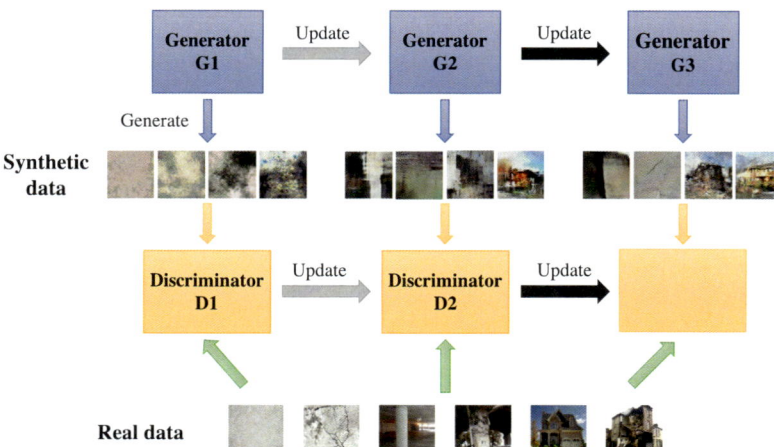

Fig. 9.1 Working mechanism of GAN in generating structural images demonstrating a procedure for GAN training

model, and the random variable, where p_z is commonly chosen as Gaussian or Uniform. $D(\cdot)$ and $G(\cdot)$ are respectively the discriminator and the generator functions with respective parameters θ_D and θ_G. It is noted that $D(x)$ is taken as a *sigmoid* function, Eq. 3.18, to output the probability of the discriminator identifying "true" data and $G(z)$ produces the synthetic data according to the selected distribution of z.

Goodfellow et al. [1] construct the value function $V(D, G) = V(\theta_D, \theta_G)$ as shown below in Eq. 9.1, where $E_{a \sim p_b}$ denotes the expectation of the data or the random variable a sampled from the distribution p_b.

$$
\begin{aligned}
V(D, G) &= E_{x \sim p_d} [\ln(D(x))] + E_{x \sim p_g} [\ln(1 - D(x))] \\
&= E_{x \sim p_d} [\ln(D(x))] + E_{z \sim p_z} [\ln(1 - D(G(z)))].
\end{aligned}
\tag{9.1}
$$

The discriminator aims to distinguish whether the fed data are real or synthetic. Thus, it is a binary classification problem and the *empirical cross-entropy*, based on Eq. 3.6, is computed by discrete samples (total number of samples $= N$) and adopted as the loss function, which is expressed as follows:

$$
L_D = -\frac{1}{N} \sum_{i=1}^{N} y_i \ln(D(x_i)) - \frac{1}{N} \sum_{i=1}^{N} (1 - y_i) \ln(1 - D(x_i)).
\tag{9.2}
$$

Label the real data, x_i, and the synthetic data, $\tilde{x}_i$, as 1 and 0, respectively, i.e., the label $y_i \in \{0, 1\}$, then Eq. 9.2 is simplified as shown in Eq. 9.3. In this equation, the empirical quantity of the loss approximately equals its theoretical value, related to Eq. 9.1, when N approaches infinity, indicating that optimizing the discriminator by estimating its parameters, $\hat{\theta}_D$, is the same as minimizing the negative of the value function, $-V(D, G)$, or maximizing the value function itself as shown in Eq. 9.4. Accordingly, the game maximizes the gain of the discriminator by distinguishing the difference between p_d and p_g as much as possible.

$$
\begin{aligned}
L_D &= -\frac{1}{N} \sum_{i=1}^{N_{\text{real}}} \ln(D(x_i)) - \frac{1}{N} \sum_{i=1}^{N_{\text{syn}}} \ln(1 - D(\tilde{x}_i)), \\
&= -C_{\text{real}} \, E_{x \sim p_d} [\ln(D(x))] - C_{\text{syn}} \, E_{x \sim p_g} [\ln(1 - D(x))],
\end{aligned}
\tag{9.3}
$$

where the constants, sometimes dropped as they do not affect the subsequent argmin operation, $C_{\text{real}} = N_{\text{real}}/N$ & $C_{\text{syn}} = N_{\text{syn}}/N$, and the number of real, N_{real}, & synthetic, N_{syn}, samples are typically taken such that $N = N_{\text{real}} + N_{\text{syn}}$.

$$
\hat{\theta}_D = \underset{\theta_D}{\operatorname{argmin}} \, L_D = \underset{D}{\operatorname{argmax}} \, V(D, G).
\tag{9.4}
$$

The generator generates the synthetic data and optimizes them aiming for high-quality synthetic data (as close as possible to real data) to fool the discriminator. Since it is a *minimax game* [3], the generator tries to minimize the gain (i.e., maximizing the value function,

$V(D, G)$) of the discriminator, defined as the generator loss function, L_G, to eventually the estimates of its parameters, $\hat{\theta}_G$, as follows:

$$L_G = \max_D V(D, G). \tag{9.5}$$

$$\hat{\theta}_G = \operatorname*{argmin}_{\theta_G} L_G = \operatorname*{argmin}_{G} \max_D V(D, G). \tag{9.6}$$

The well-defined discriminator and generator loss functions above define a minimax game that aims to optimize these functions iteratively until both converge to some local minima with the pair of parameter sets $(\hat{\theta}_D, \hat{\theta}_G)$. In practice, it is difficult to detect whether they have achieved local minima. Thus, setting a fixed number of iterations is commonly used. The training process used herein for this GAN study is summarized as follows:

Step 0: Initialize the discriminator, D, and the generator, G.
Step 1: Sample N real data points $\{x_1, x_2, \ldots, x_N\}$ using the distribution p_d.
Step 2: Sample N noise sets $\{z_1, z_2, \ldots, z_N\}$ from a selected distribution p_z.
Step 3: Update the parameters of D, i.e., $\hat{\theta}_D$, by gradient descent (Sect. 4.1.3) to minimize L_D, Eq. 9.3, using Eq. 9.7, where η is the learning rate.

$$\hat{\theta}_D \leftarrow \hat{\theta}_D + \eta \cdot \nabla_{\theta_D} \left(\frac{1}{N} \sum_{i=1}^{N} \ln\left(D\left(x_i\right)\right) + \frac{1}{N} \sum_{i=1}^{N} \ln\left(1 - D\left(G\left(z_i\right)\right)\right) \right). \tag{9.7}$$

Step 4: Update the parameters of G, i.e., $\hat{\theta}_G$, by gradient descent (Sect. 4.1.3) to minimize L_G, Eq. 9.5, using Eq. 9.8 with the same η as in Eq. 9.7.

$$\hat{\theta}_G \leftarrow \hat{\theta}_G - \eta \cdot \nabla_{\theta_G} \left(\frac{1}{N} \sum_{i=1}^{N} \ln\left(D\left(x_i\right)\right) + \frac{1}{N} \sum_{i=1}^{N} \ln\left(1 - D\left(G\left(z_i\right)\right)\right) \right). \tag{9.8}$$

Step 5: Repeat steps 1–4 until convergence is achieved or the designated number of iterations is reached.

9.2   DCGAN

In the early studies of GAN [1], a MLP, referred to as feed-forward NN introduced in Sect. 4, was used as an estimator function for the $D(\cdot)$ and $G(\cdot)$. However, this approach was criticized for its unstable performance [1, 4]. Therefore, several new GANs were developed to overcome this concern by adopting CNN, e.g., DCGAN [4]. Even though there are many

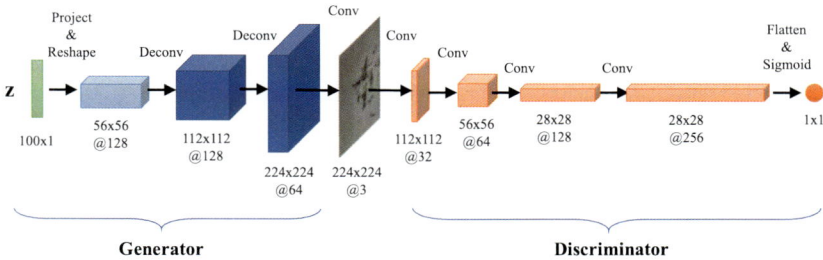

Fig. 9.2 Illustration of the DCGAN architecture

types of GANs to choose from, only DCGAN is covered in this chapter for the purpose of conducting an exploratory and educational investigation of GAN's possible usage in vision-based SHM. One example of DCGAN architecture is shown in Fig. 9.2, which is described and used in computer experiments, as discussed in Sects. 9.5 and 9.6.

The DCGAN is not the first GAN combined with deep CNN. Instead, it is defined as using a specific class of CNN architecture for both the generator and the discriminator and restricted to some constraints with the following guidelines [4]:

- Apply BatchNorm (Sect. 4.2.4) for both discriminator and generator.
- Use Conv operation to replace the deterministic spatial pooling. Specifically, apply Deconv (also known as fractional-strided or TransConv, refer to Sect. 8.1.2) in the generator for up-sampling.
- Use *LReLU*, Eq. 4.2, as an activation function in the discriminator, except for the output layer which uses the *sigmoid* function, Eq. 3.18.
- Use *ReLU*, Eq. 4.1, as an activation function in the generator, except for the output layer which uses the tanh function, Eq. 4.4.
- Eliminate the FC-layers after the last Conv layer.

Radford et al. [4] conducted extensive experiments of DCGAN on several datasets, including large-scale scene understanding (LSUN) [5], ImageNet-1k[1] [6], MNIST [7], and newly assembled Faces dataset in [4]. The experimental results indicate that the above architectural constraints, i.e., using BatchNorm, Deconv, etc., can contribute to the stable training process across a wide range of datasets and to the improved synthetic image quality.

---

[1] A labeled subset of ImageNet, which contains 1.2 million RGB images with 1,000 classes for the image classification task.

9.3   GAN Evaluation Metrics

GAN, as a DL approach, is empirically evaluated where meaningful metrics are essential. Subjective human evaluation, through the visual quality of the synthetic images compared to the real ones, is a common metric for the GAN performance evaluation. However, the lack of uniform criteria and the time-consuming processing of a large number of images in the human evaluation procedure greatly reduce the efficiency in augmented synthetic data selection and also impede the development of automated end-to-end data augmentation frameworks for real-time applications. Thus, developing efficient and quantitative metrics is an active research area of GAN.

9.3.1   Inception Score

Inception Score (IS), as an alternative to human evaluation, was first proposed in [8]. The intuition behind the *IS* is explained by the fact that a well-trained DL model which achieves high classification accuracy can be well-correlated with the human judgment of the image quality, i.e., higher *IS* implies better visual quality of the generated image. The term "Inception" comes from using a pre-trained InceptionNet [9] model as the classifier in the original study [8].

Define a set of synthetic images, namely, $\mathbf{x} = \{x_i, \ i \in \{1, 2, \ldots, N\}\}$, then the *IS* is computed as follows:

(1) Use a pre-trained inception model to determine the conditional probabilities, $p(y|\mathbf{x})$, of the class labels assigned by the network, y, to each image in $\mathbf{x}$.
(2) Compute the empirical marginal class distribution, $\hat{p}(y)$, Eq. 9.9, as the estimate of the "true" distribution, $p(y)$.
(3) Compute the *IS* using an approximate exponential of the KL divergence, using Eq. 3.25 and the notation in Sect. 3.7.3, between $p(y|\mathbf{x})$ and $\hat{p}(y)$, Eq. 9.10.

$$\hat{p}(y) = \frac{1}{N} \sum_{i=1}^{N} p(y|x_i).$$

$$(9.9)$$

$$IS(G) = \exp\left(E_{x \sim p_g}\left[D_{KL}\left(p(y|\mathbf{x}) \parallel p(y)\right)\right]\right),$$

$$\approx \exp\left(\frac{1}{N} \sum_{i=1}^{N} D_{KL}\left(p(y|x_i) \parallel \hat{p}(y)\right)\right).$$

$$(9.10)$$

If an image is real and meaningful, the objects and major targets in this image should be clear and sharp making them easily identifiable, i.e., *low entropy* of $p(y|\mathbf{x})$, call it condition "A". Moreover, realistic images should indicate a high diversity of different object categories, i.e., a *high entropy* of $p(y)$, call it condition "B". According to [10], Eq. 9.10 can be expressed as follows:

$$\ln IS(G) = H(y) - H(y|\mathbf{x}) \geq 0.0, \tag{9.11}$$

where $H(y)$ and $H(y|\mathbf{x})$ are the entropy of $p(y)$ and the conditional entropy of $p(y|\mathbf{x})$, respectively. Equation 9.11 is consistent with the information gain (i.e., mutual information) defined in Eq. 3.52. Refer to Sect. 3.9.2.1 for an in depth discussion about the entropy and the conditional entropy. Moreover, Eq. 9.11 simultaneously encodes the two conditions "A" & "B" above and it further proves that generated synthetic images are more meaningful with higher *IS* values. Note that the shown inequality in Eq. 9.11 is attributed to the fact that the conditional entropy, $H(y|\mathbf{x})$, is always less than or equal to the unconditional entropy, $H(y)$.

The upper and lower bounds of the $IS(G)$ can be easily derived from Eq. 9.11, as conducted in reference [10]. Clearly, the **lower bound** is $IS(G) \geq 1.0$, by taking the exponential of Eq. 9.11. On the other hand, the upper bound is obtained based on two facts: (i) the entropy is always positive, i.e., $[H(y) - H(y|\mathbf{x})] \leq H(y)$ and (ii) the *maximum* entropy discrete distribution is the *uniform* (rectangular) distribution, assumed to be the *least-informative* default herein when no information is given, i.e., $1/C$, where C is the number of classes, as defined in Sect. 7.1.2. Thus, the entropy can be written as $H(y) \leq -\ln(1/C)$, i.e., $H(y) \leq \ln C$. Therefore, $H(y) - H(y|\mathbf{x}) \leq H(y) \leq \ln C$, leading to the **upper bound** being $IS(G) \leq C$, also by taking the exponential of Eq. 9.11. In conclusion, the bounds on $IS(G)$ for a classification problem with C classes are expressed as follows:

$$1.0 \leq IS(G) \leq C. \tag{9.12}$$

The *IS* faces some limitations and impediments in real practice, particularly for SHM applications. In the original work [8], the classifier InceptionNet was pre-trained on ImageNet [6] (denoted as the *reference dataset*) and synthetic images were learned from CIFAR-10 [11]. Both object categories and tasks of CIFAR-10 nearly belong to a subset of ImageNet [6]. It was naturally expected that the classifier trained on the ImageNet [6] has high recognition ability on the CIFAR-10 data. However, in the domain of vision-based SHM, both the structural images with damaged patterns and the detection tasks of the scene level or more specifically the damage state classification significantly differ from those in the ImageNet [6], where the recognition ability of the pre-trained classifier can be far from that of a SHM human expert. Thus, another score should be devised for the purpose of vision-based SHM.

9.3.2   Self-inception Score

As mentioned above, the principle that a pre-trained InceptionNet classifier can replace human judgment is due to its high classification accuracy. Therefore, any other well-trained classifier with higher accuracy can be an alternative to the InceptionNet. Unlike the original

IS using ImageNet [6] for training the classifier, if the dataset used for training the GAN is well-labeled and a classifier can obtain an acceptable accuracy on this dataset, then this classifier itself can be assumed close to human judgment but possibly slightly weaker. In other words, the same data are used for training both the classifier and the GAN in recognizing (i.e., classifying) data similar to the training ones and generating synthetic images, respectively. Therefore, the *IS* computed under such a pre-trained classifier is called *Self-Inception Score* (*SIS*). This *SIS* can be thought of as a reasonable metric to overcome the drawback due to the lack of a large reference dataset.

In Eq. 9.9, the marginal class distribution is estimated by empirical distribution where its value may vary with the dataset size, N. Previous studies [8, 10] divided a large number of synthetic images into groups with different group sizes. However, they only studied the influence of group sizes without considering a new *IS* formulation. Here, a *split plan* is defined as the one that splits the image set into n_{split} groups. This is followed by determining the mean and Std of the *IS* values from all the split groups. Usually, lower mean of the *IS* and higher Std are observed with more groups, where the Std value is usually used to evaluate the variance of each group. These issues are considered in the *SIS* through averaging the *IS* values according to different split plans. Thus, the *SIS* considers local effects induced by the partial set (corresponding to different groups) of the synthetic images, which are absent if computing the *IS* from all the data. The procedure of the *SIS* computation is listed below and steps (4)–(6) are illustrated in Fig. 9.3.

(1) Partition the raw dataset into the training set and the validation set[2].
(2) Train the classifier with the training set by multiple deep NNs, e.g., VGGNet [12], InceptionNet [9], and ResNet [13] and pick the classifier with the best performance on the validation set.
(3) Train GAN with the training set and generate a large number of synthetic images.
(4) Design split plans, $n_{split} = \{1, 2, 5, \dots\}$, as shown in Fig. 9.3.
(5) Compute the mean values of the different *IS* from each split plan configuration to produce $\widehat{IS}_i$ ($i \in n_{split}$), refer to Fig. 9.3.
(6) Compute the mean and the Std of the $\widehat{IS}_i$ values in step (5) where the mean value corresponds to the required *SIS*, refer to Fig. 9.3, and Std measures the variation of the mean of *IS* among the split groups.

Although the *SIS* mitigates the limitations of the *IS* and considers the local effects, it is still a rough score. Similar to the *IS*, the *SIS* is sensitive to small weight changes of the network and synthetic images with high *SIS* may be unrealistic as observed in some adversarial synthesized samples [10].

---

[2] Here, validation means validating or verifying the generation and generalization performance of the GAN-generated data. This differs from the concept of training-validation-test dataset split in cross-validation, discussed in Sect. 3.6.

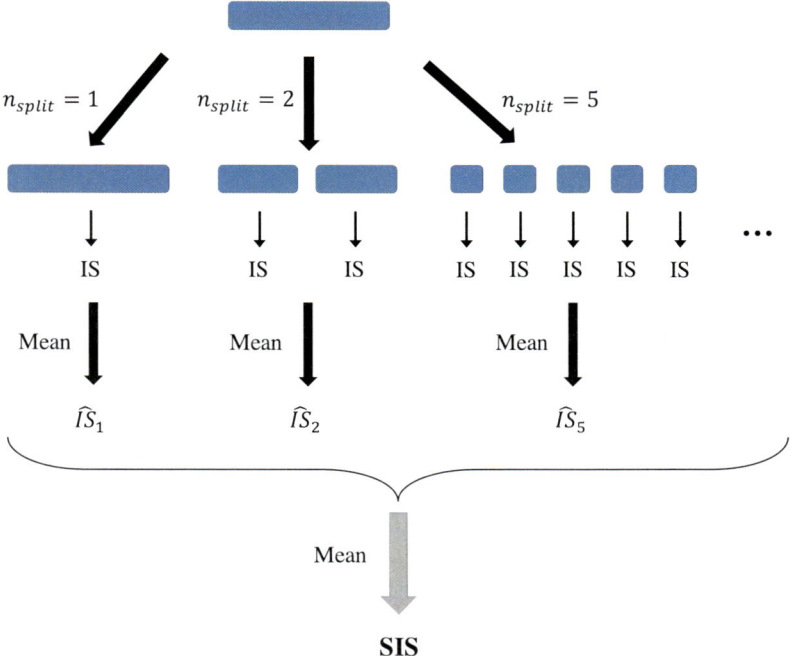

Fig. 9.3 Split plan for computing the *SIS*

9.3.3   Generalization Ability

The *SIS* considers the evaluation in terms of information gain and loss, which is limited, i.e., the image may be given a high *IS* value, even if it is classified with the wrong label. Thus, the *GA* is another independent metric to evaluate whether these synthetic images are meaningful enough to be recognized by the classifier. It is inspired by the metric of overall accuracy, Eq. 3.12, and the discriminator part of the GAN, which distinguishes the synthetic images from the real ones to improve the performance of the generator. Different from regular settings with both training and validation from real data, the *GA* measures the accuracy achieved by the classifier on the real validation set but such a classifier is only previously trained on the synthetic images generated by the GAN. For example, consider a task to classify object level images with damaged and undamaged states. Herein, synthetic images are generated and labeled first. Subsequently, they are used to train a strong classifier, e.g., VGGNet or InceptionNet. Finally, the classifier is examined using a validation set with real images only to quantitatively determine the quality, i.e., the overall accuracy, of the synthetic images used in the training. If the generated data are realistic, they will have similar representations, e.g., color and visual contents coded and stored digitally, and essential features to those of the real data, leading to a high validation accuracy on the validation set.

Moreover, from a statistical point of view, the *GA* can be thought of as a measure of the distribution difference between the synthetic training set and the real validation set in some sense. Thus, the *GA* is another reasonable and quantitative metric to evaluate the GANs.

9.4   Leaf-Bootstrapping GAN Augmentation

Compared to several datasets used in GAN-related studies [1, 4], where images share similarities with few variations, images in the ϕ-Net vary significantly due to different scales, irregularity of damage patterns, complex structural designs, and noisy backgrounds. Moreover, the size of the dataset is still small. These drawbacks may limit the GAN performance using the ϕ-Net dataset. Preliminary experiments were conducted by feeding all training images into one well-designed DCGAN and over 25,000 synthetic images were produced by a fine-tuned GAN. As expected, the quality of the images generated by the DCGAN, which was directly trained with all the training images, was not satisfactory. Although there exist several realistic synthetic images, several of other ones look like those in Fig. 9.4, which were mixtures of different levels. For example, the top four images show the sky and part of the building outlines belonging to the structural level and the bottom four images look like a wall surface belonging to the pixel level.

Statistically speaking, as a type of generative model, the ultimate goal of GAN is to learn the distribution of the data. In the case of vision-based SHM, from the perspective of the data distribution, the scene level or the damage state should have a mixed distribution by several sub-distributions with multiple peaks and ranges. For example, from the hierarchy of the ϕ-Net, discussed in Sect. 5.5 and Fig. 5.13, the structural image labeled as the damaged or undamaged state can be further separated and sorted from: (1) the pixel level with the damaged state (PD), (2) the object level with the damaged state (OD), and (3) the structural

Fig. 9.4 Unacceptable images generated by GAN trained with all the training images

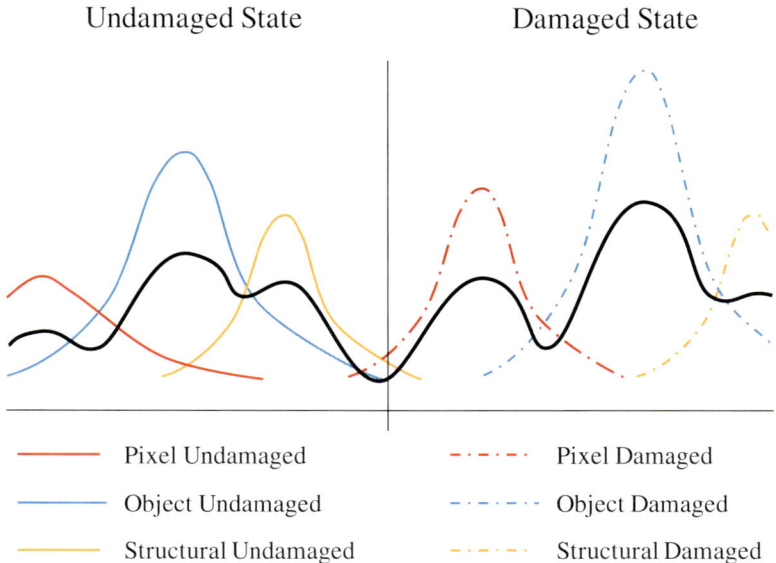

Fig. 9.5 Illustration of a mixture distribution for the structural images in a certain dimensional space. (Note: the horizontal axis represents the dimension of the features and the vertical axis represents the magnitude of the probability density function)

level with the damaged state (SD) or from (i) the pixel level with the undamaged state (PUD), (ii) the object level with the undamaged state (OUD), and (iii) the structural level with the undamaged state (SUD), respectively, refer to Fig. 9.5. Failure of directly training the GAN is due to the limited size and complexity of the dataset (without additional information to reduce the complexity) making it difficult for the original GAN to directly learn the complex mixture distribution (black curve in Fig. 9.5).

Based on domain knowledge and Bayes' theorem [14, 15], the probability distribution of the structural images can be computed from the combination of distributions of the sub-categories, i.e., the probability density of PD, OD, and SD or that of PUD, OUD, and SUD. Compared to learning a complex distribution directly, it is much easier to start from learning its sub-distributions and then learn the aggregated distribution, which shares a similar idea with the Gaussian mixture models [14, 15]. Inspired by this concept and combining it with the hierarchy ϕ-Net framework, a new training method for GAN is proposed, namely, leaf-bootstrapping (LB). Its name originates from making use of the *leaf node distribution* and generating data from a well-learned distribution, where the method also adopts the idea of *bootstrapping* [16] that samples data from a learned distribution, refer to Sect. 3.9.3 for detailed discussion about the bootstrap aggregation.

In most application scenarios, using the whole ϕ-Net framework is time-consuming and inefficient. Instead, the simplified hierarchy ϕ-Net using shallow representations, Fig. 5.14, can be considered and these shallow representations are expected to work well with the LB

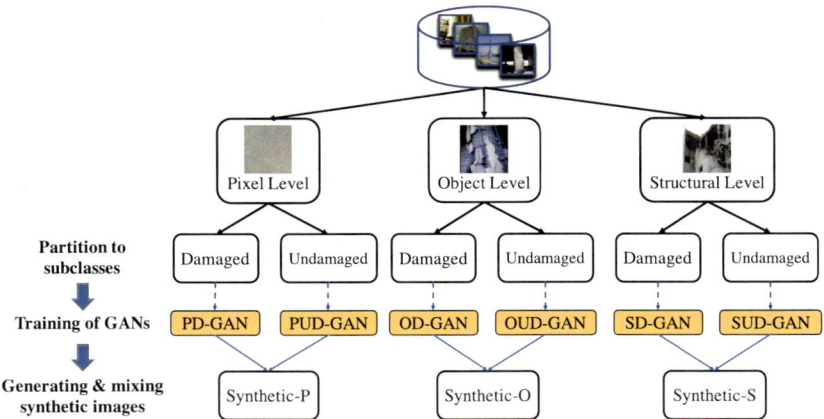

Fig. 9.6 LB for synthetic scene level using ϕ-Net depth-2 representation

method for GAN training. For example, while adopting depth-2 representation (Fig. 5.14), instead of generating synthetic images for the damaged state or the scene level directly, GAN models now can be trained on their sub-classes from the leaf nodes. Accordingly, six GAN models are trained for PUD, PD, OUD, OD, SUD, and SD. Based on each sub-class, the class-specified synthetic images are produced, e.g., synthetic PD images are sampled from the GAN trained on the PD images. Then, the synthetic images are reassembled according to their root tasks, e.g., for the synthetic pixel level images used in the scene level task, synthetic images of PD and PUD are mixed, Fig. 9.6. Similarly, in the damage state task, synthetic damaged state images are mixed from synthetic PD, OD, and SD. Moreover, the LB method can be easily extended to a depth-3 framework (Fig. 5.14), which requires mixing deeper nodes prior to their parent nodes, i.e., for access to PD, synthetic images of non-spalling and spalling need to be assembled first. In summary, the LB method through using a simplified depth-2 or depth-3 framework benefits the GANs training and samples images of higher quality through minimizing the inner class (if a class has several sub-classes) variations by simply pre-separating these inner classes into sub-classes based on domain knowledge.

As mentioned above, since the direct usage of the GAN did not work, in the following section with respect to structural image generation, the LB method is adopted in the GAN training. It should be noted that, in the following paragraphs, the GAN model using the LB method is denoted as LB-GAN for short. This method was proposed in 2019 [17] and its results on several computer experiments are discussed in the following section.

9.5   Synthetic Structural Image Generation

In the previous section, it has been shown that simply direct usage of GAN does not work for generating reasonable structural images. Thus, it is important to determine whether GAN can effectively work in vision-based SHM when using an appropriate training method. In this study of synthetic image generation in Structural Engineering applications, two common scenarios were considered, (i) scene level identification and (ii) damage state check, which are actually two basic ϕ-Net benchmark classification tasks. Moreover, besides simply validating the effectiveness of the GAN performance using the LB method, several specific questions are sought as follows:

1. Does GAN using the LB method with depth-2 representation work better than the directly trained one?
2. Does a cleaner and smaller-scale dataset generate synthetic images of better image quality?
3. Can a conventional DA method further improve the GAN performance?
4. Can deeper tree representations such as depth-3 outperform the shallower ones such as depth-2, refer to Sect. 5.5?

For answering each question, the same procedure and evaluation technique were conducted. The considered evaluation metrics are: (1) synthetic image quality by human evaluation, (2) GAN performance by the *SIS*, and (3) GAN *GA*.

9.5.1 ϕ-Net-GAN Dataset

All labeled structural images are directly collected from the ϕ-Net dataset. Because GAN is known to be inferior due to its unstable training process [1, 4] and since this is the study to explore GAN in a non-benchmark dataset, a rigorous image selection procedure and distribution design were conducted as data preprocessing. Thus, all raw data were well-selected from the dataset with multiple labels of the scene level and the damage state with relatively high quality and less ambiguity. Subsequently, the images were re-scaled to size 224×224. Finally, 6,821 images, in total, were used as the entire dataset for this case study, which is named ϕ-Net-GAN for short. Reasonably balanced label ratios for both the scene level and the damage state were achieved in the design of this dataset as shown in Fig. 9.7.

A split ratio very close to 4:1 was adopted to separate the training and test datasets, where similar label distributions were kept in both datasets. The amounts of data for both datasets are listed in Table 9.1, which is useful for the classification case study to be discussed in the following sections. It is noted that only data in the training dataset were used in training the GAN model. For the *GA* metric computation, the test dataset is also used as validation dataset here.

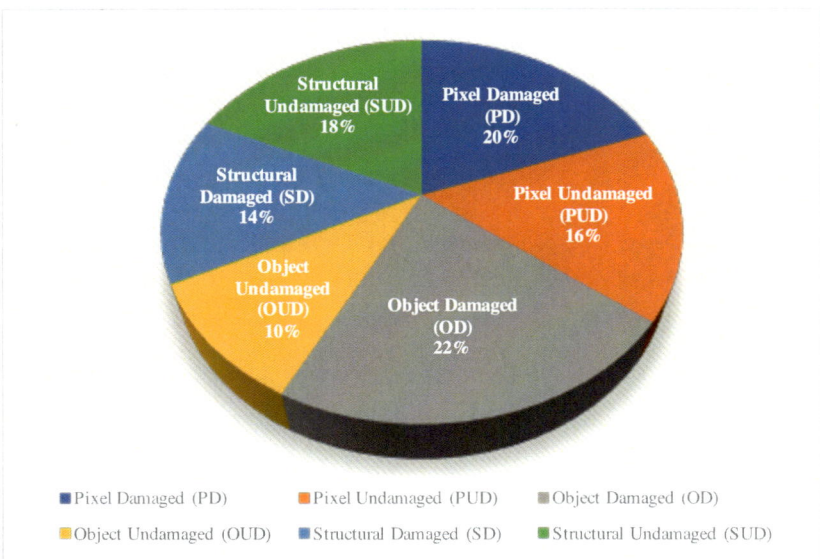

Fig. 9.7 Label ratio of the ϕ-Net-GAN

Table 9.1 Amounts of raw data in training and test datasets for six categories

Dataset	PUD	PD	OUD	OD	SUD	SD
Training	865	1,096	555	1,107	992	771
Test	218	278	137	286	247	193
Ratio	3.97:1	3.94:1	4.05:1	3.87:1	4.02:1	3.99:1

In order to address the above mentioned four questions, the following four training datasets were designed considering two scenarios:

1. <u>D2-All</u>: All raw training images with depth-2 representation.
2. <u>D2-Clean</u>: Select up to 128 images per class from D2-All with less noise for PD, PUD, OD, OUD, SD & SUD.
3. <u>D2-Clean & Aug</u>: Augment D2-Clean up to 1,024 images per class using the conventional DA method.
4. <u>D3-Clean & Aug</u>: Select up to 64 & 128 images per class for depth-3 & depth-2 representations, respectively, according to the configuration in Fig. 5.14, then perform DA up to 1,024 images per class for PD, PUD, OD, OUD, SD & SUD.

It is noted that while in image selection for D3-Clean&Aug using depth-3 representation, considering the complexity of the visual patterns in the object level, the images were split

Table 9.2 Configuration of DCGAN generator (N: # data fed into the network)

Layer	Filter Size (#)	Activation	Shape	Notes
Input	–	–	(N, 100, 1)	Noise generated from Normal distribution
FC-layer	–	*ReLU*	(N, 401408)	401408 $= 56 \times 56 \times 128$
Reshape	–	–	(N, 56, 56, 128)	–
BatchNorm	–	–	(N, 56, 56, 128)	Momentum = 0.8
Deconv	3×3 (128)	*ReLU*	(N,112, 112, 128)	Up-sampling
BatchNorm	–	–	(N, 112, 112, 128)	Momentum = 0.8
Deconv	3×3 (64)	*ReLU*	(N, 224, 224, 64)	Up-sampling
BatchNorm	–	–	(N, 224, 224, 64)	Momentum = 0.8
Conv	3×3 (3)	tanh	(N, 224, 224, 3)	–

into "beam/column" and "wall" for both undamaged and damaged states, as an extension of the representation shown in Fig. 5.14.

9.5.2   Network Configuration

The size (i.e., resolution) of the generated synthetic images from the original DCGAN is equal to or smaller than 64×64, which is much smaller than expected. It is pointed out, in Chap. 6, that several deep CNN models, e.g., VGG-type and ResNet, achieved stable and relatively optimal performance in structural damage detection using 224×224 as the network input size, where the raw images were re-scaled to match this size. Therefore, a specific DCGAN configuration using 224×224 as the network input size for the discriminator and output size for the generator was designed for this study. In other words, the synthetic image size herein is 224×224, an improvement over the previous work documented in [4]. Following the constraints of the DCGAN in Sect. 9.2, the configurations of the discriminator and the generator are listed in Tables 9.2 and 9.3, respectively, for the network shown in Fig. 9.2.

The common issue in GAN training is that hyper-parameters for DCGAN are quite sensitive. After extensive fine-tuning operations, it was found that using Adam optimization [18] with a learning rate of 10^{-4} or 5×10^{-5}, $\beta_1 = 0.5$, and $\beta_2 = 0.999$ was stable and achieved good image quality. Moreover, for the considered non-benchmark dataset, it is difficult to obtain Nash equilibrium. Thus, in this study, 40,000 maximum training iterations were considered and the best GAN model was selected with the best image quality by human

Table 9.3 Configuration of DCGAN discriminator (N: # data fed into the network)

Layer	Filter Size (#)	Activation	Shape	Notes (α: $-$ive slope coef. in *LReLU*)
Input	–	–	(N, 224, 224, 3)	–
Conv	3×3 (32)	*LReLU*	(N, 112, 112, 32)	Conv stride = 2; $\alpha = 0.2$
Dropout	–	–	(N, 112, 112, 32)	Dropout rate = 0.25
Conv	3×3 (64)	*LReLU*	(N, 56, 56, 64)	Conv stride = 2; $\alpha = 0.2$
Dropout	–	–	(N, 56, 56, 64)	Dropout rate = 0.25
BatchNorm	–	–	(N, 56, 56, 64)	Momentum = 0.8
Conv	3×3 (128)	*LReLU*	(N, 28, 28, 128)	Conv stride = 2; $\alpha = 0.2$
Dropout	–	–	(N, 28, 28, 128)	Dropout rate = 0.25
BatchNorm	–	–	(N, 28, 28, 128)	Momentum = 0.8
Conv	3×3 (256)	*LReLU*	(N, 28, 28, 256)	Conv stride = 1; $\alpha = 0.2$
Dropout	–	–	(N, 28, 28, 256)	Dropout rate = 0.25
Flatten	–	–	(N, 200704)	$200704 = 28 \times 28 \times 256$
FC-layer	–	*sigmoid*	(N, 1)	–

expert judgment. Since the LB method was adopted, multiple GAN models were trained and most GAN models were found to achieve the best performance at about 25,000 iterations. Similar to the training process of the original GAN (Sect. 9.1), the computational cost for each iteration including the gradient updating steps (Eqs. 9.7 and 9.8) depends on the batch size $2N$. In this study, a batch size of 16 was used, which corresponded to an average time of 0.12 seconds per iteration in the GAN training process. In other words, for the previously mentioned maximum of 40,000 iterations, the approximate training time for one GAN model was about 80 minutes. As for the synthetic image generating time, an average of only 0.003 seconds per image (without the file-saving operation) was achieved.

As suggested in [8], *IS* becomes suitable when using a large number of samples (at least 50,000 in total and 5,000 for a single class). The considered two scenarios were classification problems with few classes and for computational efficiency, 5,000 images per class for six leaf nodes (PD, PUD, OD, OUD, SD & SUD) were generated to form a set of 30,000 synthetic images. For depth-3 representation, 2,500 images per class were generated for sub-classes of damaged and undamaged states ("SP & NSP for PD", "Beam/Column & Wall for both OD & OUD", and "Non-collapse & Global collapse for SD"), which also summed up to 30,000 images together with the 5,000 images per the three classes of no damage, i.e., PUD, OUD, and SUD. In order to satisfy at least 5,000 images for each class, only three split plans, n_{split}={1, 2, 5}, were adopted. Therefore, for a total of 30,000 synthetic images, 30,000, 15,000 & 6,000 were the size of each split group according to these three split plans.

9.5.3   Human Evaluation

After extensive experiments and reassembling back to the six leaf tasks in depth-2 representation, several samples of the synthetic images are shown in Fig. 9.8 for four different cases.[3] In this figure, Synthetic-D2-All (Syn-D2-All for short in the sequel) means synthetic images generated by GAN trained from D2-All dataset. Other columns of synthetic images in Fig. 9.8 follow a similar nomenclature. Starting with the case "Syn-D2-All", it is shown that the GAN already learned some basic patterns such as cracks, spalling, shape of column or wall, etc., but the images look blurry and lack the details. Comparing the results for the different leaf nodes, the image quality and training difficulties vary from one to another, especially at the object level. Even though the outlines of the structural components can be roughly recognized, most images contain noise, such as the green parts in the generated two images of the first row and first & second columns of OUD, which make them unrealistic. This can be explained by the fact that the raw object level images contain many irrelevant parts, which confused the GAN. Unlike the above two image examples, synthetic PUD and

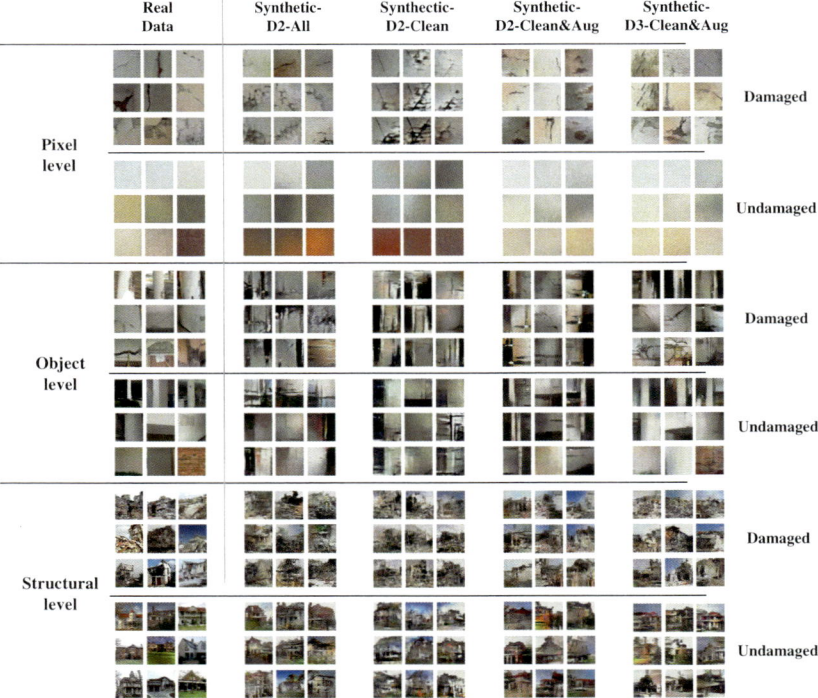

Fig. 9.8   Real images versus synthetic images generated from four cases

---

[3] The high-resolution images can be found at http://stairlab.berkeley.edu/data/.

SUD images are more realistic to the human eyes, especially when evaluated in relatively low resolution. Thus, filtering and de-noising may be necessary.

For the case "Syn-D2-Clean", after data cleaning, the generated images show more details, such as reasonable damage patterns and outlines of the structural components. In the PD examples, cracks are much sharper and clearer than in the previous case. Another important observation is that cracks occur in different locations with different slopes, which is an evidence that the GAN can learn well the visual features of the cracks, but it may lack the physical meaning. In the OD and OUD examples, the boundaries and lighting make the structural components appear more in 3D and the noise level slightly decreases. However, due to a significant decrease in the amount of data per class, more challenges are faced during the training process, such as the losses for both the discriminator and the generator not changing for the first several hundreds of iterations, leading to only generating some noisy images. Therefore, to overcome these issues, multiple trials for training are necessary benefiting from the randomness of the CNN parameters and the input latent variables, refer to Sect. 2.4. Another problem is the *mode collapse* [4], where only a few types of images are generated by the generator and these ones lack variety, i.e., most images are similar with limited changes, such as most images in the PD have a similar texture.

An augmented version of the dataset "D2-Clean" is "D2-Clean&Aug". The trained GAN with the latter slightly resolved the above issues and achieved some improvements in the image quality of the case "Syn-D2-Clean&Aug", e.g., spalling patterns in PD are more realistic. In addition, both cracking and spalling patterns are observed in OD. On the structural level, less noise and sharper boundaries are obtained. However, undesired shifting patterns are learned by the GAN and observed in the structural level images due to the affine DA, refer to Sect. 6.1. Therefore, translation, i.e., shift, operations in augmentation should be limited to a smaller range in future studies.

The case "Syn-D3-Clean&Aug" achieves the best performance, compared to the previous cases. In the pixel level examples, GAN can generate more realistic cracking and spalling patterns benefiting from further splitting into two sub-classes, namely, non-spalling (NSP) and spalling (SP). In the object level examples, more complex and detailed structural components can be generated, such as the good quality single column images (first row images of both OD & OUD), wall with corner (the image in second row and third column of OD), wall with different colors such as white on top and pink on bottom (two images in third row and first & second columns of OD), brick texture of wall (the image in third row and third column of OUD), and appearance of plants (the image in third row and first column of OUD). In the structural level examples, due to further splitting in the SD node, the global collapse (first row images of SD) and the non-collapse (third row images of SD) become more distinguishable with sharper outlines. Because of no further splitting in the SUD node, the synthetic SUD images, in this case, are similar to those of the "Syn-D2-Clean&Aug" case.

In all cases, the synthetic pixel level images describe a small range of a structural component with simple and few patterns, the synthetic object level images show the outlines of the structural components with adequate lighting and shadow effects, and the synthetic structural level images present realistic shapes of building structures with some details, e.g., windows and doors. Furthermore, for the classification, with the above mentioned sharp differences, the synthetic images are expected to distinguish between the three different levels in the scene identification. For the damage state check, differences between damaged and undamaged states are obvious, where the overall views of the synthetic images of the undamaged state are smooth and aligned, while those of the damaged state are cruder and disorganized. Especially in SD, debris from damaged structures makes the scene chaotic which describes the structural collapse features well. However, when compared to the real images with a 224×224 resolution, the synthetic images in all cases still look somewhat blurry, which is the common issue observed in most GAN applications. This drawback can be attributed to: (1) the averaging effect of GAN, where the objective functions (e.g., Eq. 9.1) are expressed in the sense of *expectations* and (2) using a relatively large image size to reconstruct (i.e., generate) with only a small amount of training data. Moreover, all synthetic images still lack some natural details, such as dirt, decoration, etc. This is particularly the case for PUD where all synthetic images look like a pure color patch. However, this issue does not hamper the classification for the damage state check, within the context of SHM, which is validated and discussed in the classification experiments in Sect. 9.5.5.

9.5.4   Scene Level Identification

9.5.4.1 *SIS* Metric

In the scene level identification, 30,000 synthetic images were generated with 10,000 per class (total number of classes, $C = 3$) for pixel (P), object (O), and structural (S) levels. To calculate the *SIS*, a strong classifier (VGG-19) was pre-trained on the real training images with about 98% validation accuracy, which is expected to have strong recognition properties close to the human judgment or even better. Therefore, the *SIS* for the whole dataset of the real training images ("Real-All", where each class has about 1,000 real images) is considered as the upper bound instead of the higher theoretical bound (computed as the number of classes, refer to Eq. 9.12, where $C = 3$ in this scenario). This is a compromise due to the lack of a large reference dataset, unlike the study in [10] where ImageNet-1k with 1,000 classes [6] is used. However, the lower bound is the same as the *IS*, i.e., 1.0 [10] and Eq. 9.12. From Table 9.4, it is observed that *SIS* for the "Real-All" is the highest with a relatively small Std, consistent with the expectations. As expected, all values of *SIS* in Table 9.4 are within the theoretical bounds of 1.0 to 3 defined by Eq. 9.12 for $IS(G)$ and accordingly the bounds of the *SIS*. However, it is noted that because of the averaging of the *IS* values according to the specified split plan, Fig. 9.3, the upper bound of *SIS* is expected to be lower than that of *IS*. In other words, the upper bound of *SIS* in this case of the 3-class scene level identification is somewhere between 2.063 and 3, refer to Table 9.4 and Eq. 9.12, respectively, which is indeed the case for all considered datasets in Table 9.4.

Table 9.4 SIS and GA for synthetic images in the scene level identification

Dataset	Real-All	Syn-D2-All	Syn-D2-Clean	Syn-D2-Clean&Aug	Syn-D3-Clean&Aug
SIS ± Std	2.063 ± 0.149	1.840 ± 0.164	1.817 ± 0.175	1.879 ± 0.163	1.903 ± 0.164
GA[a]	97.70%	85.14%	84.92%	82.41%	83.52%

[a] For "Real-All", this metric is actually the validation overall accuracy, Sect. 3.4.1

To explore the GAN performance, the synthetic images were generated based on different settings. The "Syn-D2-All" represents the synthetic images generated from the GAN learned from the "Real-All" dataset. Its SIS is relatively high, which has only $(2.063 - 1.840)/2.063) = 10.8\%$ reduction compared to the real images, Table 9.4. This result demonstrates that the synthetic images are of good quality in terms of the scene level. However, there are some images that share features from two scene levels, P & O, making it difficult to classify them accurately. Thus, a higher conditional entropy, $H(y|\mathbf{x})$, of $p(y|\mathbf{x})$ is obtained, causing the reduction of $IS(G)$ according to Eq. 9.11 and accordingly the reduction of the SIS. To explore this issue, the cleaned dataset "D2-Clean" was used to generate synthetic images "Syn-D2-Clean" leading to a lower SIS compared to "Syn-D2-All". Two reasons can explain this observation: (1) the cleaned dataset is too small, which makes it difficult for the GAN to capture the precise distribution and (2) the reduced noise by the cleaning operation leads to a decrease in the conditional entropy $H(y|\mathbf{x})$, which is expected to increase the $IS(G)$ and accordingly the SIS. However, this increase cannot compensate for the decrease of the entropy, $H(y)$, of $p(y)$ due to losing data variety by ruling out a certain number of data points. In other words, according to Eq. 9.11, the decrease of the entropy of $p(y)$ dominates the decrease of the conditional entropy of $p(y|\mathbf{x})$, which causes the final decrease of the $IS(G)$ and accordingly the decrease of the SIS for "Syn-D2-Clean" compared to "Syn-D2-All".

The "Syn-D2-Clean" problem discussed above was addressed by employing the conventional affine DA, refer to Sect. 6.1, to increase the dataset from 128 to 1,024 per class leading to the increase of the SIS from 1.817 to 1.879 for "Syn-D2-Clean&Aug", which is even higher than that for "Syn-D2-All" (i.e., 1.840), refer to Table 9.4. Moreover, for the case "Syn-D3-Clean&Aug", the SIS was further improved reaching 1.903 by benefiting from the clearer and higher quality images in dataset "D3-Clean&Aug" even though some complex and noisy images were sacrificed during the data cleaning. Comparing the SIS from the case "Syn-D3-Clean&Aug" (1.903) with its upper bound (2.063) from "Real-All", the difference is only $2.063 - 1.903 = 0.16$ (i.e., $0.16/2.063 = 7.76\%$ of the upper bound), which is relatively small considering the range and the Std of the SIS. This result demonstrates the effectiveness, i.e., achieving higher SIS in the previously discussed synthetic settings, of the proposed LB-GAN method with DA [17] in the scenario of the 3-class scene level classification.

9.5.4.2 *GA* Metric

Using the real training dataset, 97.7% validation *overall accuracy*, refer to Sect. 3.4.1, was achieved, which is the upper bound of the *GA*. Similar calculations were conducted for the synthetic images. From Table 9.4, "Syn-D2-All" has the highest validation overall accuracy (85.14%) using the synthetic training dataset (which is the *GA* metric). This performance is because "D2-All" dataset has the most informative distribution including complex and noisy data, making the generated synthetic images the most generalized ones. Since the cleaned dataset "D2-Clean" loses some variety, the *GA* (84.92%) is slightly lower for "Syn-D2-Clean". Similarly, when going deeper for depth-3 representation, the dataset loses more information. Thus, "Syn-D3-Clean&Aug" has an even lower *GA* (83.52%). One interesting observation is that when applying DA in training, "Syn-D2-Clean&Aug", the *GA* declined by $84.92 - 82.41 = 2.51\%$ compared to the case without DA, i.e., "Syn-D2-Clean". This is explained by the fact that DA on a cleaned dataset puts too much emphasis on the cleaned data, which ignores some complex and abstract patterns from the eliminated images, leading to this observed loss of generality.

To better explore the validation accuracy, the CMs are evaluated, as shown in Fig. 9.9. From Figs. 9.9b–d, it is observed that the classifier predicted many object level images (O) as non-object (pixel & structural levels) labels (P or S). Because there are several sub-classes in the O level, i.e., wall, column & beam and when generating synthetic images only using depth-2 representation, GAN merged the features from these different sub-classes resulting in a larger difference between the real and the synthetic images. On the other hand, when using depth-3 representation (Fig. 9.9e), the wall, column & beam sub-classes are separated, making the O level performance much better than that of depth-2. As a trade-off, when splitting the dataset into more granular classes, some sub-classes in the O level become similar to those in the P level (e.g., some wall images look like those in the P level). As a result, some P level images are mistakenly predicted as label O. Moreover, the O level is the intermediate level between the P & S levels. Thus, it shares some common features with the other two levels, which may confuse the classifier.

In a broader view, *GA* values obtained from all the synthetic cases are close to each other (average *GA* from all four synthetic datasets is about 84%) with an average difference of 13.7% from the *GA* of the ground truth, i.e., "Real-All". Aside for the reasons stated in the previous two paragraphs, this persistent accuracy gap is attributed to the distribution difference between the assembled set of synthetic images and that of the real images. Therefore, better GAN models and more reasonable assembling techniques should be investigated in future studies for further improvements.

9.5.5 Damage State Detection

9.5.5.1 *SIS* Metric

Similar to the scene level classification, VGG-19 was pre-trained on the whole real training images and achieved 96% validation accuracy. The same dataset settings were used as the

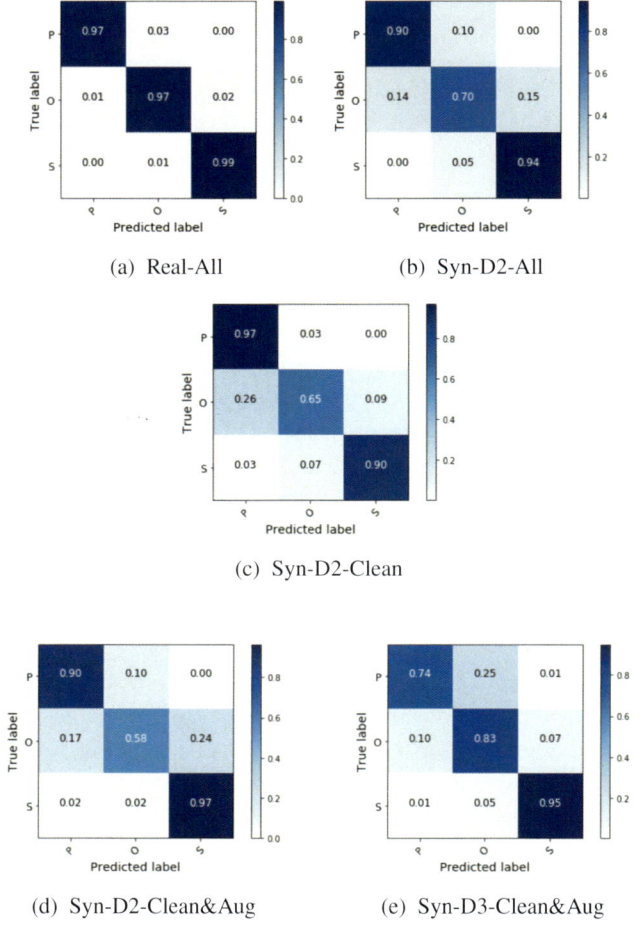

(a) Real-All   (b) Syn-D2-All

(c) Syn-D2-Clean

(d) Syn-D2-Clean&Aug   (e) Syn-D3-Clean&Aug

Fig. 9.9 CMs of the *GA* in the scene level identification

previous experiments and the results are listed in Table 9.5. Apparently, "Real-All" achieved the highest *SIS* (1.425), which represents the upper bound of this score in this experiment (the theoretical upper bound for *I S* is 2 for this binary task, i.e., damage state (DS) vs. undamaged state (US)). Among the synthetic datasets, results from "Syn-D2-All" have the highest *SIS* (1.387) without further improvement of it with dataset cleaning, augmentation, or use of deeper representation. Different from the scene level classification where the synthetic images of P, O & S are quite different from each other, in the present task, several noisy areas, e.g., green and blurry regions in the US images (refer to the two images of the first row and first & second columns of Object level, Undamaged, "Syn-D2-All", Fig. 9.8) may be mistaken (even by human evaluation) as DS patterns by the classifier. Due to losing several images, results from the cleaned dataset are inferior to those using all raw data. On the other

Table 9.5 *SIS* for synthetic images in the damage state detection task

Dataset	Real-All	Syn-D2-All	Syn-D2-Clean	Syn-D2-Clean&Aug	Syn-D3-Clean&Aug
SIS ± Std	1.425 ± 0.135	1.387 ± 0.098	1.244 ± 0.104	1.381 ± 0.091	1.381 ± 0.132

hand, results from using DA and using deeper representation are better than the data cleaning where their scores are very similar (1.381 vs. 1.387) to those of "Syn-D2-All" and also close to the upper bound score (1.425). Since several assumptions were made in computing this upper bound of this *SIS* and also in the pre-trained classifier, the score may not represent the real situation. This is consistent with concerns expressed in [10], where unreasonable adversarial images may have higher scores than realistic ones. Thus, as a metric within the perspective view of information, the performance of "Syn-D2-All", "Syn-D2-Clean&Aug", and "Syn-D3-Clean&Aug" cases are considered acceptable under this *SIS* metric.

9.5.5.2 *GA* Metric

From Table 9.6, *GA* values of the P level are generally better than those of the O & S levels in most synthetic cases because of the more uniform, relatively simpler, and less noisy patterns in the P level images. It is noted that the three columns of results in Table 9.6 separate the values of *GA* for the three scene levels by determining the validation overall accuracy to classify US versus DS from the real or synthetic images. Consistent with human evaluation, images of the O & S levels are less clear and many US images may be mistakenly classified as DS, which confuses the classifier in this damage detection task. An exception is the "Syn-D2-Clean" case, where the *GA* of the S level is only 0.03% higher than that of the P level. Similar to the scene level identification, Sect. 9.5.4, after cleaning, the accuracy declines in "Syn-D2-Clean" compared to "Syn-D2-All" due to information loss from removing some data. On the contrary, DA and deeper representation resolve this issue. However, poor results are observed for the S level of "Syn-D3-Clean&Aug", with a 20% reduction of

Table 9.6 *GA* for synthetic images in the damage state detection task

Dataset	GA		
	PD versus PUD (%)	OD versus OUD (%)	SD versus SUD (%)
Real-All[a]	94.76	93.85	97.5
Syn-D2-All	87.90	77.07	81.82
Syn-D2-Clean	78.83	70.92	78.86
Syn-D2-Clean&Aug	89.31	75.65	86.14
Syn-D3-Clean&Aug	88.10	75.89	66.14

[a]It is actually the validation overall accuracy, Sect. 3.4.1

GA compared to "Syn-D2-Clean&Aug", indicating the instability of using a deeper representation for the damage state detection task, which also stems from greatly reducing the data variety via cleaning and selecting from more sub-classes of depth-3. Compared with *GA* from "Real-All", the synthetic P level images show only a 5.45% reduction under the best case, namely, "Syn-D2-Clean&Aug", indicating a small difference in the distribution of the synthetic image dataset compared with that of the real dataset. This is indicative of the effective performance enhancement of DL models using GANs. As for the O & S levels, there exist some discrepancies between the synthetic dataset and the real one. Thus, further improvements are needed through future research endeavors.

9.5.6   Discussion

In general, synthetic images generated from four cases indicate the promising results of using DCGAN with the LB method, which is beneficial to increasing the variety of datasets. As for visual quality from human judgment, using depth-2 representation with all raw images can achieve a stable baseline quality. Applying augmentation and using deeper representation lead to improvement with clear and sharp structural patterns. Moreover, as complementary metrics to the human evaluation, results of *SIS* and *GA* are shown to be reasonable for GAN evaluation. Although the *IS* has several criticisms and drawbacks [10], due to the high accuracy of the pre-trained classifier with limited label categories, the *SIS* is shown to have a stable performance consistent with the observation that an image with better visual quality achieves a higher score. As for the *GA*, the trend indicates that with deeper representation and more data-cleaning operations, better visual qualities are obtained but some information is lost leading to a possible decrease of *GA*, which is a trade-off between quality and generalization. Using deeper representation leads to training more GAN models for new split leaf nodes, which introduces a large computational cost. Thus, how deep one needs to go depends on the complexity of the detection tasks at hand, e.g., depth-2 and depth-3 representations are sufficient for the two tasks, discussed in Sects. 9.5.4 and 9.5.5.

As indicated in many previous studies, explaining the physics behind the outcome is one of the hardest problems in the current GAN research area. In this study, damage patterns in synthetic images are generated from some learned distributions and they may indeed violate the laws of physics. However, compared to previous studies, the physical meanings have been assigned, to some extent, in the training process which somewhat relieves the potential for this physical violation issue via adding domain knowledge to the training pipeline, making use of the ϕ-Net framework. In this manner, random structural images are filtered into more fine sub-categories and this guides the GAN to generate images representing specific types of structural images, which helps to reduce the complexity of the problem and to control the variance of the image quality. From the results presented for the P level and the DS of the "Syn-D2-Clean" case in Fig. 9.8, the GAN models learned to generate synthetic images comparable to a combination of several basic crack features, i.e., diagonal cracks with different angles, widths, locations, and even interaction with spalling. This is a promising integration of domain knowledge with GAN models, suppressing the possibility of randomly generated images. Moreover, the generated images in this way are judged to be realistic compared with damaged real RC structures.

Similar to the conventional DA, the scope of using synthetic images, discussed herein, is only for the classification problem. However, finding the relationship between the synthetic damage images and the laws of physics is marked as an important future extension. In addition, previous work has shown the potential of GANs to solve this issue. According to Radford et al. [4], by manipulating the initial value of the code z in the generator (Sect. 9.1 and Fig. 9.2), the content of the synthetic image can be controlled. This observation shows the potential for future studies, where a well-learned relationship of the code z based on the laws of physics can help the GAN model to have a better understanding to generate images without violating these laws of physics. Accordingly, the image content can be controlled, e.g., with the desired crack morphology in terms of angle, damage level, spalling pattern, etc.

9.6   Synthetic Data Augmentation for Classification

In the previous section, GAN was first validated and shown to be able to generate reasonable structural images under certain settings. In this section, a second question arises, namely, whether these synthetic images can alleviate the data deficit problem while using the DL approach, e.g., in classification tasks.

One straightforward application is to aggregate synthetic images into the real dataset to enhance the classifier performance in classification problems. However, preliminary investigation demonstrated that such aggregation may render worse performance due to three issues, as follows:

- Relatively lower quality of the synthetic images compared to the real ones;
- Possible distribution difference between the dataset of the generated images and the real dataset; and
- Feeding mixed data including synthetic ones into the classifier forms a series pipeline by the GAN model and the classifier, introducing additional parameters compared to the original classifier, which may exacerbate the over-fitting issues.

Therefore, special training process and algorithms should be developed and investigated when aggregating synthetic images into the real dataset. This includes how to use synthetic data to improve the classifier performance under certain conditions, which needs to be explored as conducted in the following experiments. According to the results presented in the previous sections, the validation overall accuracy of "Real-All" trained by classical DL architectures already achieved very high values of 94.8–97.7%, where further improvement will not be so evident nor practical. Because for such a high level of accuracy, sometimes the so-called improvement is just one form of over-fitting to the current dataset. Moreover, some recent studies validated the effectiveness of using synthetic images in low-data regimes [8, 19–21]. Thus, in this section, a special case under the condition of a complex detection problem, low-data regime, and restricted computational power was designed and discussed. This special case was achieved by satisfying the following conditions:

1. Instead of independently conducting simple detection tasks (scene level & damage state classification), considering both label information formed a more complex 6-class classification problem (PUD, PD, OUD, OD, SUD & SD).
2. The "D2-Clean" dataset was treated as a training dataset with a decreased number of images in training from 5,457 to 768 to increase the difficulty of the task, but the validation dataset remained the same as that used for GA computation.
3. A weaker classifier was designed and treated as a baseline model, where the term "weaker" was represented by using a few layers and a few parameters in the CNN to cope with possible limited computational resources.

Moreover, such a complex 6-class classification problem is a meaningful practical application, which represents a complex environment and a more autonomous setting.

9.6.1   Synthetic Data Fine-Tuning

Inspired by previous studies [20, 22], instead of directly applying more complex SSL [8, 23] techniques, a fine-tuning pipeline, namely, SDF, was proposed analogous to the previously discussed TL. In the SDF, a weak classifier is firstly pre-trained on the generated synthetic images and then fine-tuned by real ones, refer to Fig. 9.10.

Fig. 9.10 Illustration of the SDF pipeline

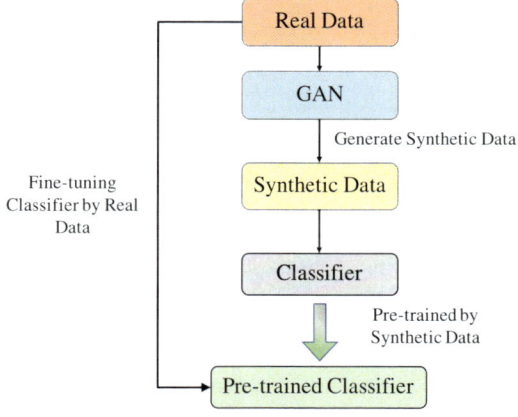

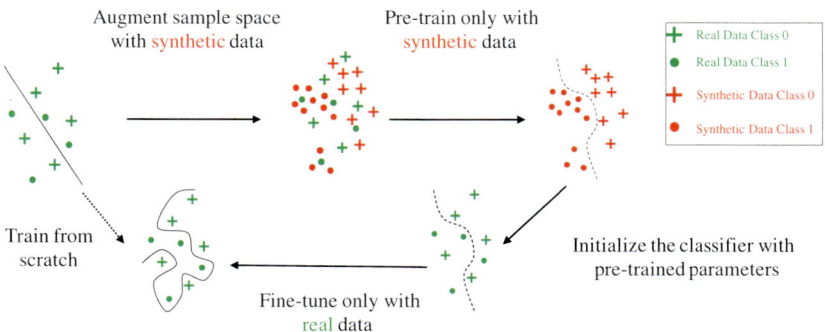

Fig. 9.11 Two learning modes of conventional training and SDF

Using a weak or non-classic CNN classifier, restricted by the available computational power, usually means that such a classifier does not have pre-trained parameters from ImageNet [6] or other datasets of large sizes in the source domain. Instead, its initialization fully depends on random, Gaussian, or other initialization approaches, e.g., Xavier initialization [24], which may be poor due to a large random parameter space. Conceptually, it is difficult to learn well the classification decision boundary by directly training from such initialization as illustrated in Fig. 9.11. When considering the SDF pipeline, regardless of the correctness of the generated synthetic images, these images can be thought of as being generated from a similar or enlarged sample space, which originates from raw real data. Therefore, it is believed that the classifier when pre-trained from such enlarged space can provide a better initialization for a fine-tuning procedure with the real data.

9.6.2   Experimental Setup

In order to comprehensively answer the question of whether direct synthetic data aggregation and SDF pipeline can enhance the classifier's performance, Sect. 9.5, a comparative experiment, with the following 10 cases, was conducted:

1. Train on "D2-Clean",
2. Train on "D2-Clean", but using DA in the classifier training,
3. Train on "D2-Clean" aggregated with "Syn-D2-Clean",
4. Same as Case 3 but using DA in the classifier training,
5. Train on "D2-Clean" aggregated with "Syn-D2-Clean&Aug",
6. Same as Case 5 but using DA in the classifier training,
7. Pre-train on "Syn-D2-Clean" and fine-tune on "D2-Clean",
8. Pre-train on "Syn-D2-Clean" and fine-tune on "D2-Clean" but using DA in the classifier training,

9. Pre-train on "Syn-D2-Clean&Aug" and fine-tune on "D2-Clean", and
10. Pre-train on "Syn-D2-Clean&Aug" and fine-tune on "D2-Clean" but using DA in the classifier training.

Cases 1 and 2 provide the baseline performance of training on a small-scale dataset with and without using the DA technique. Cases 3–6 represent directly aggregating synthetic data into the real dataset and cases 7–10 represent the SDF pipeline proposed above. It is noted that in cases 3–6, for simplicity, a 1 : 1 real-to-synthetic ratio was used. In other words, the real dataset was mixed with a similar amount of synthetic data was randomly generated, where in each case, 10 runs (10 times training and validation) were performed to be able to compute the variance of the validation accuracy, as discussed in Sect. 9.6.3.

The weak classifier designed in this experiment was a shallow CNN constructed by only 2 Conv layers with 16 3×3 filters followed by one max pooling layer with a 2×2 window size and stride $S = 2$. Finally, it used one FC-layer with 128 neurons activated by $ReLU$ and one $softmax$ layer whose output size equals the number of classes, i.e., 6 in this experiment. For a fair comparison, Adam optimization [18] with a learning rate of 10^{-4}, $\beta_1 = 0.9$, and $\beta_2 = 0.999$ was applied for all cases. In the SDF pipeline, the pre-training process took very few epochs and stopped when the validation accuracy no longer increased or continued to decrease in the following steps. Usually, the fine-tuning procedure took around 50 epochs until the results converged.

9.6.3   Results and Discussion

The results are listed in Table 9.7, where some important observations are summarized in this section. First, the two baseline results in cases 1 and 2 show that the conventional DA improved the accuracy by only 4%. From the comparison between cases 1–6, it is obvious that direct synthetic data aggregation rendered a worse performance than the baseline and DA operation led such an aggregation into an unstable direction, where the variance in terms of Std of the validation overall accuracy became larger in cases 4–6. It should be pointed out that in these cases, the large variance does not mean that both extremely high and low values of the validation accuracy were achieved among multiple runs. Instead, it was observed that most results from the conducted 10 runs were close to the mean, but only 2 or 3 sets of results achieved relatively low accuracy, e.g., 34% in case 5, leading to these larger Std values for cases 4 to 6. This can be partially explained by the fact that the conventional DA operation over-emphasized the lower-quality synthetic data (by augmenting more of such data) and diluted the contribution of the real data. Thus, simply mixing synthetic data into the real dataset did not work in this experiment. However, the idea of synthetic aggregation may work under some special training pipelines and algorithms, as reported in some previous studies [8, 19, 21]. From cases 7 to 10, the proposed SDF method can obviously improve the performance, i.e., with different approaches of the pre-training,

Table 9.7 Validation accuracy and comparisons for the cases of the experiment with and without synthetic data aggregation based on the SDF proposed method

Case	Description	Accuracy (%)	Run-time (min)
1	Train on D2-Clean	49.6 ± 0.2	6
2	Train on D2-Clean with DA	53.5 ± 0.3	9
3	Train on D2-Clean + Syn-D2-Clean	48.9 ± 0.6	7
4	Train on D2-Clean + Syn-D2-Clean with DA	49.6 ± 3.4	17
5	Train on D2-Clean + Syn-D2-Clean&Aug	47.1 ± 6.5	7
6	Train on D2-Clean + Syn-D2-Clean&Aug with DA	42.3 ± 6.6	17
7	Pre-train on Syn-D2-Clean & fine-tune on D2-Clean	55.2 ± 0.2	10
8	Pre-train on Syn-D2-Clean & fine-tune on D2-Clean with DA	54.2 ± 0.3	15
9	Pre-train on Syn-D2-Clean&Aug & fine-tune on D2-Clean	54.9 ± 0.1	10
10	Pre-train on Syn-D2-Clean&Aug & fine-tune on D2-Clean with DA	56.4 ± 0.1	15

the SDF shows accuracy improvement (54.2–56.4%) compared to that using only training on "D2-Clean" without any augmentation (49.6%), i.e., 4.6–6.8% enhancements. This is higher than the 3.9% enhancement by the conventional DA (53.5%). Here, it is restated that the parameters of the classifier in cases 1 to 6 were initialized from a random or Gaussian space and the fine-tuning pipeline by pre-training the model based on the synthetic images achieved up to 6.8% accuracy improvement, which is remarkable in this experiment.

Based on this experiment, the appropriate usage of synthetic images can improve the classification performance of a weak classifier under the low-data regime and limited computational resources. While adopting the SDF pipeline, the weak classifier was firstly pre-trained on the synthetic images for better initialization of the subsequent fine-tuning process, Fig. 9.10. This SDF pipeline is shown to be even more effective than the conventional DA and is thought to be useful and applicable in many practical scenarios. However, as a trade-off, compared to the conventional DA, the SDF pipeline is inefficient in two ways: (1) it is more time-consuming and (2) it leads to insufficient usage of the generated data. From Table 9.7, regardless of the hours of the computational cost in the GAN training,[4] the computational cost in the SDF pipeline without (cases 7 and 9) and with (cases 8 and 10) adoption of the DA in training the weak classifier is nearly double and triple of that in the baseline (case 1), respectively. However, the conventional DA (case 2) only takes 1.5 times that in the baseline. On the other hand, in the SDF pipeline, thousands of synthetic data were generated,

---

[4] This is because once the GAN is trained, the generating and sampling process for DA is fast and can be utilized repeatedly without retraining the GAN.

but they were only utilized in the pre-training process once and then discarded. Considering the high cost in training the GAN model and generating & storing the synthetic images, such a low data utilization rate is thought of as a waste of resources. Therefore, considering recent works documented in [21, 22], more efficient GANs and training pipelines should be studied further in the future.

9.7   Summary

In this chapter, besides the fundamentals of the GAN, two types of GAN-based case studies were conducted, namely, synthetic structural image generation and synthetic data augmentation in classification problems, where two tasks (scene level identification and damage state detection) were considered. The following key conclusions are inferred from all the conducted experiments:

- DCGAN is a stable GAN model that can generate reasonable structural images (by human evaluation) with the LB method applied for all levels even when only using a depth-2 representation framework without data cleaning. This LB-GAN outperforms the direct training schema of GAN.
- Using a cleaned dataset produces synthetic images with clearer objects but increases difficulties in the training process and mode collapse issues are more likely to occur.
- Conventional DA and deeper representation improve visual quality in most cases.
- Consistent with human observations, better image quality leads to a higher *SIS* value. On the other hand, *GA* is a trade-off between image quality & generalization.
- The SDF pipeline using synthetic data can enhance the performance of a weak classifier under the low-data regime and limited computational resources.

The following items are suggested as future extensions:

- There are many types of GAN models, e.g., WGAN [25], Conditional GAN [26], etc. Therefore, the selection of the GAN model as a hyper-parameter is worth exploring. Moreover, the methods and results presented in this chapter can be considered as benchmarks and references for future comparative studies.
- There are many evaluation metrics to be explored in the future, e.g., Wasserstein distance to apply WGAN and Fréchet inception distance (FID) proposed in [27].
- While adopting synthetic data aggregation for classification problems, the SDF pipeline is time-consuming and inefficient for using the generated data compared to conventional DA. More exploration is expected by using the techniques of SSL [8, 28], as explored in Chap. 10.
- More data aggregation applications, other than classification, need to be explored, e.g., damage localization, semantic segmentation, etc.

- The major drawback of applying the current GANs in engineering applications, e.g., damage classification, is that the generated synthetic data fully depend on the optimization by minimizing the difference between the distributions of the synthetic and real data. This minimization may not fully comply with the laws of physics of the problem at hand. Thus, future studies should consider encoding physics principles and information into the GAN model formulation.

9.8   Exercises

1. Provide at least 5 types of transformation or processing operations commonly utilized in conventional DA.
2. Explain the fundamental concept of GANs and how they differ from the traditional ML models.
3. Describe the roles of the generator and discriminator in a GAN and how they interact during the training process.
4. Conduct a literature review and describe at least 3 variants or extensions of GANs (e.g., Conditional GANs [26], WGANs [25], and CycleGAN [29]). Explain how these variants address specific limitations of the traditional GANs.
5. Discuss potential applications where different variants of GANs are particularly effective and explain the reasons behind such effectiveness as it pertains to these applications and the considered variants of GANs.
6. Explain why in most scenarios, directly mixing GAN-generated synthetic data with the original real data for a classification problem may not work, i.e., does not improve the test accuracy.

References

1. I. Goodfellow et al., Generative adversarial nets, in *Advances in Neural Information Processing Systems* (2014), pp. 2672–2680
2. I. Goodfellow, *NIPS 2016 Tutorial: Generative Adversarial Networks* (2016). arXiv:1701.00160
3. D.-Z. Du, P.M. Pardalos, *Minimax and Applications*, vol. 4. (Springer Science & Business Media, 2013)
4. A. Radford, L. Metz, S. Chintala, *Unsupervised Representation Learning with Deep Convolutional Generative Adversarial Networks* (2015). arXiv:1511.06434
5. F. Yu et al., *Lsun: Construction of a Large-scale Image Dataset Using Deep Learning with Humans in the Loop* (2015). arXiv:1506.03365
6. J. Deng et al., Imagenet: A large-scale hierarchical image database, in *2009 IEEE Conference on Computer Vision and Pattern Recognition* (2009), pp. 248–255
7. Y. LeCun et al., Gradient-based learning applied to document recognition. Proc. IEEE **86**(11) 2278–2324 (1998)

8. T. Salimans et al., Improved techniques for training gans, in *Advances in Neural Information Processing Systems* (2016), pp. 2234–2242
9. C. Szegedy et al., Going deeper with convolutions, in *Proceedings of the IEEE Conference on Computer Vision and Pattern Recognition* (2015), pp. 1–9
10. S. Barratt, R. Sharma, *A Note on the Inception Score* (2018). arXiv:1801.01973
11. A. Krizhevsky, Learning Multiple Layers of Features from Tiny Images. In: Technical Report TR-2009 (2009)
12. K. Simonyan, A. Zisserman, Very Deep Convolutional Networks for Large-scale Image Recognition (2014). arXiv:1409.1556
13. K. He et al., Deep residual learning for image recognition, in *Proceedings of the IEEE Conference on Computer Vision and Pattern Recognition* (2016), pp. 770–778
14. C.M. Bishop, *Pattern Recognition and Machine Learning* (Springer, New York, 2006)
15. H. Li, *Statistical Learning Methods (in Chinese)* (Tsinghua University Press, 2012)
16. C.F. Mooney et al., *Bootstrapping: A Nonparametric Approach to Statistical Inference*, vol. 95 (Sage, 1993)
17. Y. Gao, B. Kong, K.M. Mosalam, Deep leaf-bootstrapping generative adversarial network for structural image data augmentation. Comput. Aided Civil Infrastruct. Eng. **34**(9), 755–773 (2019)
18. D.P. Kingma, J. Ba, *Adam: A Method for Stochastic Optimization* (2014). arXiv:1412.6980
19. A. Antoniou, A. Storkey, H. Edwards, *Data Augmentation Generative Adversarial Networks* (2017). arXiv:1711.04340
20. M. Frid-Adar, Synthetic data augmentation using GAN for improved liver lesion classification, in *IEEE 15th International Symposium on Biomedical Imaging (ISBI 2018)*, vol. 2018 (IEEE, 2018), pp. 289–293
21. X. Zhang et al., Dada: Deep adversarial data augmentation for extremely low data regime classification, in *ICASSP 2019-2019 IEEE International Conference on Acoustics, Speech and Signal Processing (ICASSP)* (IEEE, 2019), pp. 2807–2811
22. L. Perez, J. Wang, *The Effectiveness of Data Augmentation in Image Classification Using Deep Learning* (2017). arXiv:1712.04621
23. Z. Zheng, L. Zheng, Y. Yang, Unlabeled samples generated by gan improve the person re-identification baseline in vitro, in *Proceedings of the IEEE International Conference on Computer Vision* (2017), pp. 3754–3762
24. X. Glorot, Y. Bengio., Understanding the difficulty of training deep feedforward neural networks, in *Proceedings of the Thirteenth International Conference on Artificial Intelligence and Statistics* (2010), pp. 249–256
25. M. Arjovsky, S. Chintala, L. Bottou, *Wasserstein Gan* (2017). arXiv:1701.07875
26. M. Mirza, S. Osindero, *Conditional Generative Adversarial Nets* (2014). arXiv:1411.1784
27. M. Heusel et al., Gans trained by a two time-scale update rule converge to a local nash equilibrium, in *Advances in Neural Information Processing Systems* (2017), pp. 6626–6637
28. J.T. Springenberg, *Unsupervised and Semi-Supervised Learning with Categorical Generative Adversarial Networks* (2015). arXiv:1511.06390
29. J.-Y. Zhu et al., Unpaired image-to-image translation using cycleconsistent adversarial networks, in *Proceedings of the IEEE International Conference on Computer Vision* (2017), pp. 2223–2232

Semi-Supervised Learning

10

Previous chapters demonstrate the effectiveness of ML and DL under the supervised learning setting, where all training data are well-labeled, refer to Sect. 3.1. In practice, due to costly labeling efforts, unlabeled data are more accessible. Therefore, a new mechanism, namely, SSL is developed, which utilizes both labeled and unlabeled data for training.

In this chapter, introducing SSL to GAN is specifically discussed, where the GAN is converted into a semi-supervised variant. It is believed that the SSL mechanism can more thoroughly exploit the features of the unlabeled data and simultaneously increase the model's synthetic data generation and classification capabilities. Furthermore, besides considering the restrictions of the low-data regime and limited computational power, imbalanced class and label issues are explored. To better address the imbalanced-class issues, a balanced-batch sampling technique is adopted in the training process and the whole model is accordingly named BSS-GAN, as introduced in Sect. 2.4.

10.1   SSL Mechanism

SSL is an ML paradigm that combines both supervised and unsupervised learning techniques. As introduced in Sect. 3.1.1, in supervised learning, a model is trained on a labeled dataset where each data point is associated with a corresponding label. On the other hand, unsupervised learning involves working with unlabeled data to find patterns or relationships within the data.

SSL strategically utilizes the benefits presented by both labeled and unlabeled data. In many real-world situations, obtaining labeled data can be costly, time-consuming, or a truly unfeasible endeavor, especially for segmentation tasks as introduced in Chap. 8. However, unlabeled data tend to be more plentiful and accessible. In this context, SSL

© The Author(s), under exclusive license to Springer Nature Switzerland AG 2024   279
K. M. Mosalam and Y. Gao, *Artificial Intelligence in Vision-Based Structural Health Monitoring*, Synthesis Lectures on Mechanical Engineering,
https://doi.org/10.1007/978-3-031-52407-3_10

emerges as a methodology that capitalizes on both data types, culminating in an enhanced model performance.

SSL relies on certain assumptions about the data and the relationships between labeled and unlabeled parts. Accordingly, it is usually not possible to generalize from a finite training dataset to infinite unseen test datasets. These assumptions play a crucial role in the effectiveness of the SSL algorithms and methods. Here are some common assumptions in SSL:

- Smoothness: For two arbitrary input points x_1 and x_2 that are close by in the input space, X, the corresponding labels y_1 and y_2 should be the same to guarantee the *smoothness* of the data.
- Low-density separation: The decision boundaries should prioritize passing through *low-density areas.*[1]
- Clustering: If input data points form *clusters*, each corresponding to an output class, then if the points are in the same cluster, they can be considered to belong to the same class.
- Manifold: The data points observed in high-dimensional input space are usually concentrated in low-dimensional substructures, defined as *manifolds*. The input space, X, can be composed of multiple low-dimensional manifolds. Moreover, if data points are located on the same manifold, they have the same label.

SSL has been successfully applied in various domains, including NLP, CV, and speech recognition. In addition, it can be further combined with other AI technologies, e.g., generative models, to improve model performance with limited labeled data. Therefore, in the subsequent sections, a novel variant GAN mode utilizing the SSL mechanism is introduced, especially for the SHM problems under low-data and imbalanced-class regime. For more details about SSL, refer to [1, 2].

10.2   Balanced Semi-Supervised GAN

In Chap. 9, the effectiveness of GAN in generating reasonable structural images under certain training strategies and settings is validated and discussed. To overcome the limitations of directly mixing synthetic images with real ones for training, a special union training pipeline is proposed, namely, SDF. Such a training pipeline is able to enhance the classifier performance by nearly 7% over the baseline. However, as concluded, this training pipeline is still inefficient, because both the GAN model and the classifier need to be trained and large amounts of synthesized images are left valueless after the pre-training process. Moreover, these methods do not take into account the imbalanced-class issue. According to the study of [3, 4], reformulating the problem into a SSL paradigm seems to be a promising solution.

---

[1] For example, if there exists a large gap between two classes, the data density at the boundary of the two classes is low.

The original GAN is trained in an unsupervised learning manner and its discriminator only differentiates unlabeled real samples from those synthesized by the generator. Herein, a novel GAN-based classification pipeline with balanced-batch sampling and SSL, namely, the BSS-GAN, is proposed. It formulates the original GAN in the manner of SSL [5].

Unlike the traditional GAN [6], the output size, i.e., dimension, of the discriminator increases from 2 ("real" or "synthetic") to $K + 1$ so that the discriminator can classify samples from K "real classes" (samples follow the probability distribution, $p_d(x, y)$) and one "synthetic class" (generated samples follow the probability distribution, p_g). Besides the input data, x, and the corresponding labels, y, from $p_d(x, y)$, the model simultaneously learns from the unlabeled data distribution, $p_d(x)$. These characteristics illustrate the core concept of the SSL. In addition, a balanced-batch sampling approach is introduced to pursue both class balance and real-synthetic balance during training.

10.2.1 Discriminator Loss

For each input sample, x, taken from either p_d (i.e., from both $p_d(x)$ and $p_d(x, y)$) or p_g, the discriminator, as defined in Sect. 9.1, D, outputs a $(K + 1)$-dimensional predictive model conditional probability vector as follows:

$$p_m(y = i | x) = \frac{\exp (D(x)_i)}{\sum_{j=1}^{K+1} \exp (D(x)_j)}, \quad i = 1, 2, \ldots, K + 1. \tag{10.1}$$

This predictive conditional probability is for the sample coming from the i-th class. Moreover, $D(x)_1, D(x)_2, \ldots, D(x)_{K+1}$ are logits (refer to Sect. 3.7.1) output by $D(x)$ corresponding to each class. Here, $p_m(y = K + 1 | x)$ represents the predicted probability that sample x is "synthetic" and thus $D(x)$ in the first term of Eq. 9.3 is substituted by $1 - p_m(y = K + 1 | x)$. Similarly, $1 - D(x)$ in the second term of Eq. 9.3 is substituted by $p_m(y = K + 1 | G(z))$. Both terms in Eq. 9.3 are negated to form a minimization problem of D. The unsupervised (corresponding to subscript "US") loss without using any label information of the K real classes is expressed as follows:

$$L_{US}^{(D)} = - C_{real} \, E_{x \sim p_d} \left[\ln \left(1 - p_m(y = K + 1 | x)\right) \right] \\ - C_{syn} \, E_{x \sim p_g} \left[\ln \left(p_m(y = K + 1 | x)\right) \right], \tag{10.2}$$

where, as in Sect. 9.1, the constants, which maybe dropped, $C_{real} = N_{real}/N$ & $C_{syn} = N_{syn}/N$, and the number of real, N_{real}, & synthetic, N_{syn}, samples are typically taken such that the total # of data fed into the network is $N = N_{real} + N_{syn}$. Moreover, $E_{a \sim p_b}$ denotes the expectation of the data or the random variable a sampled from the distribution p_b. Therefore, both labeled and unlabeled data can be used for unsupervised feature learning in Eq. 10.2.

For the real labeled data, the supervised (corresponding to subscript "S") discriminator loss is the cross-entropy between the real labeled data distribution, $p_d(x, y)$, and the model's predicted label distribution for the K real classes, given the real input samples, x, i.e., $p_m(y|x, y < K + 1)$. This loss is expressed as follows:

$$L_S^{(D)} = -E_{x,y \sim p_d(x,y)} \left[\ln(p_m(y|x, y < K + 1)) \right]. \tag{10.3}$$

Finally, the total discriminator loss is expressed as follows:

$$L^{(D)} = L_{US}^{(D)} + L_S^{(D)}. \tag{10.4}$$

10.2.2 Generator Loss

The generator's objective is to "weaken" the discriminator's performance. However, the original formulation of the generator loss [6] usually does not perform well. This is because the generator's gradient vanishes (refer to Sect. 4.2.4) when the discriminator has high confidence in distinguishing the generated samples from the real ones, i.e., when $D(G(z)) \rightarrow 0.0$, where $G(z)$ is as defined in Sect. 9.1. Therefore, the generator loss used in the formulation below refers to the heuristic, H, loss [7] as follows:

$$L_H^{(G)} = -\frac{1}{2} E_{z \sim p_z} \ln D(G(z)). \tag{10.5}$$

Instead of minimizing the expected log-probability of the discriminator being correct, the generator now maximizes the expected log-probability of the discriminator making a mistake, i.e., assigning a real label to a generated sample, $G(z)$. In the sequel, the constant multiplier in Eq. 10.5, i.e., $1/2$, is dropped and $E_{z \sim p_z} \ln D(G(z))$ is substituted by $E_{z \sim p_z} \ln [1 - p_m(y = K + 1|G(z))]$ to accommodate the $(K + 1)$-dimensional discriminator output. Accordingly, Eq. 10.5 becomes,

$$L_H^{(G)} = -E_{z \sim p_z} \ln [1 - p_m(y = K + 1|G(z))]. \tag{10.6}$$

Feature matching (FM) is a technique that prevents over-training of the generator and increases the stability of the GAN [5]. FM requires the generator to produce samples that result in features on an *intermediate layer* of the discriminator network similar to those of the real samples. Thus, the generator loss, considering FM, is re-formulated as follows:

$$L_{FM}^{(G)} = \left\| E_{x \sim p_d} f(x) - E_{z \sim p_z} f(G(z)) \right\|_2^2, \tag{10.7}$$

. where $\| \bullet \|_2$ is the $\mathcal{L}_2$ norm of "$\bullet$". Moreover, $f(x)$ and $f(G(z))$ are feature representations after the activation function of an intermediate layer of the discriminator for real sample x and generated sample G(z), respectively, which are defined herein by the *ReLU* (Eq. 4.1) activation on the flattened output of the last Conv layer of the discriminator network. Eq. 10.7 aims to minimize the $\mathcal{L}_2$ norm of the difference between the average feature representations (as extracted by the discriminator) of the real data and the generated data. By doing so, the generator is encouraged to produce samples that are similar to the real data not just in appearance but also in terms of their feature representations as perceived by the discriminator. This can lead to improved stability during training and to higher quality of the generated samples.

Finally, combining $L_H^{(G)}$ and $L_{FM}^{(G)}$, the total generator loss is expressed as follows:

$$L^{(G)} = L_H^{(G)} + L_{FM}^{(G)}. \tag{10.8}$$

Besides the loss formulations (Eqs. 10.2, 10.3 and 10.6) based on the cross-entropy used herein, other loss functions, e.g., WGAN loss [8] can be modified to fit the purpose of semi-supervised GAN training, which is worth future investigations.

10.2.3 Balanced-Batch Sampling

The class-imbalance issue not only affects the classification performance, but also deteriorates the perceptual quality and diversity of the generated samples [9]. During the conventional training (i.e., updating) procedure of a DL classifier or GAN, MBGD (refer to Sect. 4.1.3) is commonly adopted [10]. Due to computational limitations, instead of using *all* the data at once, the DL network is only fed with *one small batch* containing m data samples randomly selected from the full dataset of size N, where m is called the batch size and $m < N$. Statistically, if the dataset is imbalanced, the batch is also imbalanced, which eliminates neither the performance bias of the classifier nor the training instability of the GAN.

In this section, the *balanced-batch sampling* in training is introduced. As its name suggests, while forming the batch, the same amount of data is randomly sampled from each class. For balanced-batch sampling in GAN training, two types of balances are maintained in a given batch: (1) balance among all K real classes in the labeled (corresponding to superscript "l") data and (2) balance between any particular real class and the $K + 1$-th synthetic class. For (1), n^l real labeled samples are randomly selected from K real classes, where $n_1^l, n_2^l, \ldots, n_K^l$ are the numbers of data samples from each class, i.e., $n^l = \sum_{k=1}^{K} n_k^l$, expressed as follows:

$$n_1^l = n_2^l = \cdots = n_K^l = n^l / K. \tag{10.9}$$

For (2), the total number of generated (corresponding to superscript "g") samples, n^g, matches any of the sub-batches from a certain real class, i.e., $n^g = n^l / K = n_k^l, \forall k \in \{1, 2, \ldots, K\}$. In other words, each of the $K + 1$ classes contributes to the whole batch by a sub-batch of the same size to one another, as follows:

$$m = n^l + n^g = (K+1)\, n^l / K. \tag{10.10}$$

The BSS-GAN can utilize the unlabeled (corresponding to superscript "*ul*") data in feature learning, Eqs. 10.2 and 10.7. If additional unlabeled data are available, in any given batch, the ratio of unlabeled samples to any single-class sub-batch is controlled by a hyper-parameter, c, i.e.,

$$n^{ul} / n^l = c/K, \tag{10.11}$$

which modifies Eq. 10.10 to a more general form. As an example and to keep a smaller batch size for simulating limitations on the available computational resources, $c = 1$ is used here. Therefore, the overall size of each batch is expressed as follows:

$$m = n^l + n^g + n^{ul} = n + n^g = (K+2)\, n^l / K, \tag{10.12}$$

where $n = n^l + n^{ul}$.

10.2.4 BSS-GAN Pipeline

By formulating the GAN in a SSL setting and using the balanced-batch sampling technique, the integrated model BSS-GAN is established. This model builds an end-to-end pipeline for both synthetic image generation and classifier training. It is expected to have a stable and less biased performance under highly imbalanced datasets. Moreover, unlike training a supervised learning-based DL classifier, here the unlabeled data are put into use. One example of using BSS-GAN for concrete crack detection is illustrated in Fig. 10.1.

For a training batch size, m, the detailed training process of the BSS-GAN is listed in the following steps:

Step 0 Initialize the discriminator, D, and the generator, G, with their parameters θ_D and θ_G, respectively.

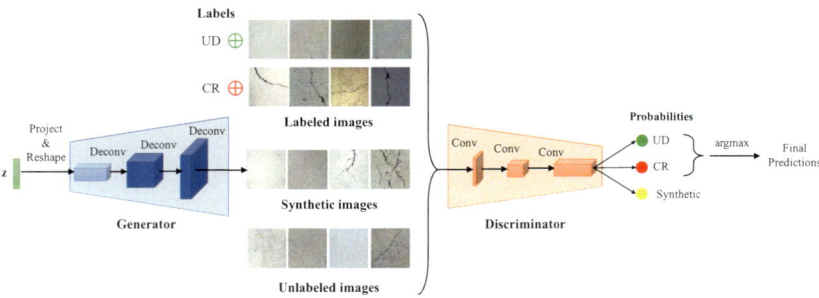

Fig. 10.1 BSS-GAN pipeline for concrete crack detection example with labels UD: undamaged and CR: cracked

Step 1 A subset of the batch represented by real data (labeled & unlabeled) is formed, $B_r = B_r^l \cup B_r^{ul} = \left\{ \left(x_1^l, y_1^l \right), \left(x_2^l, y_2^l \right), \ldots, \left(x_{n^l}^l, y_{n^l}^l \right) \right\} \cup \left\{ x_1^{ul}, x_2^{ul}, \ldots, x_{n^{ul}}^{ul} \right\}$, where n^l data-label pairs are equally sampled from the K real classes and class label $y_i^l \in \{1, 2, \ldots, K\}$.

Step 2 Random noise vectors, $z = \{z_1, z_2, \ldots, z_{n^g}\}$, are sampled from the noise prior distribution, $p_g(z)$, where z is fed to G to generate n^g synthetic samples, $B_g = \{G(z_1), G(z_2), \ldots, G(z_{n^g})\}$, to complete the final subset of the batch.

Step 3 Feed B_r to D & for $x_i \in B_r, i \in \{1, 2, \ldots, n^l + n^{ul}\}$, D outputs the probability vector $u_i = [p_m(y = 1|x_i), \ldots, p_m(y = K + 1|x_i)]^T$ of $(K + 1)$-dimension, where $u_i^{K+1} = p_m(y = K + 1|x_i)$.

Step 4 Feed B_r^l to D & for $\left(x_i^l, y_i^l \right) \in B_r^l, i \in \{1, 2, \ldots, n^l\}$, D outputs the probability vector $v_i = \left[p_m \left(y = 1|x_i^l \right), \ldots, p_m \left(y = K|x_i^l \right) \right]^T$ of $(K + 1)$-dimension and only the first K dimensions are considered, where $v_i \left(y_i^l \right) = p_m(y = y_i^l|x_i^l)$ represents the probability of the model for predicting class y_i^l.

Step 5 Feed B_g to D & for $G(z_i) \in B_g, i \in \{1, 2, \ldots, n^g\}$, D outputs the probability vector $w_i = [p_m(y = 1|G(z_i)), \ldots, p_m(y = K + 1|G(z_i))]^T$ of $(K + 1)$-dimension, where $w_i^{K+1} = p_m(y = K + 1|G(z_i))$.

Step 6 Compute the discriminator loss, $L^{(D)}$, as follows:

$$L^{(D)} = -\frac{1}{n} \sum_{i=1}^{n} \ln \left(1 - u_i^{K+1} \right) - \frac{1}{n^l} \sum_{i=1}^{n^l} \ln \left(v_i \left(y_i^l \right) \right) - \frac{1}{n^g} \sum_{i=1}^{n^g} \ln \left(w_i^{K+1} \right).$$

(10.13)

Step 7 Compute the generator loss, $L^{(G)}$, as follows:

$$L^{(G)} = \left\| \frac{1}{n} \sum_{i=1}^{n} f(x_i) - \frac{1}{n^g} \sum_{i=1}^{n^g} f(G(z_i)) \right\|_2^2 - \frac{1}{n^g} \sum_{i=1}^{n^g} \ln \left(1 - w_i^{K+1} \right).$$

(10.14)

Step 8 Optimize and update the network parameters, θ_D & θ_G, using the learning rate, η, and gradients of the loss functions, $\nabla_{\theta_D} L^{(D)}$ & $\nabla_{\theta_G} L^{(G)}$, as follows:

$$\theta_D \leftarrow \theta_D - \eta \nabla_{\theta_D} L^{(D)} \quad \& \quad \theta_G \leftarrow \theta_G - \eta \nabla_{\theta_G} L^{(G)}.$$

(10.15)

Step 9 Repeat steps (1) to (8) until convergence is achieved or the designated maximum number of iterations[2] is reached.

---

[2] In ML and as briefly mentioned in Sect. 4.1.3, typically, a single *epoch* has too much data to be sent to the computer at once. Therefore, each epoch is divided into smaller *batches*, where the batch size is the total number of training data points within a particular batch. Here, one iteration means updating a batch once, and the total number of *iterations* refers to the total number of batches required to complete.

In the above algorithm, the balanced-batch sampling (steps 1 to 5) is only adopted in training. Once the BSS-GAN is trained, when referencing or predicting new data for classification purposes, all new data are fed into D only, and the predicted class is the one with the highest predictive probability among the first K entries of the D's output. Similarly, for synthetic data generation, the noise vector, z, is sampled and fed into G, which then outputs the synthetic data. Moreover, BSS-GAN is an *online* DAmethod, where only n^g synthetic data are generated and utilized during each training batch, and can be discarded when the training moves to the next batch. The total amount of the generated data depends on the number of training batches. For example, when training moves to the e-th epoch, the total amount of the generated data, $N_e^g = n^g \cdot e \cdot \lceil N^l / n^l \rceil$, where N^l is the total amount of the labeled data and $\lceil \bullet \rceil$ is the ceiling operation for "$\bullet$", refer to Sect. 4.1.3. It is noted that $n^g \ll N_e^g$ because usually the multiplier, $e \gg 1$, and $N^l \gg n^l$. This makes the BSS-GAN, being an online DA method, an efficient method in terms of the needed computing resources for data storage during the training.

10.3 Case Study: RC Damage Detection

10.3.1 Low-data and Imbalanced-class Problem

In this section, the low-data and imbalanced-class issue in one common task in vision-based SHM, namely, RC damage detection, is investigated. Three key classes: (i) undamaged state (UD), Fig. 10.2a, (ii) cracked (CR), Fig. 10.2b, and (iii) spalling (SP), Fig. 10.2c (more details in Sect. 5.2.2), are considered, describing three damage levels in the order of increasing corrosion risk to embedded reinforcing bars. In real-world applications, the class ratios of UD:CR and UD:SP are usually high. To simulate such imbalances in a realistic SHM data collection, an empirical ratio of 32:2:1 (UD:CR:SP) is selected for experimental purposes, where SP is treated as less frequent than CR. It is noted that the main purpose of this case study is to have a solution to circumvent the high imbalance issue in practice. Thus, slightly imbalanced or even balanced class ratios are not discussed herein. Three major validation scenarios are designed:

1. Binary crack detection with UD:CR class ratio of 16:1,
2. Binary spalling detection with UD:SP class ratio of 32:1, and
3. Ternary damage pattern classification with UD:CR:SP class ratio of 32:2:1.

Scenarios (1) & (2) simulate the real-world binary damage detection in vision-based SHM, where the number of "undamaged" cases (UD) far exceeds that of "damaged" cases (CR or SP). On the other hand, scenario (3) integrates the two damage cases CR & SP into a comprehensive but more complex 3-class classification, which aims to evaluate the DL model when used in the context of an imbalanced multi-class problem.

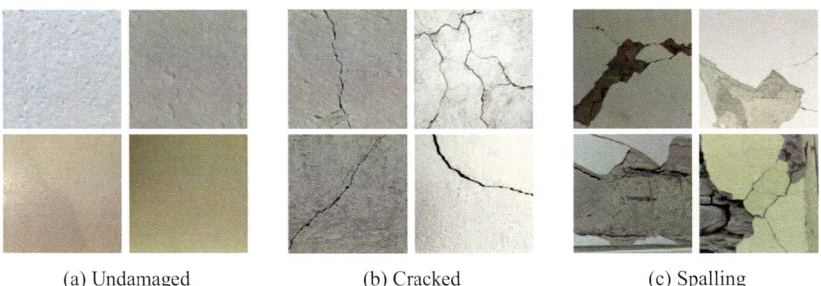

(a) Undamaged   (b) Cracked   (c) Spalling

Fig. 10.2 Sample images of three classes

For a comparative study, in each scenario, six pipelines are configured and compared as follows:

- A baseline shallow CNN classifier (BSL),
- BSL with under-sampling the majority-class data to restore class balance (BUS),
- BSL with over-sampling the minority-class data by conventional DA, e.g., flip, translation, and rotation (BOS-DA),
- BSL with over-sampling the minority-class data by ordinary GAN-generated data (BOS-GAN),
- BSL adopting the SDF training pipeline (BSL-SDF), and
- BSS-GAN.

In BUS, BOS-DA & BOS-GAN pipelines, an equal class ratio is maintained after under/over-sampling. The performance of each case is evaluated by appropriate metrics, e.g., recall, CM, and F_β score, which are covered in detail in Sect. 3.4. Beyond these metrics, more intuitions are discussed in terms of (a) synthetic image quality of ordinary GAN and BSS-GAN and (b) effectiveness of unsupervised feature learning using different amounts of unlabeled data in the BSS-GAN.

10.3.2 Datasets

For generality, the images in this study were obtained from two open-source structural image datasets: ϕ-Net [11] (refer to Sect. 5.4) and SDNET2018 [12]. The structural images cover scenarios ranging from undamaged to extreme cracking or spalling. The images were further processed for the experiments as follows:

1. Clean the dataset and select pixel level (close-up) images with high visual quality.
2. Select and store the images with labels UD, CR, and SP to build the full dataset.

Table 10.1 Label statistics of the datasets

Dataset	UD	CR	SP	Class ratio
Training	9,600	600	300	32:2:1
Test	4,800	300	150	
Total	14,400	900	450	

3. Rescale the images to a uniform size of 128×128 pixels using bicubic resampling (e.g., bicubic interpolation) [13].

In this way, a dataset with a total of $N = 15,750$ real images was constructed including 14,400 UD, 900 CR, and 450 SP images, Table 10.1. In addition, a 2:1 split ratio of training:test was applied, compared to the more common 3:1 or 4:1 ratios. This led to fewer training data, which also helped to simulate the data deficiency. On the other hand, in this manner, there were sufficient test data (especially minority-class data) to be appropriately evaluated using the proposed BSS-GAN.

Compared with the ϕ-Net benchmark dataset (refer to Chap. 5) and general CV applications, this dataset contains imbalanced low-data. In addition, the dataset is publicly available upon request from http://stairlab.berkeley.edu/data/. To investigate the contribution of unlabeled data in the BSS-GAN, an imbalanced "hybrid" dataset for concrete crack detection (UD vs. CR) was formed with 20% of the labeled training data $(0.2 \times (9,600 + 600) = 2,040)$ in the original crack detection, which simulated a more severe low-data scenario. The remaining training data $(0.8 \times (9,600 + 600) = 8,160)$ were treated as unlabeled. Both labeled and unlabeled sub-sets of the data remained at UD:CR = 16:1 class ratio.

10.3.3 Network Configurations

In previous studies, the network design is sophisticated and dependent on the tasks at hand and the used datasets. To avoid loss of generality while considering the possible limitations of the computing resources in practical engineering applications, the BSL is designed as a general multi-layer CNN discriminative classifier with a simple network configuration and hyper-parameter tuning, Table 10.2. For a fair comparison, the BSS-GAN's discriminator uses the same architecture as the BSL, e.g., the same numbers of layers and filters, except for the output dimension ($C = K + 1$ in BSS-GAN as opposed to $C = K$ in the BSL). According to [14], *LReLU* [15] is used as the activation function with a negative slope coefficient $\alpha = 0.2$, and BatchNorm [16] layers are inserted after the intermediate Conv layers with a momentum $= 0.8$. To avoid over-fitting, a 0.25 dropout rate is also applied.

The generator of the BSS-GAN is configured based on previous studies [14, 17], Table 10.3. Due to the small image size, 128×128, a conventional 100-dimensional noise

Table 10.2 Configurations of the BSL or the discriminator of the BSS-GAN

Layer	Filter size (#)	Activation	Shape	Notes
Input	–	–	$(N, 128, 128, 3)$	Input RGB images of size 128×128
Conv	3×3 (32)	LReLU	$(N, 64, 64, 32)$	Stride = 2, $\alpha = 0.2$
Dropout	–	–	$(N, 64, 64, 32)$	Dropout rate = 0.25
Conv	3×3 (64)	LReLU	$(N, 32, 32, 64)$	Stride = 2, $\alpha = 0.2$
BatchNorm	–	–	$(N, 32, 32, 64)$	Momentum = 0.8
Dropout	–	–	$(N, 32, 32, 64)$	Dropout rate = 0.25
Conv	3×3 (64)	LReLU	$(N, 32, 32, 64)$	Stride = 1, $\alpha = 0.2$
Flatten	–	–	$(N, 65536)$	$65536 = 32 \times 32 \times 64$
FC-layer	–	softmax	(N, C)	BSS-GAN: $C = K + 1$; BSL: $C = K$

Table 10.3 Configuration of the generator of the BSS-GAN

Layer	Filter size (#)	Activation	Shape	Notes
Input	–	–	$(N, 100)$	Noise $\sim$ Gaussian distribution
FC-layer	–	ReLU	$(N, 131072)$	$131072 = 32 \times 32 \times 128$
Reshape	–	–	$(N, 32, 32, 128)$	–
Deconv	3×3 (128)	ReLU	$(N, 64, 64, 64)$	Stride = 2
BatchNorm	–	–	$(N, 64, 64, 64)$	Momentum = 0.8
Deconv	3×3 (64)	ReLU	$(N, 128, 128, 3)$	Stride = 2
BatchNorm	–	–	$(N, 128, 128, 3)$	Momentum = 0.8
Deconv	3×3 (3)	tanh	$(N, 128, 128, 3)$	Stride = 1

vector is randomly generated from the Gaussian distribution as the input to the generator without dropout [18]. BatchNorm layers with a 0.8 momentum are added after the Deconv layers except for the final layer. For the other two GAN-based pipelines (BOS-GAN & BSL-SDF), the GAN portions are consistent with that of the BSS-GAN. The exceptions are the loss functions for the generator and discriminator, which are taken as the ones used in the original GAN [6] based on the value function, Eq. 9.1.

10.3.4 Experimental Setup

In the three major scenarios introduced in Sect. 10.3.1, all data were labeled. For the first five pipelines, also listed in Sect. 10.3.1, a batch size of $m = 60$ was used. For the BSS-

GAN (sixth pipeline), the number of labeled real data was maintained at 60. However, for the remaining batch data, based on the balanced-batch sampling, the numbers varied for different cases. For the two binary cases and the one ternary case, the total batch size was $m = 60 + 60/2 = 90$ and $m = 60 + 60/3 = 80$, respectively. While discussing the effectiveness of using different amounts of unlabeled data in binary crack detection, the batch size was $m = 60 + 60/2 \times 2 = 120$. All six pipelines were trained for 300 epochs, and results were saved for each epoch.

Regarding the performance evaluation, the classification (overall) accuracy (Eq. 3.12) is not informative for the imbalanced classification problems, where merely guessing all samples from the majority class may yield a misleadingly high accuracy. Thus, in this case study, several more appropriate metrics are used: (1) CM along with TPR (or recall), TNR & precision and (2) F_β score. The details of these metrics can be found in Sect. 3.4. In the binary tasks, the best model was selected by the highest TPR with over 90% TNR. For the ternary class task, the best model was selected by the highest recall of SP with over 90% recall of UD. The optimizer was Adam [19] with an initial learning rate of 2×10^{-5}. All computations were conducted on the TensorFlow platform and performed on CyberpowerPC with a single GPU (CPU: Intel Core i7-8700K@3.7 GHz 6 Core, RAM: 32 GB & GPU: Nvidia Geforce RTX 2080Ti).

10.4   Results and Analysis

10.4.1 Scenario 1: Crack Detection

From Table 10.4, the accuracy values for all six pipelines are higher than 90%, which is deceivingly promising. However, simply predicting all images as UD can easily lead to a $16/17 = 94.1\%$ overall accuracy under the UD:CR=16:1 class ratio. More focus should be placed on the TPR & TNR, representing the accuracy of detecting CR & UD, respectively. The resulting TPR & TNR are listed in Table 10.4 and along the diagonal cells of the CMs in Fig. 10.3.

Table 10.4 Classification performance for the crack detection (%)

Pipeline	TPR	TNR	F_2	F_5	Accuracy
BSL	31.0	99.4	35.2	31.7	95.4
BUS	46.0	97.0	46.6	46.1	94.0
BOS-DA	45.7	99.4	50.1	46.5	96.2
BOS-GAN	29.3	99.5	33.6	30.1	95.4
BSL-SDF	49.0	92.8	43.4	47.8	90.2
BSS-GAN	89.3	92.2	72.8	85.6	92.1

Bold # indicates a much better result

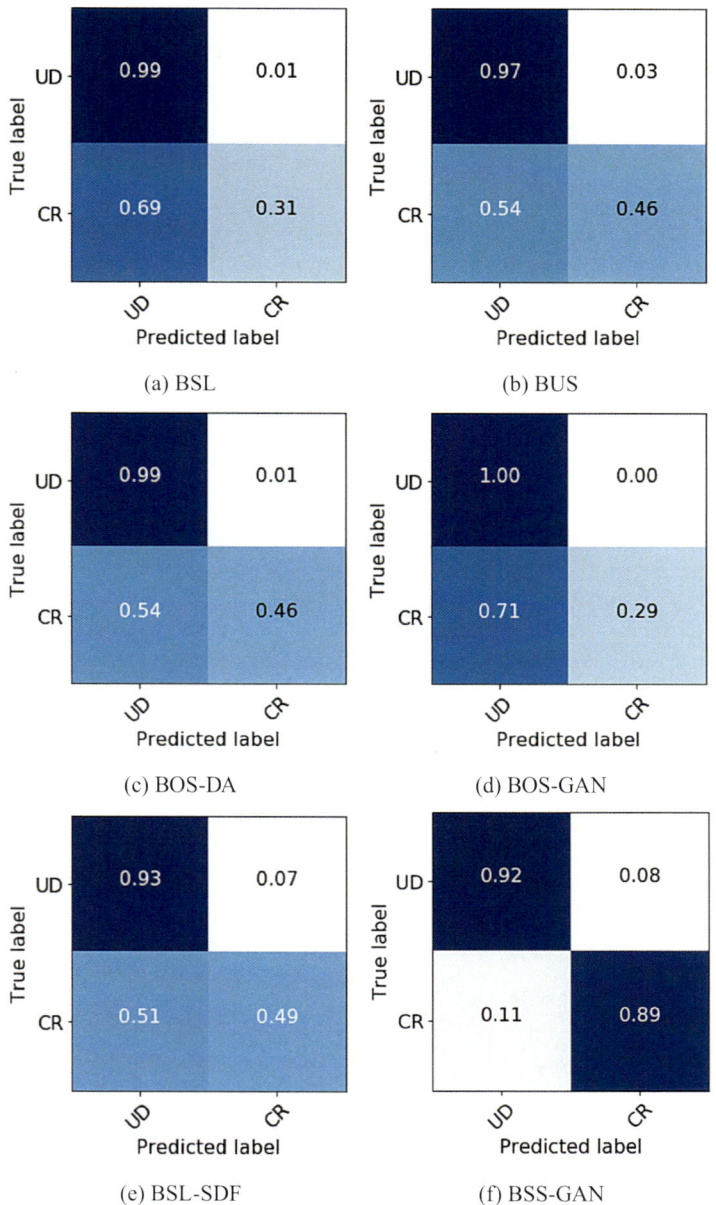

Fig. 10.3 Normalized CMs for the crack detection

Starting from the low TPR of the BSL, it can be inferred that a shallow DL model can easily become biased due to extreme class imbalance (16:1). The BUS's under-sampling worked to some degree, as it improved the TPR from 31% to 46% without significantly compromising the TNR. Similarly, BOS-DA's over-sampling helped increase the TPR to 45.7% without a drop in the TNR, yet its TPR (along with that of the BUS) is still unsatisfactory. BOS-GAN had the worst TPR (lower than the BSL), which conforms with the observations in [17] of the low performance of directly mixing synthetic data into the pipeline. BOS-GAN is extremely biased, and it is more prone to mis-predicting data as UD, causing a meaninglessly high TNR. Three factors lead to this poor and biased behavior: (i) the risk of introducing extra parameters, which may exacerbate over-fitting as mentioned in Sect. 9.6, (ii) manually selecting augmented images is subjective, and (iii) some GAN-generated images may be "adversarial" images [20], i.e., intended for the sole purpose of confusing the neural network, resulting in the misclassification of a given input. For factor (iii), although the GAN-generated images might be realistic-looking to human eyes, small feature perturbations undetectable by humans within these images might cause the classifier to make false predictions. Lastly, in the case of the BSL-SDF pipeline, it obtained a comparable performance to BUS and BOS-DA with a small sacrifice in TNR to make up for the $49 - 46 = 3\%$ enhancement in TPR. BSL-SDF is still biased, for which we can infer that even though SDF improved the model initialization, it did not help solve or even alleviate the class imbalance issue.

In general, the five pipelines above are unsatisfactory for crack detection under a UD:CR = 16:1 class ratio. These five pipelines have TPR below 50% and misleadingly high TNR, implying severe biases toward the UD class. Accordingly, these pipelines can only detect less than 50% of all cracked structures or components, which is unacceptable in SHM practice. On the contrary, the BSS-GAN model not only maintained an equally good TNR (over 92%) as others, but its TPR was also substantially higher (over 89%), indicating a nearly unbiased performance. Moreover, BSS-GAN is efficient in training, and it has an advantage over other models. compared to BUS, BSS-GAN can utilize all accessible (labeled or unlabeled) data, which provide additional information. On the other hand, compared with BOS-DA, BOS-GAN & BSL-SDF, the DA process is hidden (i.e., online, as discussed in Sect. 10.2.4) within the architecture of the BSS-GAN and it is part of the training process, with no extra data storage or need for multi-step training used in SDF.

The F_2 & F_5 scores (i.e., $\beta = 2$ and 5, respectively, of the F_β score, Sect. 3.4.4) were computed to take the recall (i.e., TPR) and precision into account. As aforementioned, the selection of β usually depends on its real-world interpretation. To avoid loss of generality, F_β scores (Eq. 3.17) with varying β values are plotted in Fig. 10.4. It is observed that starting from $\beta = 1$ (weighting recall and precision equally), BSS-GAN and BSL-SDF have increasing trends while BSL, BOS-DA, and BOS-GAN show decreasing trends. In addition, the F_β values converge to recall scores (TPR) as β increases. In SHM, it is more crucial to reduce FN than FP, so a large β is preferred. From Fig. 10.4, as β becomes larger, F_β

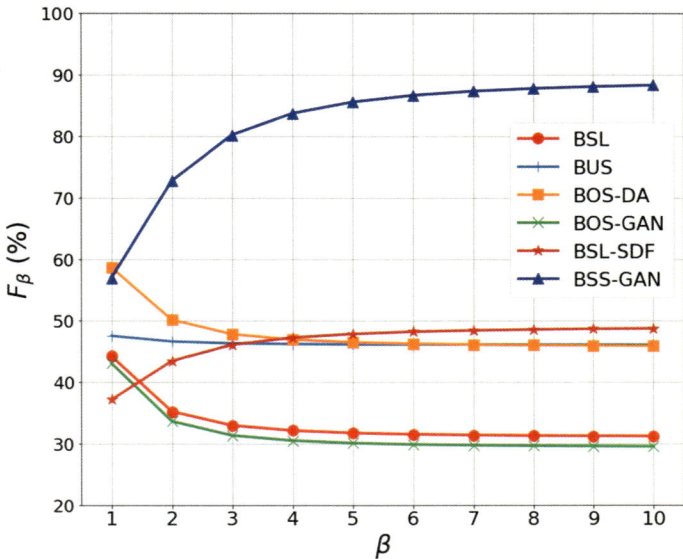

Fig. 10.4 F_β with varying β for the crack detection

from BSS-GAN exceeds those of other models with growing differences, suggesting its superiority in crack detection problems with high class imbalance.

10.4.2 Scenario 2: Spalling Detection

Even though the imbalanced class ratio (UD:SP = 32:1) in this scenario is twice that of the crack detection task, from Table 10.5 and Fig. 10.5, the performance of all pipelines surpasses that of scenario 1. This observation can be partially explained by different degrees

Table 10.5 Classification performance for the spalling detection (%)

Pipeline	TPR	TNR	F_2	F_5	Accuracy
BSL	62.0	99.8	66.1	62.8	98.6
BUS	84.7	97.1	73.2	82.2	96.7
BOS-DA	83.3	99.9	85.5	83.7	99.4
BOS-GAN	64.0	99.9	68.6	64.8	99.0
BSL-SDF	84.0	94.1	62.3	78.7	93.8
BSS-GAN	**98.0**	95.8	77.6	**93.3**	95.9

Bold # indicates a much better result

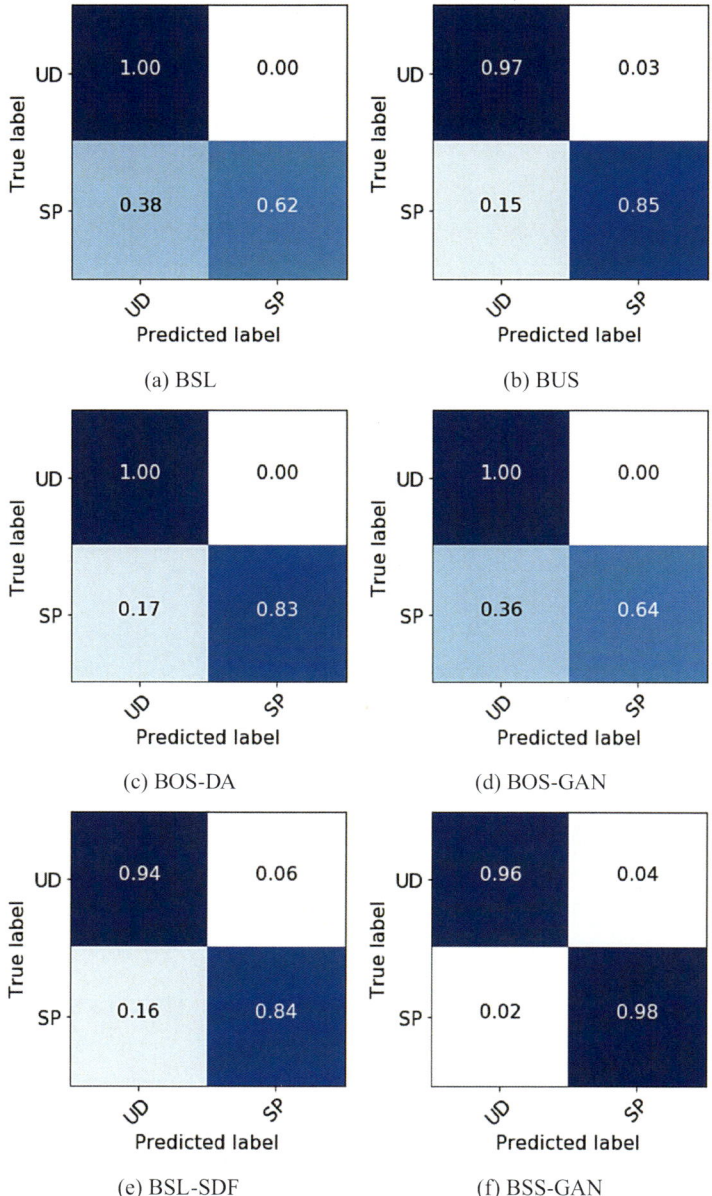

(a) BSL

(b) BUS

(c) BOS-DA

(d) BOS-GAN

(e) BSL-SDF

(f) BSS-GAN

Fig. 10.5 Normalized CMs for the spalling detection

of visual pattern similarity among UD, CR, and SP. CR images (Fig. 10.2b) are similar to UD images (Fig. 10.2a) except for the appearance of surface fissures or fine cracks. On the contrary, SP images (Fig. 10.2c) are more dissimilar, where the areas of spalling break the surface patterns in both color and texture, making the SP features more distinguishable. In this scenario, all pipelines obtained satisfactory TNR and over 60% TPR values. As in scenario 1, BSL and BOS-GAN had the lowest TPR, which again showed the ineffectiveness of directly mixing synthetic data into the pipeline. BUS, BOS-DA, and BSL-SDF achieved similar performance, with TPR values reaching nearly 84%. BSS-GAN achieved the highest TPR (consistent with its crack detection performance in scenario 1). It not only maintained a high TNR (95.8%) as with other pipelines, but also significantly improved the TPR from BSL (62.0%) to a surprising 98.0%, which is nearly 14% higher than those of BUS, BOS-DA, and BSL-SDF. Clearly, the BSS-GAN outperformed the other five pipelines.

In spalling detection, due to high class imbalance, a small decrease in TNR will over-emphasize the increase of FP, leading to a higher F_2 score. For example, a mere $99.9 - 95.8 = 4.1\%$ drop in TNR from BOS-DA to BSS-GAN makes F_2 of BOS-DA higher by $85.5 - 77.6 = 7.9\%$, although the TPR of BOS-DA is $98.0 - 83.3 = 14.7\%$ lower than that of BSS-GAN. Under the UD:SP = 32:1 class ratio, the F_2 score does not place enough emphasis on recall (i.e., TPR). Thus, a larger $\beta = 5$ is more informative. According to Fig. 10.6, both BSL and BOS-GAN share similar values and trends, while BUS, BOS-DA, and BSL-SDF converge to the same value after $\beta = 5$. As β increases, especially beyond $\beta = 5$, BSS-GAN significantly outperforms the other five pipelines.

Fig. 10.6 F_β with varying β for the spalling detection

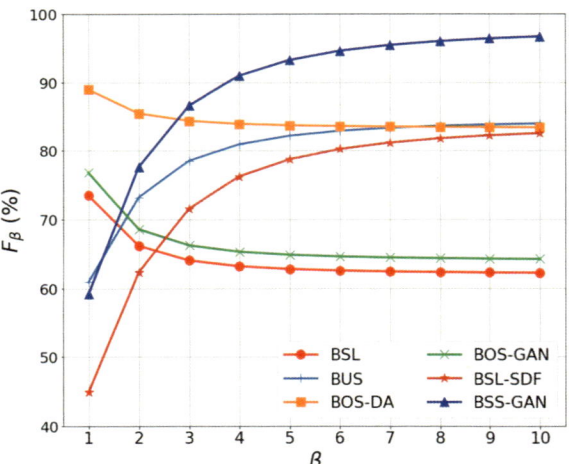

10.4.3 Scenario 3: Damage Pattern Recognition

In this scenario, a more complex multi-class classification was investigated, where the imbalanced ratio for UD:CR:SP is 32:2:1. The performance results are listed in Table 10.6 and along the diagonal cells of the CMs in Fig. 10.7. As with the two binary scenarios 1 & 2, the BSL was biased in favor of UD with low recall values (around 30%) for both minority classes CR & SP. BOS-GAN was the second worst pipeline in terms of class recall (around 31% & 61% for CR & SP, respectively). BUS, BOS-DA, and BSL-SDF did not perform well either, as their improvements in SP recall were merely sacrifices of the CR recall. For example, BOS-DA's SP recall reached 90%, but its CR recall remained low at 29%.

In general, the first five pipelines share a common drawback, namely, the bias against CR, as characterized by their low CR recall values. This issue is attributed to the high visual similarity between UD and CR. GAN-based DA worsens this issue by generating images with blended features between UD and CR. On the contrary, the BSS-GAN pipeline outperformed the other five pipelines with 91%, 70%, and 94% recalls for the respective UD, CR, and SP classes. Additionally, the BSS-GAN only misclassified 6% of the SP images as CR, and no SP images were misidentified as UD. Accordingly, the BSS-GAN is much more reliable in detecting severer damage.

10.4.4 Investigation of Failure Cases

For more generality, the failure cases of the ternary damage pattern recognition task are investigated. Selected examples of misclassified images are presented in Fig. 10.8.

From the CM in Fig. 10.7f and among the misclassified images, it is shown that the ratios of UD images misclassified as CR (5%) or SP (4%) and SP images misclassified as CR (6%) are small and comparable to each other, while the ratios of CR images misclassified as UD (17%) or SP (13%) are much larger. For misclassified UD images, wall texture and color contrast are the main contributors to the inaccuracies. For example, the BSS-GAN

Table 10.6 Classification performance for the damage pattern recognition (%)

Pipeline	Accuracy	Recall		
		UD	CR	SP
BSL	93.6	99.5	28.7	32.0
BUS	91.7	95.4	42.6	69.3
BOS-DA	95.2	99.3	29.0	90.0
BOS-GAN	88.3	92.7	31.3	60.7
BSL-SDF	89.6	93.4	35.0	73.3
BSS-GAN	89.7	90.9	**70.0**	**94.0**

Bold # indicates a much better result

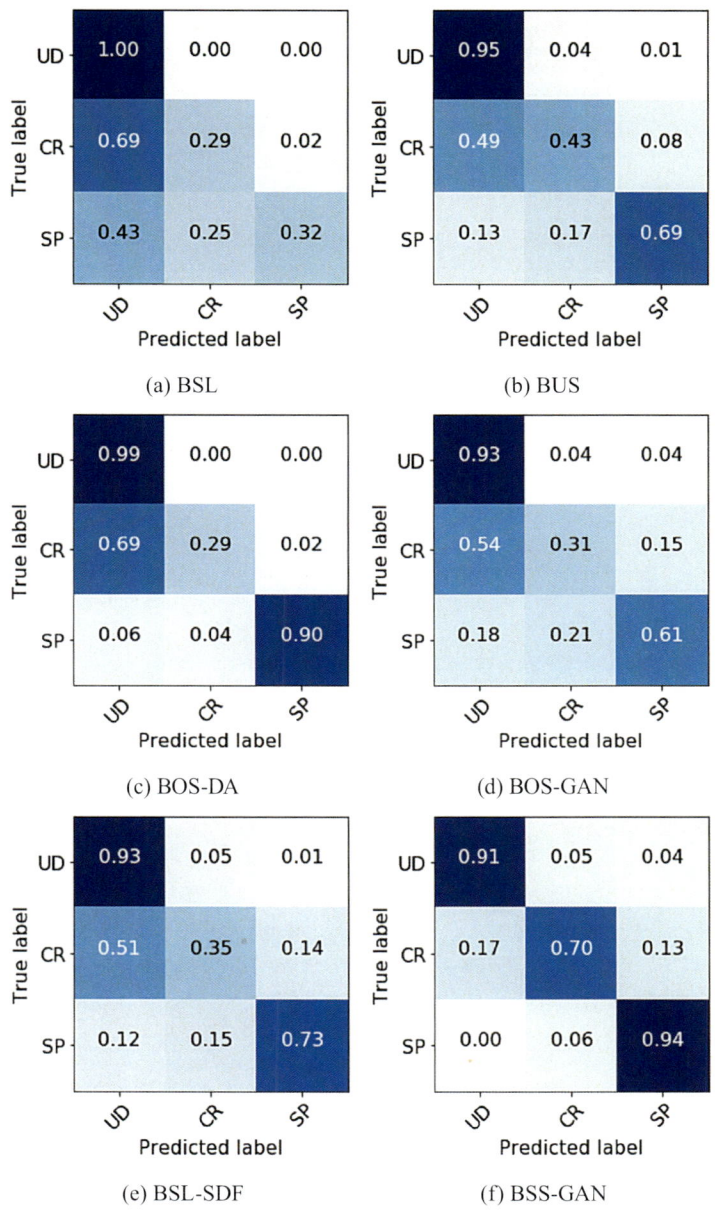

Fig. 10.7 Normalized CMs for the damage pattern recognition

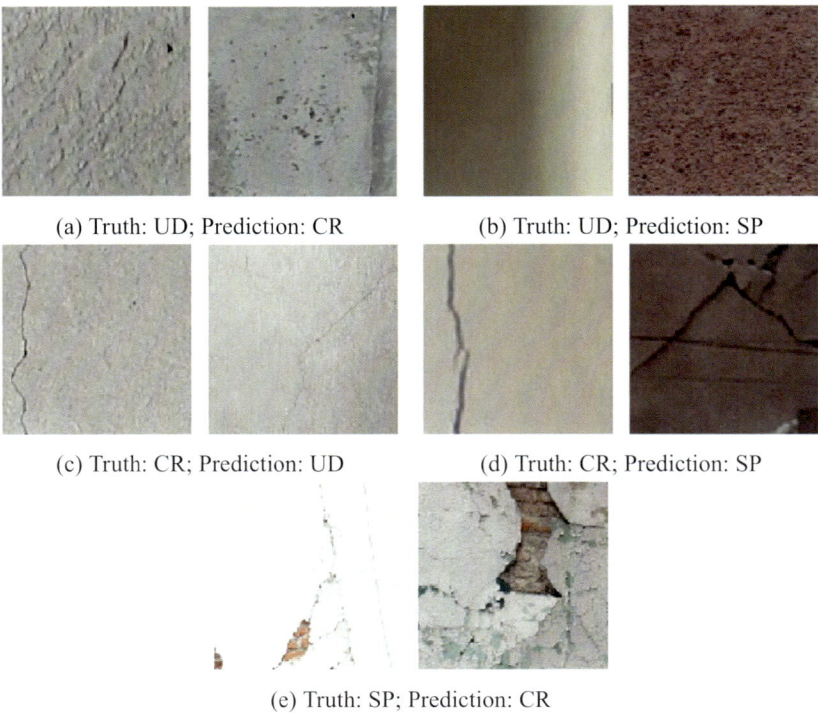

(a) Truth: UD; Prediction: CR   (b) Truth: UD; Prediction: SP

(c) Truth: CR; Prediction: UD   (d) Truth: CR; Prediction: SP

(e) Truth: SP; Prediction: CR

Fig. 10.8 Sample images of BSS-GAN's failure cases

model tends to misclassify rougher wall textures, Figs. 10.8a and 10.8b, as damaged (CR or SP) and color contrasts of large patches, Fig. 10.8b, as SP. For misclassified SP images, no images are classified as UD, but some images of mixed damage types with both CR & SP features (usually labeled as SP), Fig. 10.8e, are misjudged by the model as CR.

The CR images, containing features in between UD & SP, are more likely to be misclassified than UD or SP images, Figs. 10.8c and 10.8d. In other words, the CR samples lie at the intersection of UD & SP in the *feature manifold* and are particularly hard to be distinguished by the BSS-GAN model. Physical attributes such as crack sizes in terms of their widths directly affect the model's judgment on whether an image is more likely to be misclassified as UD or SP.

10.4.5 Investigation of Synthetic Image Quality

In this section, synthetic images generated from well-trained BSS-GAN models in the above scenarios are presented and, when possible, also compared with those generated by the

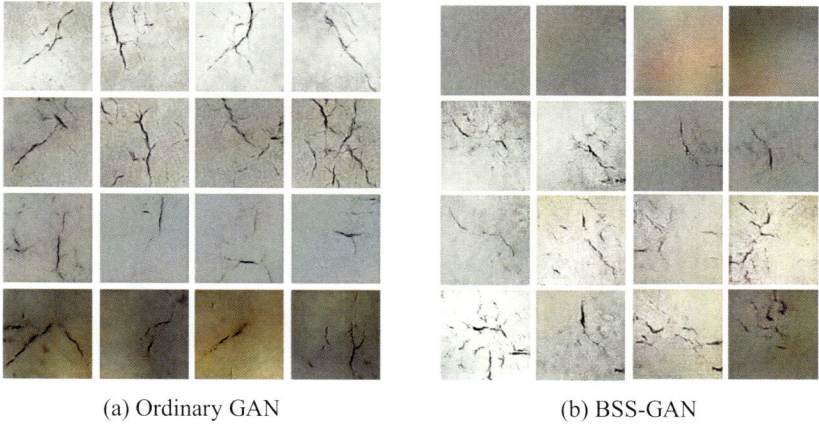

(a) Ordinary GAN (b) BSS-GAN

Fig. 10.9 Sample synthetic images for the crack detection

ordinary GAN used by BOS-GAN and BSL-SDF pipelines, Figs. 10.9 to 10.11. It is noted that the generator in the BSS-GAN was trained using mixed-class images instead of class-specific images (minority class) as in BOS-GAN and BSL-SDF. Therefore, the BSS-GAN learned a mixed distribution of UD, CR & SP, refer to Fig. 9.5. For example, for crack detection, BSS-GAN generated both synthetic UD and CR images, Fig. 10.9b.

Overall, there is no obvious mode collapse issue (i.e., when the generator only produces limited varieties of images, also refer to Sect. 9.5.3) in either the ordinary GAN or BSS-GAN. Besides basic visual features like textures and colors, the generator in both models can generate images with various more complex features, e.g., crack orientation, location & width, and spalling shape, location, and area. As mentioned in Chap. 9, structural images have complex and mixed distributions, which make it difficult for GAN to generate clear and class-discriminative[3] images. However, conditioning operation (i.e., considering class information related to a specific class of images) makes the ordinary GAN capable of generating higher-quality images toward that class. For example, when training the GANs of BOS-GAN and BSL-SDF pipelines for crack detection, only feeding minority-class (CR) images can be viewed as one type of conditioning operation, which significantly reduces the data distribution complexity. This cleaner and simpler sub-distribution boosts the learning process, and avoids the model being trapped in certain local saddle points leading to mode collapse. Thus, in Fig. 10.9a, the synthetic images show very realistic visual qualities along with variety. On the other hand, the generator in the BSS-GAN was trained with all images in an unsupervised manner. Thus, it had to learn a mixed distribution from both UD and CR. As a result, the synthetic images generated by BSS-GAN have features of UD, CR, or even

---

[3] Discriminative implies being supportive of the classification tasks to predict the class label of an input based on some features by determining the decision boundary between classes, instead of modeling the distribution of the data.

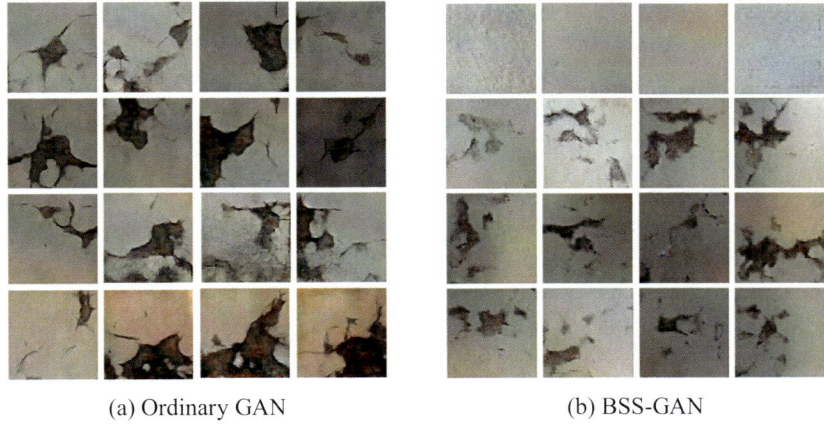

(a) Ordinary GAN   (b) BSS-GAN

Fig. 10.10 Sample synthetic images for the spalling detection

the intermediate (mixed) state. The ordinary GAN using all images without any conditioning operations was unable to generate reasonable structural images and occasionally led to mode collapse issues similar to the failure cases in Sect. 9.5.3 and other failure cases can be found in [17]. However, the balanced-batch sampling mechanism strictly enforces equal occurrences of each class within one training batch, which makes the model less biased to certain classes and thus reduces the mode collapse issue. To show this, in Fig. 10.9b, synthetic images in the first row are smooth and resemble UD, while the remaining images resemble CR, but are somewhat blurry, as expected, because of the learned mixed distribution (Fig. 9.5).

Another possible explanation for the differences in image quality is the used loss function. Unlike ordinary GAN, the loss function in BSS-GAN focuses more on classification than the quality of the generated images. This is reflected by the supervised cross-entropy loss (Eq. 10.3). On the contrary, the ordinary GAN only utilizes the unsupervised loss (Eq. 10.2), which is more about feature learning than classification. These different training objectives influence the performance of the generator even though the same network architecture was used. It is thus inferred that BSS-GAN trades off its generator performance for more improvement in the discriminator's classification capabilities.

In summary, the ordinary GAN with class conditioning is able to generate higher-quality images than BSS-GAN by human judgment under the 128×128 pixel resolution. However, according to [5, 17], sometimes the realistic-looking generated images may not align well with the correct features for the classification, e.g., an image is recognized as CR by the model, but it looks like UD by a human, such as the generated samples in the first row and first three columns in Fig. 10.11. Thus, if the synthetic image selection is performed through human interaction, i.e., manually selecting high-quality synthetic images by human visual inspection and assigning them with class labels to be subsequently mixed with a dataset of real images for training or pre-training purposes, using these synthetic data may worsen

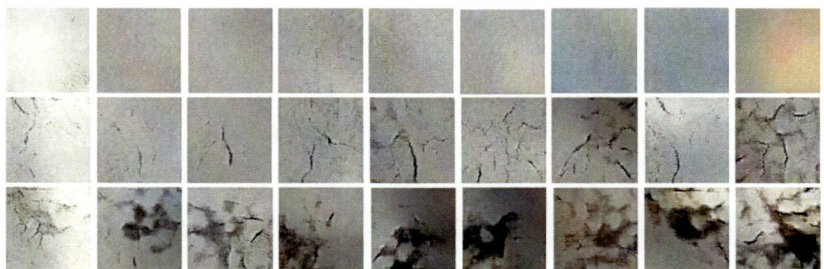

Fig. 10.11 Sample synthetic images for the damage pattern recognition generated by the BSS-GAN trained model

the classifier's performance as shown in BOS-GAN & BSL-SDF. On the contrary, for BSS-GAN, the steps from synthetic image generation to feature learning and classification are automatically and implicitly embedded in the training process, which is characterized by the game-theoretic (refer to Sect. 2.4) competition between D & G, and no human interaction is required. This important characteristic helps the BSS-GAN learn meaningful representations better and improves both training efficiency and discriminative performance, as supported by the results in Figs. 10.3f, 10.5f and 10.7f.

10.4.6 Impact of Unsupervised Feature Learning

In this section, both BSL and BSS-GAN were initially trained using only 2,040 labeled samples the training dataset of UD & CR classes only (20% of the training dataset of 10,200 images, refer to Table 10.1). More unlabeled data were progressively added to subsequent BSS-GAN trials (0% for BSS-GAN-0, 50% for BSS-GAN-50 & 100% for BSS-GAN-100 of the remaining 8,160 samples). The results of the four cases are listed in Table 10.7 and along the diagonal cells of the CMs in Fig. 10.12.

According to the results in Table 10.7 and Fig. 10.12, initially given 20% of training data with imbalanced classes, BSL was biased with a TPR of only 29.0%. Although the TPR of BSS-GAN-0 dropped to 60.7% compared to using a fully labeled dataset, i.e., 89.3%

Table 10.7 Classification performance in the study of unlabeled data utilization (%)

Pipeline	Unlabeled data	TPR	TNR	F_5	Accuracy
BSL	–	29.0	99.2	29.7	95.1
BSS-GAN-0	–	60.7	95.0	59.7	93.0
BSS-GAN-50	4,080	65.0	93.2	63.2	91.5
BSS-GAN-100	8,160	**72.3**	92.8	**70.0**	91.6

Bold # indicates a much better result

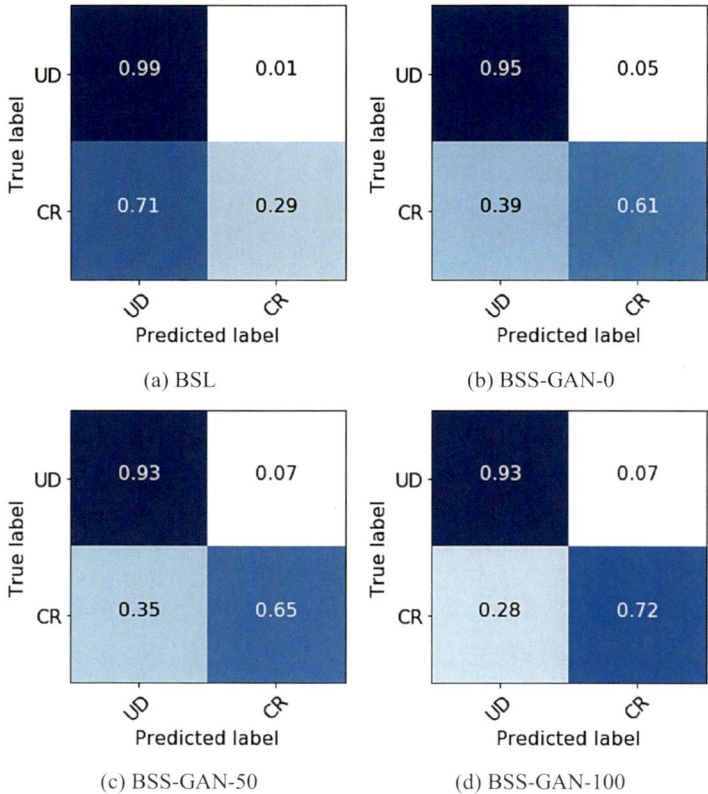

Fig. 10.12 Normalized CMs of four cases under a reduced-scale training labeled dataset with additional unlabeled data

from Table 10.4, it was still far less biased than BSL in this case of using reduced-scale training labeled dataset. Moreover, the BSL cannot improve beyond this point, as it can only learn from the labeled data. However, as we introduced more unlabeled data to BSS-GAN (Figs. 10.12c and 10.12d), the TPR of BSS-GAN improved by $65.0 - 60.7 = 4.3\%$ and $72.3 - 60.7 = 11.6\%$ by supplementing 50% and 100% of the unlabeled data, respectively, compared to BSS-GAN-0. Although the supplementary data do not provide label information, under the SSL setting, BSS-GAN can still utilize information from the unlabeled samples. During balanced-batch sampling, the number of unlabeled data fed to each batch stays consistent with that of a single-class sub-batch, i.e., $n^{ul} = n^l / K$, refer to Sect. 10.2 and Eq. 10.11 where $c = 1$. As a result, even as more unlabeled data are introduced, they do not overwhelm the labeled samples during batch-by-batch training. Thus, the unlabeled data, once handled appropriately in each training batch, are able to supplement the learned features and improve the classifier's performance.

One seeming caveat observed from the results in Table 10.7 and Fig. 10.12 is the decreasing TNR of the BSS-GAN as more unlabeled data are supplemented. However, the upward trend of the F_5 score suggests that the BSS-GAN models trained with more unlabeled data are better, which is based on our interest where high recall is prioritized over precision. It shows once more that the overall accuracy is a deceptive measure as the BSL achieved a 95.1% accuracy compared with BSS-GAN-100, which achieved 91.6% (with 8,160 unlabeled samples), yet the BSL's performance is the worst overall in terms of the more important metrics of TPR and F_5 score, especially for SHM applications.

10.4.7 Concluding Remarks

The following key conclusions are drawn from the case study:

✓ In general, BSS-GAN outperformed conventional methods in both binary crack and spalling detection under low-data and imbalanced-class settings. It achieved a significant improvement in TPR by reducing FN with only a slight decrease in TNR. BSS-GAN achieved better F_β, e.g., F_5, scores with more weights on recall (i.e., TPR) over precision.

✓ Over-sampling the minority class by GAN-generated images (BOS-GAN) led to worse performance than the baseline (BSL) in spalling detection tasks. This is caused by: (1) the introduction of extra parameters, (2) the subjective manual synthetic image selection, and (3) the generation of "adversarial" images. These factors caused unstable training behavior and exacerbated BOS-GAN's bias in favor of the majority class (UD).

✓ BUS, BOS-DA & BSL-SDF had similar but limited improvements over BSL, which are not satisfactory in practice. Their flaws include: in BUS, under-sampling eliminated a large portion of the labeled majority-class data, causing information loss, in BOS-DA, the conventional DA failed to increase feature variety, and in BSL-SDF, the model did not sufficiently address the imbalanced-class issue although it improved the parameter initialization.

✓ In the three-class classification, all pipelines except BSS-GAN were prone to predicting CR as UD (leading to low CR recalls). On the contrary, BSS-GAN obtained a promising CR recall of about 70%, while maintaining a high SP recall of 94% and a good UD recall of 91%. This further indicates the stable and high potential of the BSS-GAN for handling imbalanced multi-class tasks.

✓ BSS-GAN generated images of all classes without mode collapse, because it learned from a mixed-class distribution with balanced-batch sampling. If only considering the generated image quality by human visual judgment, the ordinary GAN generator used in BOS-GAN and BSL-SDF was slightly better. It is inferred that the improvement of the BSS-GAN's discriminator weakened its generator, but the generator was yet able to generate realistic images for the classifier to accurately learn new features from these images.

✓ When labeled data have limited availability, the semi-supervised setting of the BSS-GAN allows it to utilize unlabeled data. With a proper ratio of unlabeled data placed into each training batch, BSS-GAN is able to capture meaningful information from these unlabeled data.

10.5 Exercises

1. List and discuss at least 3 scenarios having imbalanced-class label issues in practical SHM or Civil Engineering projects.
2. Explain why the overall accuracy (Eq. 3.12) may not be an appropriate evaluation metric in an imbalanced-class label classification problem.
3. Conduct literature review and comparisons of at least 3 classical SSL methods and discuss their algorithms.

References

1. X.J. Zhu, Semi-supervised learning literature survey (2005)
2. J.E. Van Engelen, H.H. Hoos, A survey on semi-supervised learning. Mach. Learn. **109**(2), 373–440 (2020)
3. A. Madani et al., Chest x-ray generation and data augmentation for cardiovascular abnormality classification, in *Medical Imaging 2018: Image Processing*, vol. 10574 (International Society for Optics and Photonics, 2018), p. 105741M
4. A. Madani, Semi-supervised learning with generative adversarial networks for chest x-ray classification with ability of data domain adaptation, in *IEEE 15th International Symposium on Biomedical Imaging (ISBI 2018)*, vol. 2018 (IEEE, 2018), pp. 1038–1042
5. T. Salimans et al., Improved techniques for training gans, in *Advances in Neural Information Processing Systems* (2016), pp. 2234–2242
6. I. Goodfellow et al., Generative adversarial nets, in *Advances in Neural Information Processing Systems* (2014), pp. 2672–2680
7. I. Goodfellow, NIPS 2016 tutorial: generative adversarial networks (2016). arXiv:1701.00160
8. M. Arjovsky, S. Chintala, L. Bottou, Wasserstein gan (2017). arXiv:1701.07875
9. G. Mariani et al., Bagan: data augmentation with balancing gan (2018). arXiv:1803.09655
10. I. Goodfellow, Y. Bengio, A. Courville, *Deep Learning* (MIT Press, Cambridge, 2016)
11. Y. Gao, K.M. Mosalam, PEER Hub ImageNet: a large-scale multiattribute benchmark data set of structural images. J. Struct. Eng. **146**(10), 04020198 (2020)
12. S. Dorafshan, R.J. Thomas, M. Maguire, Comparison of deep convolutional neural networks and edge detectors for image-based crack detection in concrete. Constr. Build. Mater. **186**, 1031–1045 (2018)
13. S. Gao, V. Gruev, Bilinear and bicubic interpolation methods for division of focal plane polarimeters. Opt. Express **19**(27), 26161–26173 (2011)
14. A. Radford, L. Metz, S. Chintala, Unsupervised representation learning with deep convolutional generative adversarial networks (2015). arXiv:1511.06434

15. A.L. Maas, A.Y. Hannun, A.Y. Ng, Rectifier nonlinearities improve neural network acoustic models. In: Proc. icml. Vol. 30. 1. 2013, p. 3
16. S. Ioffe, C. Szegedy, Batch normalization: accelerating deep network training by reducing internal covariate shift (2015). arXiv:1502.03167
17. Y. Gao, B. Kong, K.M. Mosalam, Deep leaf-bootstrapping generative adversarial network for structural image data augmentation. Comput. Aided Civ. Infrastruct. Eng. **34**(9), 755–773 (2019)
18. N. Srivastava et al., Dropout: a simple way to prevent neural networks from overfitting. J. Mach. Learn. Res. **15**(1), 1929–1958 (2014)
19. D.P. Kingma, J. Ba, Adam: a method for stochastic optimization (2014). arXiv:1412.6980
20. I.J. Goodfellow, J. Shlens, C. Szegedy, Explaining and harnessing adversarial examples (2014). arXiv:1412.6572

Active Learning

Manually labeling data is time-consuming. Therefore, it is usually difficult to obtain a large amount of labeled data compared to having unlabeled data. If the SSL method is employed, both labeled and unlabeled data can be utilized simultaneously, which can improve the AI model performance to some extent, as demonstrated with the BSS-GAN in Chap. 10.

In this Chapter, another technique, namely, AL, is introduced to complement the SSL. While applying AL in classification problems, a group of unlabeled data that are recognized by the classifier with a low confidence level (i.e., low probability of predicting a specific class) is first queried. Subsequently, these data are labeled by human annotators and then fed into the original classifier for performance improvement. In addition, autoencoder (AE)-type network, e.g., stacked convolutional AE (SCAE) [1], is known as a good feature extractor in unlabeled data, which helps the model better utilize the unlabeled data [2]. Ultimately, SSL, AL, and SCAE are fused into a new training paradigm, namely, SCAE-based AL, which is further validated through a case study on RC bridge crack identification.

11.1 Active Learning Mechanism

The core idea of AL revolves around leveraging an existing trained model's ability to generate pseudo-labels for new, unlabeled data. These pseudo-labels are deemed "pseudo" because they are not ground truth labels provided by human annotators; instead, they are inferred from the model's predictions. Subsequently, these new data-label pairs are added to the training dataset for the next round of training. If the confidence levels (i.e., probabilities of the predictions generated by the model) for certain images are low, it is inferred that it is hard to recognize the real conditions and distinguish the classes by this model. In such

© The Author(s), under exclusive license to Springer Nature Switzerland AG 2024   307
K. M. Mosalam and Y. Gao, *Artificial Intelligence in Vision-Based Structural Health Monitoring*, Synthesis Lectures on Mechanical Engineering,
https://doi.org/10.1007/978-3-031-52407-3_11

instances, those images should be subjected to manual labeling. Before going into details of AL, self-training (Self-T) is introduced first.

11.1.1 Self-training

Self-T [3] is one early realization of AL. It does not rely on complex assumptions and its algorithm structure is straightforward. For a K-class classification problem, it is supposed that there are N^l pairs of labeled data $\{(X^l, Y^l)\}$, and N^{ul} *initially* unlabeled data $\{X^{ul}\}$, where X represents the input data or the data features in a more general way, and Y represents the data labels. An *initial* training set T is formed with all labeled data, $T = \{(X^l, Y^l)\}$. The related algorithm (Fig. 11.1) is as follows:

1. Train a classifier with all labeled data in T.
2. Use the trained classifier to predict the unlabeled dataset $\{X^{ul}\}$, and generate the K-dimension output, $\{O^{ul}\}$.
3. Select N_{hc} images with high confidence, $\{X^{hc}\}$, from $\{X^{ul}\}$, and assign them the pseudo-label, $\{O^{hc}\}$, based on their outputs, $\{O^{ul}\}$, from step 2.
4. Remove $\{X^{hc}\}$ from the unlabeled set, $\{X^{ul}\}$, and add them to the labeled data, $\{(X^l, Y^l)\}$, to *update* T as follows:

$$\{X^{ul}\} \leftarrow \{X^{ul}\} - \{X^{hc}\}. \tag{11.1}$$

$$T \leftarrow \{(X^l, Y^l)\} \bigcup \{(X^{hc}, O^{hc})\}. \tag{11.2}$$

5. Retrain the classifier with the updated training set T.
6. Repeat steps 2 to 5 until all unlabeled data are labeled.

It is noted that in step 3, a threshold, $1.0 \le \epsilon_{cl} \le 0.0$, for filtering data with confidence level is predefined. If the probability of predicting a specific class is higher than ϵ_{cl}, the input data is defined as having high confidence, and vice versa. The value of ϵ_{cl} is defined by the user based on demands, e.g., the higher the value of ϵ_{cl}, the stricter the criterion for data filtering, and the more manual labeling work is expected.

11.1.2 AL

Although Self-T can somewhat improve the baseline performance utilizing new unlabeled data, the improvement may be limited because the pre-trained model has limited recognition ability leading to labeling errors [4], i.e., the generated pseudo-label with high confidence can be incorrect. These errors make the model unable to update its parameters, and may even

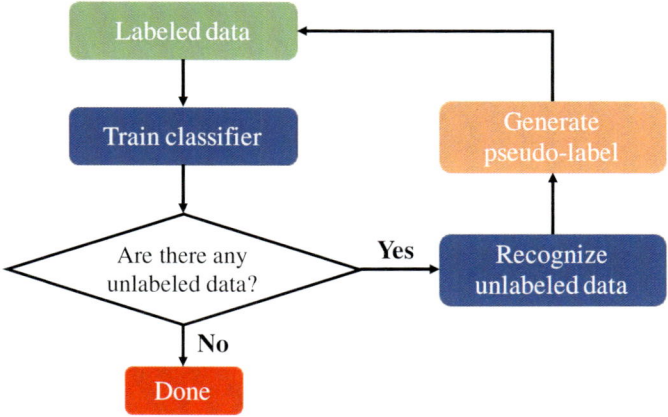

Fig. 11.1 Procedure of Self-T

lead to over-fitting, and ultimately deteriorating the model performance. Therefore, if manual labeling is conducted on those hard-to-distinguish images, the model can correct the wrong parameters and improve its recognition performance. AL plays such a role. For example, in the crack detection problem, if the confidence levels of predicting "non-cracked" (none of the cracks occur on the surface of the structural component) and "cracked" (cracks occur on the surface of the structural component) are close for images recognized by the model, it is inferred that it is hard to recognize the real conditions and automatically and autonomously distinguish the classes by this model. Under such circumstances, those images should be subjected to manual labeling.

Based on the defined K-class classification in Sect. 11.1.1, the procedure of AL (Fig. 11.2) is as follows:

1. Train a classifier with all labeled data in T.
2. Randomly separate the *initially* unlabeled dataset, $\{X^{ul}\}$, to M batches, i.e., $\{X^{ul}\}_1$, $\{X^{ul}\}_2, \ldots, \{X^{ul}\}_M$.
3. Randomly select one batch $\{X^{ul}\}_m$, $m \in \{1, 2, \ldots, M\}$, use the trained classifier to predict the labels for all unlabeled data in this m-th batch, and then select the N_d most difficult images, $\{X^{rl}\}$, in this batch with low confidence levels for manual labeling. Obtain the real labels of these N_d images manually, and form the data-label pairs as $\{(X^{rl}, Y^{rl})\}$.
4. To increase the number of $\{(X^{rl}, Y^{rl})\}$ pairs, offline DA (refer to Sect. 6.1) can be conducted to generate more data-label pairs, resulting in a larger data-label subset, denoted $\{(\hat{X}^{rl}, \hat{Y}^{rl})\}$.
5. Remove the selected batch, $\{X^{ul}\}_m$, from the unlabeled dataset, $\{X^{ul}\}$, and add $\{(\hat{X}^{rl}, \hat{Y}^{rl})\}$ to the labeled data, $\{(X^l, Y^l)\}$, to *update* T as follows:

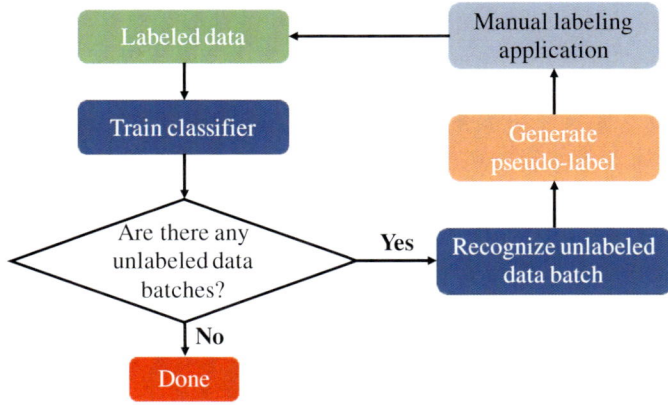

Fig. 11.2 Procedure of AL

$$\left\{X^{ul}\right\} \leftarrow \left\{X^{ul}\right\} - \left\{X^{ul}\right\}_m . \tag{11.3}$$

$$T \leftarrow \left\{(X^l, Y^l)\right\} \bigcup \left\{\left(\hat{X}^{rl}, \hat{Y}^{rl}\right)\right\}. \tag{11.4}$$

6. Re-train the classifier with the updated dataset T.
7. Repeat steps 3 to 6 until all M unlabeled batches are removed from $\left\{X^{ul}\right\}$.

11.2   SCAE-Based AL

11.2.1 Stacked Convolutional Autoencoder

In order to further utilize informative features in the data while conducting AL, SCAE is adopted into the AL loop. The SCAE is one type of convolutional AE (CAE) structure, where several CAEs can be stacked to form a deep hierarchy [1]. As illustrated in Fig. 11.3, a SCAE consists of several layers of CAE, where the output of each layer is connected to the input of the successive layer. Through *nonlinear dimensional reduction*, high-dimensional data are mapped to the low-dimensional space, and thus the hidden features are extracted. No labeling information is required, so SCAE can utilize abundant unlabeled data to obtain a good feature extraction ability. Subsequently, through *feature fusion*, the features extracted by SCAE from the data without considering labeling information are combined with those extracted by the original classifier from the labeled data.

Suppose the input image, I, has a shape of $(H \times W \times C)$, where H, W, and C are its height, width, and number of color channels, respectively. The first AE layer, $AE^{(1)}$, generates the feature F_1 with the shape (H_1, W_1, C_1) and the output of the corresponding DE layer, $DE^{(1)}$, has the same shape of the input I. Subsequently, using F_1 as the input

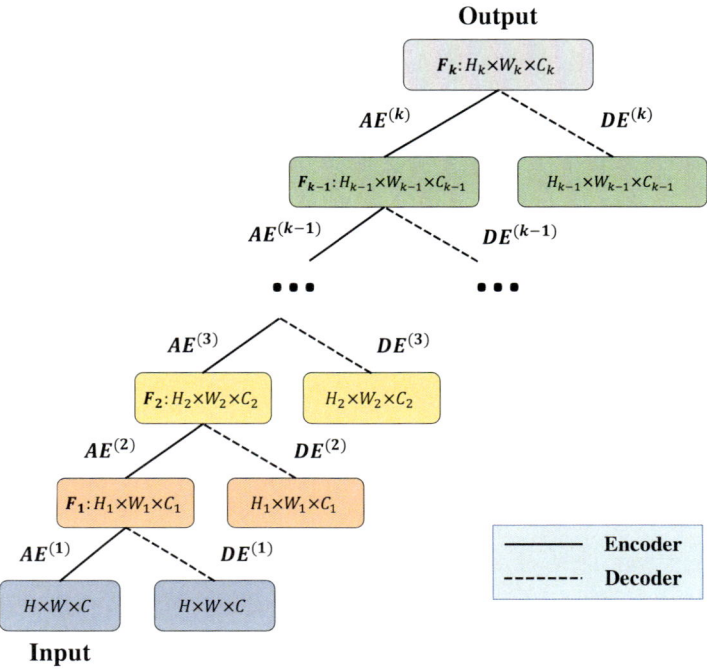

Fig. 11.3 Illustration of stacking structure of the SCAE

to the second AE layer, $AE^{(2)}$, the second layer feature, F_2, is extracted with the shape (H_2, W_2, C_2) and the second DE layer, $DE^{(2)}$, maps F_2 back to have the shape of F_1. Keep stacking until the designated depth k, and the final feature F_k is extracted by $AE^{(k)}$ with the shape (H_k, W_k, C_k). It is noted that while designing the network structure of the SCAE, the final feature F_k should share the same dimension with the features F_{clf} extracted by the original classifier (step 1 in the AL procedure), which is expected to be a NN-like model.

11.2.2 Feature Fusion with AL

One of the basic assumptions of *feature fusion* is that different feature extraction methods have different representative abilities. Through a parallel combination of multiple features from different sources, a relatively comprehensive feature group can be obtained, where features extracted by different methods can complement each other. Therefore, after feature fusion, the performance of the model is expected to be improved. In this chapter, SCAE with a feature fusion technique is applied along with AL. Figure 11.4 illustrates that the updated labeled dataset T (step 5 in the AL procedure) is used for feature fusion of the features, F_{clf}, which are extracted by the original classifier using labeling information and features, F_k, which are, on the other hand, extracted by the SCAE without labeling information,

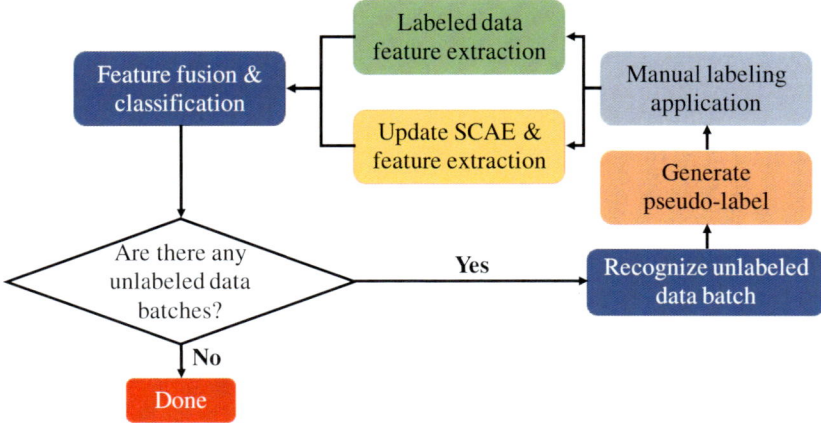

Fig. 11.4 Procedure of SCAE-based AL

i.e., using SSL. Both F_{clf} and F_k are concatenated to the shape $(H_k, W_k, 2C_k)$ firstly, and then GAP (refer to Sect. 4.3.2 and references [5, 6]) is applied to reduce its dimension to a $(1 \times 1 \times C_k)$ fusion feature vector, F_{fu}. Subsequently, F_{fu} is fully connected to the output layer with the number of neurons that equals the number of total classes. The procedure is summarized as follows (note that steps 1 to 5 are identical to those in the procedure shown in Sect. 11.1.2):

6. Update SCAE using unlabeled data $\{X^{ul}\}_m$, and extract the features F_k from the updated labeled dataset, T.
7. Re-train the classifier with the updated T, and extract the features F_{clf}.
8. Conduct feature fusion for F_k and F_{clf} to generate the fusion feature F_{fu}.
9. Train the single FC-layer with the input F_{fu}.
10. Repeat steps 3 to 9 until all unlabeled batches are removed from $\{X^{ul}\}$.

11.3   Case Study: RC Bridge Crack Identification

11.3.1 Background

Based on statistical data, by the end of 2020, there were more than 618,000 bridges in the United States, where nearly 36% of the bridges need repair work, and 7.3% of them are considered structurally deficient [7]. Similarly, there are 912,800 highway bridges in China, including 6,444 long-span bridges and 119,935 short-span bridges. Moreover, the total mileage of the highway in China is about 5.19 million km, of which the maintenance mileage of the highway is 5.14 million km, accounting for 99.0% of the total mileage [8]. The remaining 1% is for newly constructed highways with no need for immediate maintenance. Therefore, surging demands in the maintenance work of highway bridges in the United States, China, and many other countries are worldwide realities and it is a necessity to develop effective methods to evaluate the service functionalities of bridge structures. Because of cost-efficiency and easy-forming, RC is used for most bridge structures worldwide. However, damage to RC is commonly initiated on its surface in the form of cracks. The appearance of cracks not only affects the bearing capacity of the structure, but also leads to possible corrosion of the internal reinforcement, which greatly reduces the service life of the structure. Therefore, accurate identification and measurement of these cracks are of great importance during bridge health monitoring (BHM) process.

As mentioned above, for the practical scenarios, there are less labeled data annotated by human experts and more unlabeled data, which are only images or videos without annotations. Therefore, in this case study, AL techniques are applied along with SSL for a bridge crack identification (classification) problem.

11.3.2 Convolutional AL Framework

To close the gap between vision-based technologies and their applications in real-world projects, the integrated multi-step framework, namely, convolutional AL (CAL for short) is proposed herein as illustrated in Fig. 11.5. Based on findings from previous studies: (i) the training error and uncertainty can be reduced by adopting a TL technique, i.e., re-training a benchmark model through tuning its parameters has a better and more stable performance than training directly from scratch and (ii) the usage of the unlabeled images can reduce the labeling efforts and improve the model generalization ability. The working principle of the CAL framework is organized into two major steps: (1) pre-training a benchmark classification model and (2) re-training a semi-supervised AL model.

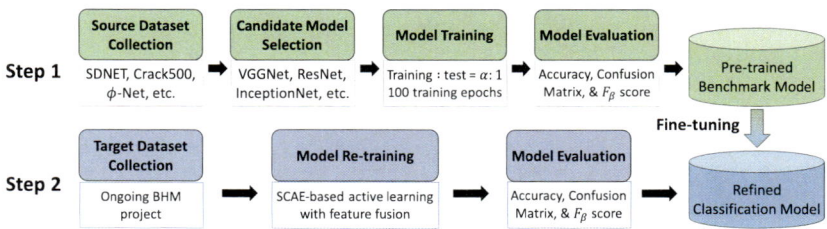

Fig. 11.5 CAL framework

11.3.2.1 Step 1: Pre-training a Benchmark Classification Model

The objective of this step is to determine a pre-trained benchmark model, which can be used in a general crack identification (classification) problem and then can be easily adjusted based on newly collected data. Firstly, a pre-training dataset is built, where data can be collected from several open-source structural and crack datasets, e.g., SDNET [9, 10], Crack500 [11], and ϕ-Net [12] (Sect. 5.4). Then, the data are split into training and test sets with a ratio of Training : Test $= \alpha : 1$ for model performance evaluation. A variety of current mainstream networks, e.g., VGGNet [13], ResNet [6], and Inception-ResNet [14], are selected as candidates for the benchmark model. Subsequently, these models are trained and tested based on the pre-training data and their performance is evaluated through multiple metrics, i.e., accuracy, CM, and F_β score (Sect. 3.4).

The accuracy (Eq. 3.12) is defined by the number of correct predictions divided by the total number of predictions made for a dataset. However, in the practice of BHM, the occurrences of damage, e.g., "cracked" label, are far less than "non-cracked" label. Thus, the model having a high accuracy may predict all the inputs as "non-cracked" resulting in a biased model, with misleading high accuracy. Thus, CM is also adopted herein. Since this case study is a binary classification problem, denote "cracked" as a positive class and "non-cracked" as a negative class. Accordingly, the CM has four possible outcomes: TP, TN, FP, and FN. For example, TP represents the scenario where the model predicts the input as the positive class "cracked", and the prediction is indeed consistent with the ground truth "cracked". Based on these outcomes, *recall* (i.e., TPR), Eq. 3.14, TNR, Eq. 3.15, and *precision*, Eq. 3.16, are computed. It is noted that recall is the proportion of images with cracks that can be recognized by the model, which reflects how comprehensive the recognition is, TPR & TNR evaluate the accuracy of detecting TP & TN outcomes, respectively, and precision is the proportion of images with cracks in the predicted images, which reflects the accuracy of recognition. In order to take into account both recall and precision of the model simultaneously, F_β score (Eq. 3.17) is also adopted. Finally, according to this comprehensive model performance evaluation, the relatively optimal model with the highest F_β score or least number of parameters, if F_β scores are close (within 1%), is selected as the benchmark model.

11.3.2.2 Step 2: Re-Training a Semi-supervised AL Model

In this step, the TL and domain adaptation techniques are adopted to transfer information and knowledge from the source domain to the target domain [15, 16]. Specifically, the benchmark model pre-trained on the source dataset in step 1 is treated as a basis for learning general damage due to cracking. This model is further re-trained on a custom target dataset to identify the damage on the specific bridge structure and components, where data are acquired from the bridge structure for the BHM project at hand or from a similar scenario to this specific project.

Due to costly manual labeling, the custom target dataset usually consists of two parts: (i) a small portion of labeled data and (ii) a larger portion of unlabeled data. The traditional supervised learning methods only work on the labeled data, neglecting the abundant information of the unlabeled data. However, herein, SCAE-based AL with feature fusion is used, which is a SSL method to utilize information from both labeled and unlabeled data simultaneously. It combines the features from the labeled data through the original classifier and from the unlabeled data through SCAE, where the ratio labeled : unlabeled data is set as $\gamma : 1$. Through the AL mechanism, a small portion of the unlabeled data, initially recognized by the classifier with less confidence, is manually labeled and added to the labeled dataset for improving the classifier's performance. In step 2, the model evaluation metrics are the same as in step 1, i.e., accuracy, CM, and F_β score. Finally, a well-tuned recognition model based on the custom dataset is obtained and directly used for the crack identification task as the intended target of the BHM project at hand. The "cracked" images recognized by this model can be further investigated for crack localization and segmentation models, as introduced in Chaps. 7 and 8.

11.3.3 Bridge Crack Dataset

In this case study, two datasets are prepared, i.e., source dataset for benchmark model pre-training, and custom target dataset for model fine-tuning and adaptation. These two datasets are described in the following two sections.

11.3.3.1 Source Dataset

The images of the source dataset were selected from four open-source datasets: Concrete Crack Images for Classification (CCIC) [17], Crack500 [11], SDNET2018 (for bridge decks, highway pavements, and walls) [10], and Concrete Crack Detection (CCD) [18]. All data have been labeled with "one-hot vector encoding", refer to Sect. 3.7.1, i.e., [1, 0] for non-cracked, and [0, 1] for cracked. A few sample images are shown in Fig. 11.6.

A total of 3,000 labeled images was selected for the source dataset, where the ratio between cracked (CR) and non-cracked (NCR) labels was 1 : 1. Subsequently, all images were re-scaled to $256 \times 256 \times 3$, and split into training and test sets with a 5 : 1 ratio (i.e., $\alpha = 5$). This led to 2,500 labeled images used to train the benchmark model and the remaining 500 labeled images used for testing its performance.

11.3.3.2 Target Dataset

In step 2, a new dataset similar to the target domain, i.e., bridge damage, was collected and partially labeled. To construct this target dataset (Fig. 11.7), a total of 175 images with a high resolution of $6{,}000 \times 4{,}000$ were collected. This dataset mainly consisted of laboratory concrete components and university campus bridges with both non-cracked (NCR) and

Dataset	CCIC	Crack500	SDNET2018			CCD
			Deck	Pavement	Wall	
Non-cracked [1,0]						
Cracked [0,1]						

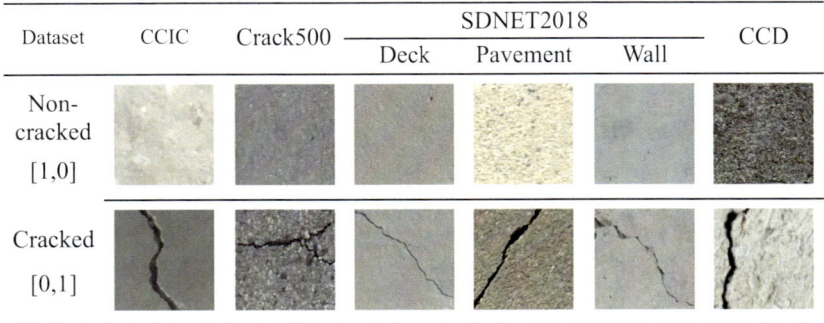

Fig. 11.6 Samples from the source dataset

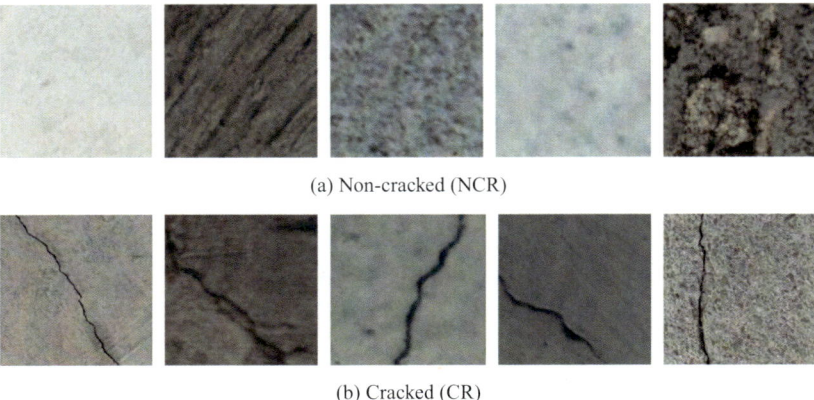

(a) Non-cracked (NCR)

(b) Cracked (CR)

Fig. 11.7 Samples from the target dataset

cracked (CR) classes. The image data acquisition equipment was Nikon D7100 camera, and all images were vertical surfaces of the objects. An offline DA was applied to all images to increase the dataset size. For a given image, a random 3,000 × 3,000 pixels region was cropped and then saved as a new image. This was repeated 5 times for each image in the dataset. Subsequently, all images were re-scaled to 256 × 256 pixels.

To validate the effectiveness of reducing the labeling efforts, no labeled training data were provided in advance and the pre-trained classifier started from step 2 of the AL procedure with all unlabeled data (i.e., $\gamma = 0$). Based on this setting, the target dataset was established, which consisted of an unlabeled training set, *Unl*, with 60,000 images, and a labeled test set, *L*, with 1,000 images, where 824 and 176 images were labeled as NCR and CR, respectively, i.e., *L* is moderately imbalanced with an NCR : CR ratio of about 1 : 4.68. Accordingly, the labeling operations were only conducted for fewer images during each training iteration.

11.3.4 Benchmark Model Pre-training

In step 1 of the CAL framework, multiple candidate DL structures were considered for more robust performance. A total of 7 network structures, i.e., VGG-11, VGG-16, ResNet-18, ResNet-26, Inception-ResNet-18, SE-ResNet-18, and SE-Inception-ResNet-18, were considered and compared. These 7 structures were variants of several widely-used network designs, and their original model details can be found in [6, 13, 14, 19]. Different from the original implementations, modifications were made to these models for: (i) adjusting the number of Conv layers and (ii) matching the dimension of the *softmax* layer to the number of classes. Twenty images were processed per batch as dictated by the available computational capacity, and a total of 100 epochs was applied, in which the optimization method to update the network parameters in backpropagation was the SGD (refer to Sect. 4.1.3). Through multiple runs for parameter tuning, the best performance was obtained for each candidate model, which reduced the uncertainty in the training process.

The performance of all candidates with 5 metrics (# of parameters, accuracy, precision, recall, and F_1 score, i.e., F_β, $\beta = 1$, which is selected due to a balanced 1 : 1 class ratio) are listed in Table 11.1. According to the results, using a deeper network structure did not necessarily improve the model recognition performance, e.g., compared to VGG-11 and ResNet-18, recall values of VGG-16 and ResNet-26 slightly decreased. This is mainly attributed to the use of a small-scale dataset for only two classes identified where the deeper network introduced more complexity and over-fitting issues than the shallow one. From the comprehensive model performance evaluation for all 7 models, SE-ResNet-18 had the least # of parameters (13 million) and the best accuracy with the highest F_1 score and the second highest precision and recall. Thus, SE-ResNet-18 was selected as the benchmark model for the subsequent step. The configuration of SE-ResNet-18 is listed in Table 11.2, where, e.g.,

Table 11.1 Pre-training recognition performance of 7 candidate models

Network	Parameters (#)	Accuracy (%)	Precision (%)	Recall (%)	F_5 score
VGG-11	160M	87.6	86.7	88.8	0.877
VGG-16	165M	92.2	96.1	88.0	0.919
ResNet-18	13M	94.0	95.8	92.0	0.939
ResNet-26	19M	94.2	96.6	91.6	0.940
Inception-ResNet-18	28M	94.8	95.9	93.6	0.947
SE-ResNet-18	13M	94.8	97.5	92.0	0.947
SE-Inception-ResNet-18	28M	94.8	97.9	91.6	0.946

Table 11.2 Network configuration of the benchmark model, SE-ResNet-18

Layer	Output size	Remarks
Input	256×256	Input RGB image
Conv block 1	128×128	7×7, 64, stride 2
Max pooling	64×64	3×3, stride 2
Conv block 2	64×64	$\begin{bmatrix} 3 \times 3, 64 \\ 3 \times 3, 64 \end{bmatrix} \times 2$
Conv block 3	32×32	$\begin{bmatrix} 3 \times 3, 128 \\ 3 \times 3, 128 \end{bmatrix} \times 2$
Conv block 4	16×16	$\begin{bmatrix} 3 \times 3, 256 \\ 3 \times 3, 256 \end{bmatrix} \times 2$
Conv block 5	8×8	$\begin{bmatrix} 3 \times 3, 512 \\ 3 \times 3, 512 \end{bmatrix} \times 2$
GAP	1,000	–
FC-layer	2	Activation: *softmax*

7×7, 64 means 64 7×7 size Conv filters and, e.g., $\begin{bmatrix} 3 \times 3, 128 \\ 3 \times 3, 128 \end{bmatrix} \times 2$ represents repeating twice the 2-layer structure using 128 3×3 Conv filters. More details can be found in [19].

11.3.5 Fine-Tuning with AL

The pre-training can provide better parameter initialization, and in this step, the benchmark model was further combined with the SCAE module as introduced in Sect. 11.3.2. The configuration of the SCAE is illustrated in Fig. 11.8 and details about each AE layer are listed in Table 11.3. In order to avoid both gradient explosion and gradient vanishing issues, features were extracted layer by layer, and normalization was performed before feature stacking. From the procedure of the SCAE-based AL, the unlabeled data, X^{ul}, were separated into $M = 6$ batches. Furthermore, for simplicity, $N_d = 100$ most difficult images were selected for manual labeling in each iteration. Unlike the source dataset with a balanced 1 : 1 class ratio used in Step 1, the target dataset had a moderately imbalanced class ratio, i.e., 1 : 4.68. Thus, $\beta = 2$ was considered for F_β score evaluation, i.e., F_2 score.

After 6 rounds of iterations, the re-trained benchmark model using SCAE-based AL, i.e., SCAE-AL, obtained a stable and satisfying performance, as shown in Fig. 11.9. In addition, comparisons were made with other methods, i.e., Self-T and AL, where results are summarized in Table 11.4. Even though the benchmark obtained 90.0% accuracy on the target dataset, this is deceptive. Due to the skewed class distribution of the target dataset, the recall (i.e., TPR) value of the benchmark was just slightly higher than 60%. Considering

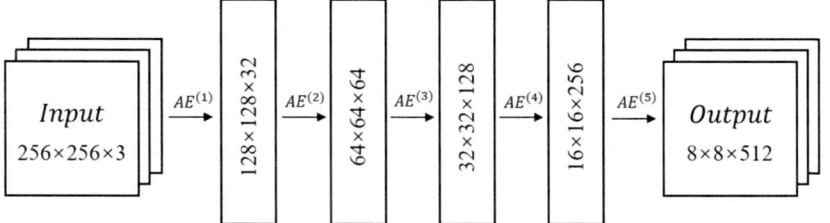

Fig. 11.8 Network configuration of the SCAE

Table 11.3 Network configuration of the $AE^{(k)}$

Layer	$AE^{(1)}$	$AE^{(k)}*$
Input	$256 \times 256 \times 3$	$m \times m \times n$
Conv layer	3×3, 32 filters, stride 2	3×3, $2n$, stride 2
Conv layer	1×1, 32 filters, stride 1	1×1, $2n$, stride 1
Output	$128 \times 128 \times 32$	$m/2 \times m/2 \times 2n$

* For $k = 2 : m = 128$ & $n = 32$.
For $k = 3 : m = 64$ & $n = 64$.
For $k = 4 : m = 32$ & $n = 128$.
For $k = 5 : m = 16$ & $n = 256$

its high TNR value (95.6%), the results indicate that the benchmark was more prone to predicting the images as non-cracked. In addition, the Self-T and AL models worked better than the benchmark, where AL was slightly better than Self-T because of the assistance of human-corrected labels. However, the improvements for each metric were limited (less than 5%) compared to the benchmark. Considering that both Self-T and AL need limited labeling efforts, i.e., not required for Self-T and only 600 ($N_d = 100$ for 6 iterations) images were manually re-labeled among 60,000 unlabeled data for AL, such a small improvement was deemed acceptable.

Adding the SCAE into the AL, the SCAE-AL model obtained significant improvements in both precision (from 75.7% to 84.1%) and recall (from 63.6% to 90.3%), which relieved the previously mentioned bias issue. This is explained by a stronger feature representation ability through feature fusion from the benchmark model and the SCAE structure. The SSL mechanism in the SCAE branch improved the utilization of the abundant unlabeled data in the target dataset, which made the SCAE able to extract deep features from the target dataset, and thus improved the generalization ability of the SCAE. Furthermore, through feature fusion, the extracted features from the benchmark and the SCAE were combined, providing more comprehensive information about the target dataset. Therefore, the SCAE-AL achieved a better generalization ability than the other methods.

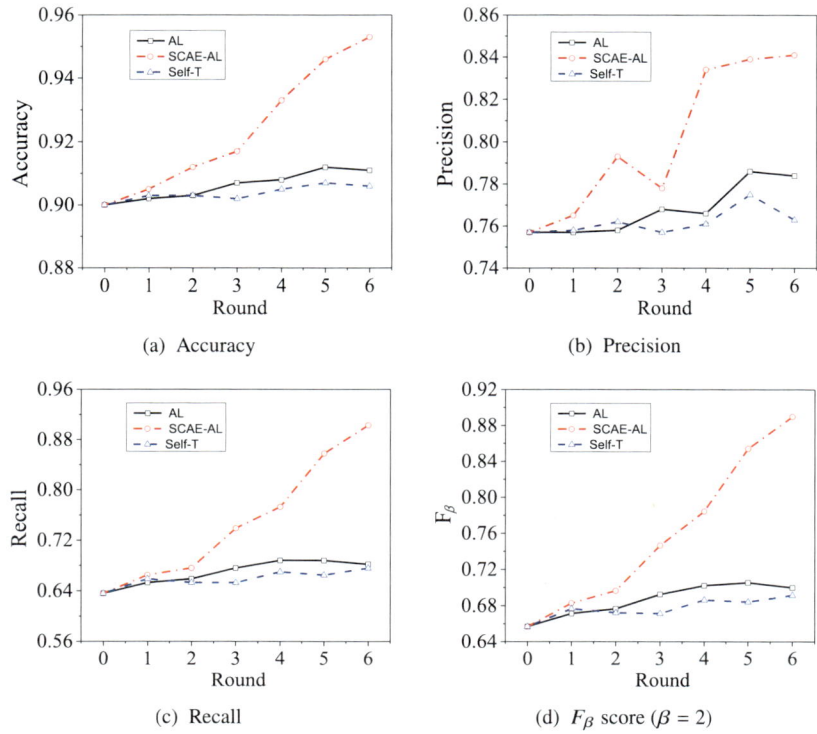

Fig. 11.9 Evaluation of the model re-training

Table 11.4 Re-training recognition performance of the benchmark model on the target dataset

Network	Accuracy (%)	Precision (%)	Recall (%)	F_2 score	TNR (%)
Benchmark	90.0	75.7	63.6	0.657	95.6
Self-T	90.6	76.3	67.6	0.692	95.5
AL	91.1	78.4	68.2	0.700	96.0
SCAE-AL	95.3	84.1	90.3	0.890	96.4

11.3.6 Summary

A multi-step crack recognition framework, namely, CAL, is established in this case study. It integrates AL to reduce the burden of labeling efforts while maintaining classification accuracy. The CAL framework contains two steps: (1) pre-training benchmark classification model and (2) re-training semi-supervised AL model.

Notably, re-trained by semi-supervised AL with limited labeling efforts (only labeled 600 among 60,000 images), the benchmark classification model achieved better generalization ability with significant accuracy improvement, e.g., F_2 score, precision, and recall increased by 23.3%, 8.4%, and 26.7%, respectively. Therefore, the results demonstrate the effectiveness and efficiency of the CAL for BHM applications.

It is important to note that the current functionality of the CAL framework is limited to classification models, and it does not extend its utility to the training of tasks involving localization and segmentation, where the Bbox and pixel-wise labeling are even more costly. Therefore, future explorations to mitigate the labeling efforts associated with these tasks are recommended.

11.4 Exercises

1. Explain the concept of AL and its significance in SHM.
2. Describe how the learning model is updated in AL based on the nature of the acquired labeled data.
3. What are the drawbacks of Self-T?
4. Provide examples or scenarios of SHM applications where AL can be applied to optimize data labeling and improve model accuracy.
5. What are the advantages of integrating SCAE with AL?

References

1. J. Masci et al., Stacked convolutional auto-encoders for hierarchical feature extraction, in *International Conference on Artificial Neural Networks* (Springer, 2011), pp. 52–59
2. P. Vincent et al., Stacked denoising autoencoders: learning useful representations in a deep network with a local denoising criterion. J. Mach. Learn. Res. **11**(12), (2010)
3. S. Clark, J.R. Curran, M. Osborne, Bootstrapping POS taggers using unlabelled data, in *Proceedings of the Seventh Conference on Natural Language Learning at HLT-NAACL 2003* (2003), pp. 49–55
4. B. Settles, Active learning literature survey. University of Wisconsin-Madison Department of Computer Sciences, 2009
5. C. Szegedy et al., Going deeper with convolutions, in *Proceedings of the IEEE Conference on Computer Vision and Pattern Recognition* (2015), pp. 1–9
6. K. He et al., Deep residual learning for image recognition, in *Proceedings of the IEEE Conference on Computer Vision and Pattern Recognition* (2016), pp. 770–778
7. American Road and Transportation Builders Association. Bridge Conditions Report 2021 (2021)
8. Ministry of transport of the People's Republic of China. Statistical bulletin on the development of transportation industry 2020 (2020)
9. S. Dorafshan, R.J. Thomas, M. Maguire, Comparison of deep convolutional neural networks and edge detectors for image-based crack detection in concrete. Constr. Build. Mater. **186**, 1031–1045 (2018)

10. S. Dorafshan, R.J. Thomas, M. Maguire, SDNET2018: an annotated image dataset for non-contact concrete crack detection using deep convolutional neural networks. Data Brief **21**, 1664–1668 (2018)

11. L. Zhang et al., Road crack detection using deep convolutional neural network, in *IEEE International Conference on Image Processing (ICIP)*, vol. 2016 (IEEE, 2016), pp. 3708–3712

12. Y. Gao, K.M. Mosalam, PEER Hub ImageNet: A large-scale multiattribute benchmark data set of structural images. J. Struct. Eng. **146**(10), 04020198 (2020)

13. K. Simonyan, A. Zisserman, Very deep convolutional networks for large-scale image recognition (2014), arXiv:1409.1556

14. C. Szegedy et al., Inception-v4, inception-resnet and the impact of residual connections on learning, in *Thirty-first AAAI Conference on Artificial Intelligence* (2017)

15. S.J. Pan, Q. Yang, A survey on transfer learning. IEEE Trans. Knowl. Data Eng. **22**(10), 1345–1359 (2009)

16. Y. Gao, K.M. Mosalam, Deep transfer learning for image-based structural damage recognition. Comput. Aided Civ. Infrastruct. Eng. **33**(9), 748–768 (2018)

17. Ç.F. Özgenel, Concrete crack images for classification. Mendeley Data **1**(1), (2018), http://dx.doi.org/10.17632/5y9wdsg2zt

18. R. Satyen, Concrete-Crack-Detection (2018), https://github.com/satyenrajpal

19. J. Hu, L. Shen, G. Sun, Squeeze-and-excitation networks, in *Proceedings of the IEEE Conference on Computer Vision and Pattern Recognition* (2018), pp. 7132–7141

The term "resiliency" in structural engineering refers to the capacity of a structure or structural system to recover from a shock caused by an extreme event or a disaster by maintaining its fundamental characteristics and swiftly recovering its functionalities to mitigate losses. In Part IV, a novel concept of "resilient AI" is introduced to establish an intelligent SHM system, which aims to enhance the robustness, efficiency, and inter-pretability of the AI-enabled SHM system to ultimately contribute to heightened resilience in the built environment. However, several crucial challenges such as multimodality of data sources, multitasking of learning the features, and the "black box" issues of the DL models remain largely unexplored.

Chapter 12 initiates a reformulation of the vision-based SHM problem into a multi-attribute multi-task setting. Each image is attributed with multiple labels describing its distinct characteristics. To enhance performance across all tasks, a novel hierarchical framework, known as the multi-attribute multi-task transformer (MAMT2) is proposed. This framework integrates multi-task TL mechanisms and employs a transformer-based network as the backbone. The performance of the framework is further studied through numerical experiments involving multiple classification, localization, and segmentation tasks.

Given the intrinsic complexity of AI, particularly in DL models, as well as the general lack of physical meaning of these DL models, they are hard to understand by a human, resulting in the "black box" issues. Chapter 13 shifts the focus toward addressing this "black box" issue in AI for SHM. It begins by introducing various classical "eXplain-able AI" (XAI) methods, namely, guided backpropagation (GBP), class activation map (CAM), gradient-weighted CAM (Grad-CAM), and Guided Grad-CAM. These methods are designed to aid humans in understanding DL network behavior by visualizing the backward signal or the activated features. The chapter then comprehensively explores and discusses the performance of these XAI methods in basic SHM vision tasks. In summary, advocating for an explainable model in SHM represents a crucial stride toward fortifying the resilience of intelligent disaster prevention systems.

Multi-task Learning

12

The previous chapters have demonstrated the feasibility and adaptability of AI technologies to recognize building and infrastructure damage via images. However, many previous studies solely work on the existence of damage in the images and directly treat the problem as a single attribute classification or separately focus on finding the location or area of the damage as a localization or segmentation problem. Abundant information in the images from multiple sources and inter-task relationships are not fully exploited. As a result, the trained AI model or system may not be robust and efficient enough to address the complex tasks, including functional recovery [1], in the real environment. In other words, such an AI system lacks the tools to address issues of direct relevance to "resiliency" when extreme events occur.

In this chapter, the vision-based SHM problem is first reformulated into a multi-attribute multi-task setting, where each image contains multiple labels to describe its characteristics. Consequently, extended from ϕ-Net, a multi-attribute multi-task dataset, namely, ϕ-NeXt, is established, which contains 37,000 pairs of multi-labeled images covering classification, localization, and segmentation tasks. To pursue better performance in all tasks, a novel hierarchical framework, namely, multi-attribute multi-task transformer (MAMT2) is proposed, which integrates multi-task TL mechanisms and adopts a transformer-based network as the backbone. Through numerical experiments, the results demonstrate the superiority of the MAMT2 in all tasks, which reveals a great potential for practical applications and future studies in both Structural Engineering and CV.

The original version of this chapter has been revised: Figure 12.1 has been corrected. A correction to this chapter can be found at https://doi.org/10.1007/978-3-031-52407-3_14

K. M. Mosalam and Y. Gao, *Artificial Intelligence in Vision-Based Structural Health Monitoring*, Synthesis Lectures on Mechanical Engineering, https://doi.org/10.1007/978-3-031-52407-3_12

12.1   Multi-attribute Multi-task SHM Problem

In Chap. 5, a general structural image detection framework, namely, ϕ-Net, is proposed. It includes several basic tasks for the purpose of automated damage assessment, where each task represents one structural attribute. According to the logic of the hierarchy framework in ϕ-Net, the structural image is processed layer by layer following tree branches, namely, Pixel, Object, and Structural. However, previous work on ϕ-Net focused on the multi-class classification problems among eight key tasks: (1) scene level, (2) damage state, (3) concrete cover spalling condition, (4) material type, (5) collapse mode, (6) component type, (7) damage level, and (8) damage type. These tasks were treated as eight independent ones, i.e., not fully exploring and utilizing the internal relationships between all attributes.

Based on the ϕ-Net, a new multi-attribute multi-task SHM problem is re-defined in this chapter. Firstly, the same eight classification tasks are maintained as the fundamental ones. Subsequently, damage localization and segmentation tasks are complemented to them as downstream tasks. Finally, these ten tasks are reorganized into a similar but simplified framework, named ϕ-NeXt, as shown in Fig. 5.15 in Sect. 5.6. The damage localization and segmentation tasks for only the concrete cover spalling, as a representative example, are presented here. Similar to the original ϕ-Net, the number of tasks of ϕ-NeXt can be expanded based on demands and more typical tasks, e.g., locating and segmenting damage patterns of reinforcing bar exposure, steel corrosion, and masonry crushing, are listed as future extensions. The definitions of the ten benchmark tasks are summarized below for completeness.

- **Task 1 Scene level**. It is defined as a 3-class classification: *Pixel level*, *Object level*, and *Structural level*, which represents distance toward the target from a close, mid-range, and far range, respectively.
- **Task 2 Damage state**. It is a binary classification: *Damaged* and *Undamaged*, which is straightforward and describes a general condition of structural or component's health. The damaged patterns include concrete cracking or spalling, reinforcing bar exposure, buckling or fracture, and masonry crushing.
- **Task 3 Spalling condition**. It is a binary classification: *Spalling* (*SP*) and *Non-spalling* (*NSP*), where spalling means the loss of concrete cover material (covering the reinforcing steel) from a structural component surface.
- **Task 4 Material type**. It is a binary case: *Steel* and *Others*, which intends to identify the construction material of the structure or component. For simplicity, all material types other than steel are grouped into the "others" class.
- **Task 5 Collapse mode**. It recognizes the severity of damage that occurred in the structural-level image: *Non-collapse*, *Partial collapse*, and *Global collapse*.
- **Task 6 Component type**. It identifies the type of structural components: *Beam*, *Column*, *Wall*, and *Others*. It is conducted if the image is for the object level.

- **Task 7 Damage level**. It is for the severity of component damage from object level images: *Undamaged*, *Minor damage*, *Moderate damage*, and *Heavy damage*.
- **Task 8 Damage type**. It describes the type of damage that occurred in structural components from object level images based on a complex, irregular, and even abstract semantic vision pattern: *Undamaged*, *Flexural damage*, *Shear damage*, and *Combined damage*.
- **Task 9 Damage localization**. It localizes the damage patterns by Bboxes. In this chapter, for brevity, only spalling areas are monitored.
- **Task 10 Damage segmentation**. It quantifies the damage by finding the whole damaged area, where each pixel has its own label and regions of pixels with the same label are grouped and segmented as one object (class). Similar to Task 9, each pixel is labeled as either *SP* or *NSP* and images are from Task 3.

Based on the ϕ-NeXt hierarchical relationships and dependency between the inter-task attributes, a structural image may go through multiple tasks where the output (i.e., label) of each task is treated as a single structural attribute. Thus, a sequential set of attributes is obtained for vision-based SHM. For details about the task definitions, e.g., additional physical interpretations, refer to Sect. 5.2.1.

12.2   Multi-attribute Dataset Development

With the defined tasks above in the ϕ-NeXt framework, a well-labeled dataset is established, which is a multi-task version of the original ϕ-Net dataset. In some scenarios, an online image may have been described with one form of labeling by different experts with respect to different attributes, e.g., expert A describes its scene level (Task 1) and expert B defines its damage state (Task 2). Therefore, a multi-attribute label merging algorithm and a dataset development procedure are introduced.

12.2.1 Data Collection and Labeling

Along with the image data in ϕ-Net, multi-class labels are provided for the first eight classification tasks. As mentioned above, spalling damage is the focus of the damage localization and segmentation tasks herein. Therefore, only spalling-related images are selected from the ϕ-Net for these two tasks. Besides, newly-collected spalling-related images are included. For Tasks 9 & 10, the labeling tool "*Labelme*" [2] is used. Finally, the corresponding Bboxes of the spalled regions and pixel-wise labeling are obtained. However, the collected and labeled data are still conducted independently, i.e., given one image, the user does not know all attributes at the same time. In other words, one image can have multiple copies in different parts of the dataset. Therefore, how to find the same image is a key goal of the developed algorithm. The Message-Digest algorithm 5 (MD5) check method [3] is a feasible solution. While adopting the MD5 check, each image is assigned a 32-character hexadecimal number, namely, the MD5 checksum value. If two images have the same MD5 checksum value, these two images are the same in the proposed label merging algorithm, refer to Algorithm 12.1, which is discussed below.

Suppose N pairs of labeled images $\left(x_i, y_i^j\right)$ are collected, where $i \in \{1, 2, \ldots, N\}$ is the image index and $j \in \{1, 2, \ldots, P + Q + R\}$ is the task index. The first P tasks are classification ones, whose labels are discrete values, and tasks from $P + 1$ to $P + Q + R$ are Q localization plus R segmentation tasks. In the computational loops, the MD5 checksum values ($v_i's$) of all images from different sources (i.e., datasets) are computed, where some images may occur multiple times in different datasets but labeled with different tasks, denoted by different $j's$ accordingly. Subsequently, all images with the same MD5 value are grouped for merging their different labels of the same task. For the classification attributes ($j \leq P$), if one attribute has a clear majority label from different sources, including having a single label from one source, this label is used; otherwise, a random selection is made, to break the tie if there is no clear majority, including having only two labels from two sources.[1] As more complex and informative attributes, i.e, the localization ($P < j \leq P + Q$) and segmentation ($P + Q < j \leq P + Q + R$) attributes, the coordinates of the Bboxes and pixel-wise labels are directly stored. If no label is found for a certain attribute, the label "Not Applicable" (NA) is assigned to represent the missing information. Finally, the merged labels for each attribute are saved to a JavaScript Object Notation (JSON) file. It is noted that this label merging algorithm not only works for existing data, but also can be easily used for newly-collected data.

For the current ϕ-NeXt, $P = 8$, $Q = R = 1$, and all labels of the different tasks are encoded to be presented numerically and accordingly computationally manipulated. For the first eight classification attributes, the labels are encoded to integers from 1 to the predefined number of classes, C, e.g., in 3-class classification, $C = 3$ and $y \in \{1, 2, 3\}$. The "NA" label is encoded as "-1", which can be easily filtered out during training and not be taken into the loss computation. Besides, one spalling localization and one spalling segmentation tasks are covered. Finally, through processing the data from the ϕ-Net and the additional images, **37,000** images with multiple labels are obtained.

---

[1] Example 1: Assume 3 image copies A, B & C with same MD5 values. Copy A has attributes: O for "object in Task 1", UD for "undamaged in Task 2" & Column for "Task 6". Copy B has attributes: O, UD & Wall for "Task 6". Copy C has attributes: P for "pixel in Task 1", UD & Column. The final attributes of the merged 3 copies become: O, UD & Column. Example 2: Assume 4 image copies A, B, C & D with same MD5 values. Copies A, B & C are the same as in Example 1. Copy D has attributes: P, UD & Column. The final attributes of the merged 4 copies become: a random choice of {O, O, P, P}, UD & Column, i.e., either O, UD & Column or P, UD & Column.

Algorithm 12.1 Multi-attribute label merging algorithm for ϕ-NeXt

---

1: **Given**: data-label pairs $(x_i, y_i^j), i \in \{1, 2, \ldots, N\}, j \in \{1, 2, \ldots, P, P+1, \ldots, P+Q+R\}$
2: **Initialize** $V = [\ ], X = [\ ], Y = [\ ], M = [\ ], k = 1$
3: **for** $i = 1, 2, \ldots, N$ **do**
4:　　Run **MD5 Check** for x_i and get v_i
5:　　**if** v_i is not in V **then**
6:　　　　**Initialize** $(P+Q+R)$-long array y_k with "NA" labels for all elements
7:　　　　$v_k = v_i, x_k = x_i$ and $y_k[j] = y_i^j$
8:　　　　Append v_k, x_k, and y_k into V, X, and Y, respectively
9:　　　　$k = k + 1$
10:　　**else**
11:　　　　**Find** l, where $y_l \in Y, v_l = v_i \in V$
12:　　　　**if** $y_l[j] =$NA **then**
13:　　　　　　**Update** $y_l[j] = y_i^j$
14:　　　　**else**
15:　　　　　　**if** $j \leq P$ **then**
16:　　　　　　　　Append (v_i, j, y_i^j) into M
17:　　　　　　　　**Update** $y_l[j] =$ Majority Voting $\forall y^j | v_l = v_i \in M$
18:　　　　　　**end if**
19:　　　　**end if**
20:　　**end if**
21: **end for**
22: **Output** X and Y

12.2.2 Dataset Split and Statistics

For benchmarking purposes, splitting training and test sets is one essential step in the label merging algorithm. However, adopting a fixed training-to-test splitting ratio is inappropriate for multi-attribute problems to avoid label imbalance issues [4]. Unlike a single-attribute task, a coarse and fixed split may disrupt the distributions of labels in the training and test sets and lead to a biased split in certain attributes, e.g., we may end up with all labels of the collapse mode attribute (for Task 5) are in the training set and accordingly images in the test set have no meaningful labels (i.e., "NA") related to Task 5. Therefore, following the recommendation from a previous study [4], a training-to-test split range is maintained between 8:1 to 9:1 across all tasks and their attributes. Moreover, the label distributions are kept identical between training and test datasets. Since most data are directly from the already built ϕ-Net, which has been assigned to either training or test set, the split ratio check is only conducted for the newly added images, which have been judged by the MD5 check, as presented in the pseudo-code in Algorithm 12.1.

12.3   Multi-attribute Multi-task Transformer Framework

To comprehensively consider the multi-attribute multi-task problems, a unified framework across classification, localization, and segmentation is proposed, namely, MAMT2, Fig. 12.1. The following paragraphs describe the details of MAMT2.

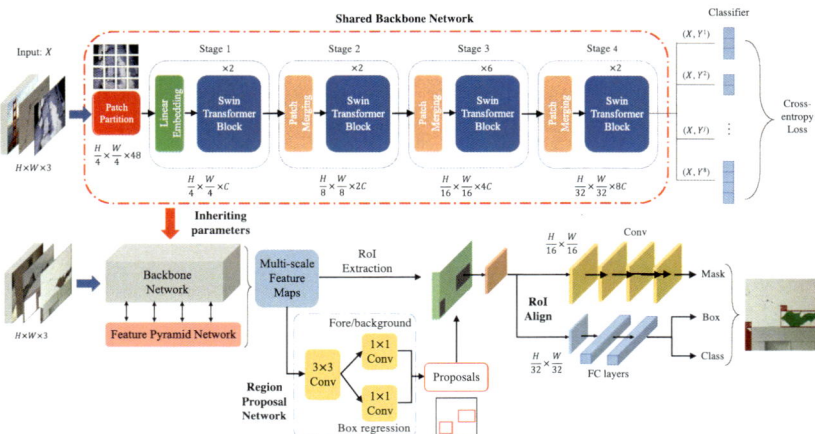

Fig. 12.1 MAMT2 framework

The realization of the classification is different from localization and segmentation, i.e., identifying discrete class labels of images is more straightforward than obtaining Bboxes and pixel-wise labels of the target objects. The localization and segmentation tasks usually require certain networks, e.g., RPN (refer to Sect. 7.4.3), to locate the RoIs first. Therefore, one of the novelties of the MAMT2 is to design the framework in a two-step manner, which conducts classification and other tasks (e.g., localization and segmentation) asynchronously and hierarchically. As illustrated in Fig. 12.1, first, the coupled multi-attribute classification problems in ϕ-NeXt are solved using a shared backbone network followed by multi-head branches, where each branch corresponds to one specific task (and its attributes), $j \in \{1, 2, \ldots, 8\}$. Each branch is constructed by a MLP classifier and the number of output neurons is equal to the number of classes in each classification task, C^j. Suppose there are $N^{(1)}$ pairs of labeled data $\left\{ \left(x_i, y_i^j \right) \right\}, i \in \{1, 2, \ldots, N^{(1)}\}$, where y is the ground truth label, i.e., $y = c, c \in \left\{ 1, 2, \ldots, C^j \right\}$ and denote the classifier prediction as $\hat{y}_i^j$. The cross-entropy loss, $L^{(1)}$, is computed as follows:

$$L^{(1)} = -\frac{1}{N^{(1)}} \sum_{i=1}^{N^{(1)}} \sum_{c=1}^{C^j} p\left(y_i^j = c\right) \ln \left[q\left(\hat{y}_i^j = c\right) \right], \tag{12.1}$$

where $p(y)$ and $q(\hat{y})$ are the probability distributions of the ground truth and the predictions, respectively.

The previous studies in [4–7] have demonstrated that the shared features among the tasks contributed to complementing the information and are useful in enhancing the accuracy and reducing the training costs. It is noted that the MAMT2 is a general framework, where the backbone can be any type of network, e.g., CNN, RNN, or Transformer. In the CV domain, CNN has served as a standard network for the past decades and most studies in the vision-based SHM area adopted CNN as their best choice. Recent studies [8, 9] show that transformer-based networks, e.g., ViT, Sect. 4.5.4, and Swin transformer, Sect. 4.5.5, started to achieve equivalent or even better performance than CNN in classification, localization, and segmentation tasks. Therefore, for the purpose of achieving a concrete benchmark performance, the Swin transformer, one of the state-of-the-art backbone networks for many CV applications [9], is selected for the MAMT2.

Based on the concept of TL, the shared backbone network used in classification can share key features and information for localization and segmentation. Therefore, the transformer backbone trained from the multi-task classification is directly inherited by the localization and segmentation tasks. To fully benefit from the multi-scale features extracted from each stage in the transformer, the backbone is further connected with a FPN, as illustrated in Fig. 4.26. The 1×1 Conv operations are performed on the feature maps from the final layer of each stage to project their depth to a uniform space, call it C_1. Additionally, these extended features from stages 3 to 1 are fused with their subsequent features (i.e., stages 4 to 2) along with up-sampling operations. Subsequently, another round of Conv operations with a different uniform space dimension, call it C_2, making use of 3×3 filters, are performed. Finally, a stack of multi-scale feature maps (a.k.a. pyramid feature maps) is generated, i.e., F_1 to F_4, Fig. 4.26. In this MAMT2 framework, both C_1 and C_2 are taken as 256.

Similar to the classical object detection framework using Faster R-CNN, these pyramid feature maps are fed into a RPN to generate a set of rectangular object proposals, which are parameterized to the same number of reference boxes, i.e., anchor boxes, refer to Sect. 7.4.3, on the raw input image. Following the same proposal selection procedure [10], multiple RoIs are determined on the feature maps for all scales, where each RoI is a rectangular window region of the extracted feature maps. To improve the efficiency of small target detection and relieve the misalignment issue in the feature maps, the pyramid RoIAlign operation[2] is performed.

Inspired by the design of Mask R-CNN [11], two branches are added to further process the feature maps of each RoI, where one branch consists of multiple Conv layers to determine the mask for each object and the other branch is constructed by FC-layers to obtain the Bbox coordinates of the object and the class label in each box, refer to the bottom right part

---

[2] The pyramid RoIAlign operation is a developed module in [11] to utilize multiple RoIAlign operations to extract proposal features at different pooling sizes. It is noted that the RoIAlign operation accurately extracts a small feature map from each RoI in detection & segmentation tasks. It properly aligns the extracted features with the input making use of bilinear interpolation [12].

of Fig. 12.1. The total loss of this step (with $N^{(2)}$ data points available for localization & segmentation), $L^{(2)}$, consists of 5 parts, Eq. 12.2: (1) classification loss of RPN, L_{cls}^{RPN}, (2) regression loss of RPN, L_{reg}^{RPN}, (3) classification loss of the prediction box, L_{cls}, (4) regression loss of the Bbox, L_{loc}, and (5) average binary cross-entropy loss of the mask, L_{mask}. Because the first 4 loss terms, L_{cls}^{RPN}, L_{reg}^{RPN}, L_{cls}, and L_{loc}, are the same as those in Eq. 7.8, Sect. 7.4.3, they are not discussed here. It is noted that the hyper-parameter, λ, used in Eq. 7.8 to balance the weight between classification and regression, is not used in Eq. 12.2.

$$L^{(2)} = L_{cls}^{\text{RPN}} + L_{reg}^{\text{RPN}} + L_{cls} + L_{loc} + L_{mask}. \tag{12.2}$$

Suppose in the i-th data point, there are I_i RoIs extracted and each RoI has a shape of $C \times m \times m$ corresponding to $m \times m$ binary-value masks of the C classes. Each pixel in the mask is either 1 or 0 representing the existence of the object or not. The mask loss, L_{mask}, is computed by an average pixel-wise binary cross-entropy loss of class $\hat{y}$ determined from the classification branch, L_{cls}, as follows:

$$L_{mask} = -\sum_{i=1}^{N^{(2)}} \sum_{j=1}^{I_i} \left[\sum_{k=1}^{N_p} PR_k \Big/ N_p \right]. \tag{12.3}$$

where $PR_k = r_k \ln \left[p\left(\hat{r}_k\right) \right] + (1 - r_k) \ln \left[1 - p\left(\hat{r}_k\right) \right]$, N_p is the total number of pixels in the mask corresponding to class $\hat{y}$, r_k represents whether the k-th pixel belongs to the class $\hat{y}$, i.e., $r_k = 1$, or not, i.e., $r_k = 0$, and $p(\hat{r}_k)$ is the probability distribution of the k-th pixel prediction.

12.4   Benchmark Experiments on ϕ-NeXt

To verify the effectiveness of the MAMT2 framework and to set the benchmark performance on the ϕ-NeXt dataset, a series of experiments including classification, localization, and segmentation were conducted and the effects of different approaches and models were compared and discussed. The implementation of these experiments was based on TensorFlow and PyTorch and performed on CyberpowerPC with a single GPU (CPU: Intel Core i7-8700K@3.7 GHz 6 Core, RAM: 32 GB, and GPU: Nvidia Geforce RTX 2080Ti).

12.4.1 Classification Experiments

In the classification experiments, six comparison cases were designed and separated into three groups: baseline (VGG-16, VGG-19, and ResNet-50) using both TL and DA,[3] HTL

---

[3] These baseline models are actually the TL-DA models defined in Sect. 6.3.5.

model, and two MTL models, where partial results of baseline models and HTL model are reported in Sects. 6.3.5 and 6.5.

To investigate the advantage of using a transformer-like network in handling the inter-relationships among tasks, the performance of MAMT2 was compared with a conventional MTL model using CNN as the backbone instead of the transformer, which is denoted as convolutional-MTL (C-MTL). Similar to MAMT2, for a batch of image samples in an iteration, the parameters of the CNN and the corresponding classifiers were trained simultaneously using labels from different tasks and were optimized by the joint cross-entropy loss, Eq. 12.1. Moreover, similar to the HTL settings, ResNet-50 was adopted for the C-MTL backbone with ImageNet [13] pre-trained parameters before training and the remaining settings were kept the same. The eight classifiers were FC-layers that had the output dimensions according to the number of classes for each of the eight considered classification tasks.

For all cases, the evaluation metric was the overall accuracy, Eq. 3.12. The test accuracy of the considered six models is listed in Table 12.1. MTL usually learns to share the representations among multiple tasks [14], which enables the model to generalize well for these tasks. From Table 12.1, it is obvious that the MTL models achieved better performance than the baseline and HTL models. Especially for hard tasks (i.e., Tasks 5 to 8), the MAMT2 and C-MTL improved the baseline performance of 74.7%, 76.9%, 72.7%, and 72.4% by 6.0%, 1.7%, 6.2%, and 2.8%, respectively. This demonstrates the state-of-the-art performance of MTL-based models over the conventional approaches. In addition, the proposed transformer-based backbone used in MAMT2 indicated superiority over the CNN-based MTL model, i.e., the MAMT2 achieved the best performance in all tasks except Task 6: component type, where the CNN backbone MTL model was only about 0.9% better in overall test accuracy. In CNN, shallow Conv layers capture local information with a relatively small receptive field, refer to Sect. 4.5. A large receptive field is obtained in the deep layers by stacking more layers. However, they lose the local information of data during forward propagation. On the contrary, through MSA, Sect. 4.5.3.3 and the hierarchical mechanism, Sect. 4.5.5.1, the MAMT2 extracts the features with a large respective field while maintaining global attention from the shallow layers to the deeper layers. These characteristics efficiently capture the inter-task relationships and utilize abundant features and information of structural images from different source domains.

In summary, from the perspective of the classification tasks in vision-based SHM, the MAMT2 framework has the following merits: (1) it achieves state-of-the-art performance in most tasks, (2) it exploits the inter-task relationships and global information in the data, and (3) it significantly enhances the training efficiency compared to other approaches. From practical considerations, test accuracy values for Tasks 1 ($\approx 94\%$) and 4 ($\approx 99\%$) are very promising for real practical applications in vision-based SHM and the remaining tasks, i.e., hard ones, are still acceptable ($> 75\%$) and can be improved with more data.

Table 12.1 Classification results (test accuracy, %)

Task	Model					
	VGG-16	VGG-19	ResNet-50	HTL	C-MTL	MAMT2
1	92.5	92.1	<u>93.0</u>	93.0	93.2	**93.5**
2	87.6	87.4	<u>88.9</u>	88.3	88.3	**89.4**
3	81.6	81.8	<u>83.0</u>	83.3	83.4	**84.6**
4	97.0	96.9	<u>98.5</u>	–	98.7	**98.8**
5	71.9	72.6	<u>74.7</u>	–	74.0	**80.7**
6	<u>76.9</u>	76.9	76.3	–	**78.6**	77.7
7	<u>72.7</u>	64.7	<u>72.7</u>	74.1	76.3	**78.9**
8	70.5	<u>72.4</u>	68.1	70.9	74.0	**75.2**

Bold # indicates the best result for each task
Underlined # indicates the best considered baseline

12.4.2 Localization and Segmentation Experiments

In the localization and segmentation, two comparison cases were considered, namely, Mask R-CNN and MAMT2. Because the ϕ-NeXt labels follow the style of MS COCO [15], as one of the most powerful and prevalent detection networks, the Mask R-CNN was selected to represent the baseline performance in this case study. In addition, since a regular image size of 224×224 was used in this experiment, the RPN in the MAMT2 was inherited from a pre-trained Mask R-CNN model based on the MS COCO dataset [15].

Based on the definitions in the MS COCO dataset and previous studies [9, 11, 15], the standard evaluation metrics for both localization and segmentation are AP and average recall (AR) with variant IoU thresholds, i.e., ϵ's, as introduced in Chaps. 7 and 8. Besides the widely-used AP^{50} metric (introduced in Sect. 7.1.2), for the benchmarking purpose, following the work in [9, 11], the standard COCO metrics include a set of AP variations, namely, AP^{75}, AP^{S}, AP^{M}, and AP^{L}, refer to Sect. 7.1.2, which were used here. While computing these three metrics, the AP values were computed 10 times using a set of IoUs from $\epsilon = 0.5$ to 0.95 with an increment of 0.05. In addition, a set of AR variations was computed, i.e., AR^{50}, AR^{75}, AR^{S}, AR^{M}, and AR^{L}. AR^{50} and AR^{75}, by definitions, were computed under $\epsilon = 0.5$ and 0.75, respectively. Similar to AP^{S}, AP^{M}, and AP^{L}, the computations of AR^{S}, AR^{M}, and AR^{L} involved the average values of recall for $\epsilon = 0.5$ to 0.95 with an increment of 0.05 by only considering the data with the three scales ground truth Bboxes used for the definitions in AP^{S}, AP^{M}, and AP^{L}, as discussed in Sect. 7.1.2. Besides the standard evaluation metrics, mIoU, Sect. 8.1.1, is computed, by averaging the IoUs for all classes, and compared for both localization and segmentation tasks. This is also widely used in previous relevant SHM studies [16, 17]. It is noted that the localization results were compared by the rectangular region (represented by the coordinates) for the ground truth and the predicted Bboxes. The localization and segmentation numerical results of all 11 performance eval-

Table 12.2 Localization and segmentation results of 11 evaluation metrics (%)

Metric	Localization		Segmentation	
	MAMT2	Mask R-CNN	MAMT2	Mask R-CNN
AP^{50}	**78.9**	72.5	**81.3**	70.3
AP^{75}	**63.6**	55.0	**59.6**	52.0
AP^{S}	**45.4**	43.2	**42.8**	37.5
AP^{M}	**63.0**	55.2	**60.9**	52.6
AP^{L}	**78.2**	59.7	**72.1**	59.4
AR^{50}	**65.1**	59.5	**61.3**	54.8
AR^{75}	**65.1**	59.5	**61.3**	54.8
AR^{S}	**53.7**	50.1	**49.8**	45.2
AR^{M}	**70.4**	66.4	**67.8**	61.6
AR^{L}	**86.3**	71.1	**80.0**	67.2
mIoU	**64.9**	57.3	**86.7**	72.9

Bold # indicates the best result for each task

uation metrics are listed in Table 12.2. Moreover, the visual results of selected 10 image samples are shown in Fig. 12.2. These results are discussed in the following two sections.

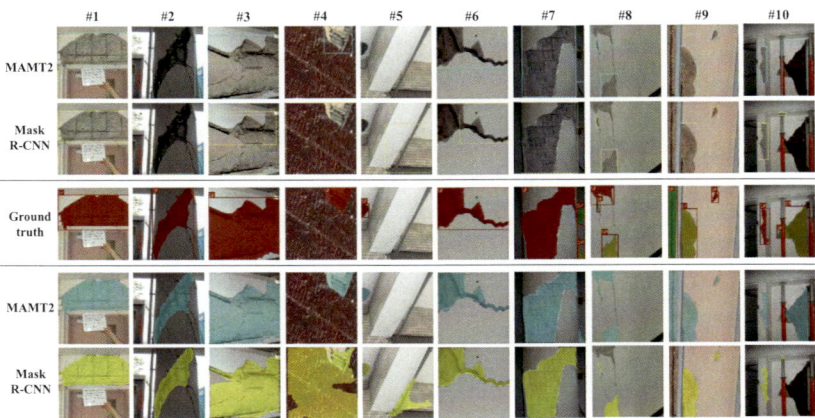

Fig. 12.2 Results of localization (top two rows) and segmentation (bottom two rows) for 10 image samples in comparison with the ground truth (middle row)

12.4.2.1 Localization

In general, the performance of the proposed MAMT2 was better than Mask R-CNN under all evaluation metrics. With respect to AP, MAMT2 achieved nearly 79% AP^{50} score and had

6.4% enhancement over Mask R-CNN under a single IoU threshold, $\epsilon = 0.5$, which is very promising compared to results reported in several benchmark experiments [9, 11]. When ϵ increased to a more strict value of 0.75, the performance of Mask R-CNN significantly decreased and the leading advantage of MAMT2 increased from 6.4% to 8.6%, which indicated a more robust performance. For the evaluation of objects in three different scales, it was found that both models were encountering difficulties in identifying small objects ($< 32 \times 32$) where MAMT2 was only leading by 2.2%. With the increase in the object size, MAMT2 obtained a significant improvement, especially for the large objects ($> 96 \times 96$), where AP^L of MAMT2 outperformed Mask R-CNN by 18.5%. As a complementary to the AP scores, the AR scores measure the assertiveness of the object detectors for a given class. Herein, the AR scores presented in Table 12.2 were computed for the specific "spalling" class of Task 3: spalling condition, Sect. 12.1. Both models were insensitive to the IoU threshold, where AR^{50} and AR^{75} scores were the same, but MAMT2 outperformed Mask R-CNN by 5.6%. Similar to the trend observed the AP scores, MAMT2 achieved greater improvement than Mask R-CNN for different scales, especially the obtained 15.2% increase in detecting large objects. For the mIoU score, MAMT2 also outperformed Mask R-CNN by 7.6%, which further demonstrated the superiority of MAMT2 from another perspective.

AP, AR, and mIoU scores, discussed above, only reflected the general conditions of both models, Mask R-CNN and MAMT2. Several examples are illustrated in Fig. 12.2 to complement the discussion where samples #1 & 2 represent the most common spalling patterns and both Mask R-CNN and MAMT2 achieved relatively satisfactory performance. Consistent with the AP and AR conclusions, in samples #4, 8, 9 & 10, Mask R-CNN missed certain spalling regions in different scales while MAMT2 accurately captured them all. Furthermore, toward some large spalling patterns, e.g., samples #3 & 6, Mask R-CNN produced duplicated Bboxes. On the contrary, MAMT2 produced a single precise Bbox, which demonstrated the robustness of MAMT2 with respect to the detection of large objects.

In conclusion, due to MAMT2's strong multi-scale ability, it can achieve better performance on objects of different sizes than classical models like Mask R-CNN. It improves the model's detection ability toward small and medium objects moderately and significantly boosts the performance on large objects. Therefore, based on this case study, the MAMT2 model increases the confidence of the researchers and engineers to accurately detect spalling areas larger than 96×96 pixels resolution for practical usage in vision-based SHM problems.

12.4.2.2 Segmentation

The segmentation results are listed in Table 12.2, which were computed by comparing the masked region pixel-wise between ground truth and predicted mask. It is noted that, for organizational purposes and similar to Fig. 2.6 in Sect. 2.3, multiple disconnected spalling areas of the ground truth were masked with different colors representing different damaged locations. The same 10 image samples used in the localization task were also considered for this segmentation task as shown in Fig. 12.2. Similar to the observations in the localization task, MAMT2 presented better performance than Mask R-CNN under all evaluation metrics and both models achieved more accurate results toward larger objects, i.e., $> 96 \times 96$ pixels resolution. It is noted that MAMT2 obtained a value of 81.3% for AP score under IoU computed with $\epsilon = 0.5$, i.e., for AP^{50}, which outperformed Mask R-CNN by 11%, and it also led Mask R-CNN by 13.8% for the mIoU score, indicating a significant improvement of MAMT2 above the current mainstream method, Mask R-CNN. Furthermore, the performance differences between MAMT2 and Mask R-CNN in detecting small and medium objects increased, where AP^S, AP^M, AR^S, and AR^M were improved by 5.3%, 8.3%, 4.6%, and 6.2%, respectively. This was also reflected in image samples #4 & #10 in Fig. 12.2, where Mask R-CNN made a wrong prediction in #4 and it missed detecting the spalled area of a small-size in the middle of image #10 and also that of a large-size on the right-hand side of image #10.

From samples #1, 2, 3 & 6, both models covered the spalled areas, but the masks generated by MAMT2 were more precise, especially near the edges of the spalling areas, e.g., the mask of Mask R-CNN was incomplete in sample #3. Besides, influenced by the localization performance, Mask R-CNN sometimes did not generate meaningful masks as shown in samples #4 & 5. Similar to the results in localization, in samples #4, 7, 8, 9 & 10, Mask R-CNN missed a few spalled regions while MAMT2 recognized all of them precisely, except missing a small spalling area in sample #7 at the right edge near the middle part of the image. Therefore, these results demonstrated the promising and robust performance of MAMT2 in spalling segmentation, especially for medium-size and large-size spalling areas.

Both MAMT2 and Mask R-CNN have multiple branches, which can output localization and segmentation simultaneously. In this case study, the inference, i.e., prediction, time per image for the two models based on Mask R-CNN) and MAMT2 were 0.094 and 0.055 s, respectively. These results demonstrate a higher computational efficiency of MAMT2, indicating its feasibility for real-time, i.e., rapid, assessment in vision-based SHM problems.

12.5 Summary

In this chapter, several typical vision-based SHM problems are first reformulated into a multi-attribute multi-task setting to describe the characteristics of the collected structural image data in a more comprehensive manner. Extended from the ϕ-Net, a more general structural image detection framework, namely, ϕ-NeXt, is developed, which introduces ten benchmark tasks covering the classification, localization, and segmentation. Accordingly, a multi-labeled dataset along with the dataset forming and label merging algorithm is established, which contains 37,000 images with multiple attributes.

To better address the multi-attribute multi-task problems, a novel hierarchical framework, namely, MAMT2, is presented. Swin transformer is adopted as the backbone, whose hierarchical mechanism and SW-MSA are intended to improve the model recognition ability.

Benefiting from the MTL mechanism, MAMT2 can utilize rich information and inter-task relationships from different sources and attributes. The joint training mechanism with a shared backbone network improves the recognition accuracy and preserves computational efficiency. In addition, through TL, the trained backbone for the classification tasks is inherited directly as the pre-trained model for the localization and segmentation tasks.

Finally, the MAMT2 framework is validated on ten benchmark tasks in ϕ-NeXt. In the classification tasks, MAMT2 is compared with baseline models (VGG-16, VGG-19, and ResNet-50) trained separately on single tasks, the HTL model whose transfer path is designed based on domain knowledge, and one MTL model using CNN (C-MTL) as the backbone. The results show that MAMT2 achieves the best performance in all tasks except Task 6: component type (although results are still comparable with other models), and both MTL-based models (MAMT2 and C-MTL) outperform the conventional methods, especially in hard tasks (i.e., Tasks 5 to 8, Sect. 12.1). In the localization and segmentation tasks, the numerical (AP and AR scores) and visual results demonstrate a more precise and robust performance of MAMT2 than the classical model, i.e., Mask R-CNN. The extensive experimental results presented in this chapter provide a benchmark reference for future relevant studies pursuing state-of-the-art performance in both Structural Engineering and general CV applications.

12.6   Exercises

1. Discuss the benefits of vision-based SHM problems within the MTL framework.
2. Investigate and provide insights into the potential reasons behind the improved performance observed when utilizing a Transformer backbone as opposed to a CNN backbone in a DL model for vision-based SHM.
3. Explain why the conventional approach of dividing data into training and test sets is ineffective when dealing with multi-attribute problems in the context of vision-based SHM.

References

1. C. Molina Hutt et al., Toward functional recovery performance in the seismic design of modern tall buildings. Earthq. Spectra **38**(1), 283–309 (2022)
2. K. Wada, Labelme: image polygonal annotation with python (2018), https://github.com/wkentaro/labelme
3. R. Rivest, The MD5 message-digest algorithm. Tech. rep. (1992)
4. Y. Gao, K.M. Mosalam, PEER Hub ImageNet: a large-scale multiattribute benchmark data set of structural images. J. Struct. Eng. **146**(10), 04020198 (2020)
5. S.J. Pan, Q. Yang, A survey on transfer learning. IEEE Trans. Knowl. Data Eng. **22**(10), 1345–1359 (2009)

6. Y. Gao, K.M. Mosalam, Deep transfer learning for image-based structural damage recognition. Comput. Aided Civ. Infrastruct. Eng. **33**(9), 748–768 (2018)

7. Y. Gao, K.M. Mosalam, Deep learning visual interpretation of structural damage images. J. Build. Eng. 105144 (2022)

8. A. Dosovitskiy et al., An image is worth 16x16 words: transformers for image recognition at scale (2020), arXiv:2010.11929

9. Z. Liu et al., Swin transformer: Hierarchical vision transformer using shifted windows, in *Proceedings of the IEEE/CVF International Conference on Computer Vision* (2021), pp. 10012–10022

10. S. Ren et al., Faster r-cnn: towards real-time object detection with region proposal networks. Adv. Neural Inf. Process. Syst. **28**, (2015)

11. K. He et al., Mask r-cnn, in *Proceedings of the IEEE International Conference on Computer Vision* (2017), pp. 2961–2969

12. E.J. Kirkland, E.J. Kirkland, Bilinear interpolation, in *Advanced Computing in Electron Microscopy* (2010), pp. 261–263

13. J. Deng et al., Imagenet: a large-scale hierarchical image database, in *2009 IEEE Conference on Computer Vision and Pattern Recognition* (2009), pp. 248–255

14. A. Argyriou, T. Evgeniou, M. Pontil, Multitask feature learning, in *Advances in Neural Information Processing Systems* (2007), pp. 41–48

15. T.-Y. Lin et al., Microsoft coco: common objects in context, in *Computer Vision–ECCV 2014: 13th European Conference, Zurich, Switzerland, September 6–12, 2014, Proceedings, Part V 13* (Springer, 2014), pp. 740–755

16. S.O. Sajedi, X. Liang, Uncertainty-assisted deep vision structural health monitoring. Comput. Aided Civ. Infrastruct. Eng. **36**(2), 126–142 (2021)

17. J. Zhao et al., Structure-PoseNet for identification of dense dynamic displacement and three-dimensional poses of structures using a monocular camera. Comput. Aided Civ. Infrastruct. Eng. **37**(6), 704–725 (2022)

Interpreting CNN in Structural Vision Tasks

13

In previous chapters, promising results have been achieved using AI, e.g., DL-based models, in vision-based SHM problems. However, the internal working principle in AI, especially for the DL model, is hard to understand by a human and is treated as a discouraging "black box", especially for the inquisitive engineers. To conquer this problem, CV researchers have been recently striving to develop visual interpretation methods, e.g., guided backpropagation (GBP) [1], class activation map (CAM) [2], gradient-weighted CAM (Grad-CAM), and Guided Grad-CAM [3], to understand the DL-based network model behavior through visualizing the backward signal (i.e., data flow) or the activated features, which is also known as XAI. As mentioned in Chap. 1, pursuing an explainable model for vision-based SHM is one necessary step for strengthening the resiliency of the intelligent disaster prevention system.

In this chapter, a comprehensive exploration of the performance of the XAI approaches in basic tasks of the vision-based SHM is presented. Before diving into the computer experiments of such an endeavour, several basics and background information about the DL visual interpretation methods are introduced first.

13.1   Explainable AI for Vision Problems

Recent research activities [1–6] in the CV domain attempt to explain and understand deep CNN through visualization by plotting the *saliency map*, presented as a heat or grey-scale depth map, where the colors and shades represent the activeness, i.e., importance, of pixels in the image toward the model's final decision. Several studies adopt either *gradient-based approaches* via manipulating the backward gradient from the output score to the input image or *feature reconstruction approaches* through weighting the Conv features to identify the most discriminative regions and then up-sampling back to the image pixel space.

K. M. Mosalam and Y. Gao, *Artificial Intelligence in Vision-Based Structural Health Monitoring*, Synthesis Lectures on Mechanical Engineering, https://doi.org/10.1007/978-3-031-52407-3_13

The difficulties in explaining deep CNN originate from the nature of its recognition ability. While learning complex patterns from input data, a deep CNN determines a set of optimal parameters in the heavily parametrized network, where the learning process is numerically performed by updating forward activated information signals and backward gradients to minimize the loss of the objective function. This intensively numerical approach makes it hard to directly perceive the discrimination principle of the network and to ultimately understand how it works rendering the working principle of deep CNNs a "black box". This situation is not only encountered in vision-based SHM, but also faced by other applications such as medical imaging [7] and self-driving cars [8], where the explanation of AI (or more specifically deep CNNs) is equally as important as acquiring accurate recognition results.

In recent years, the first efforts were made by Zeiler and Fergus [5] to develop a Deconv approach using a multi-layer deconvolution network (DeconvNet) to project feature activation back to the image pixel space. The results, i.e., saliency maps, are reconstructed from restoring the gradient through a backward pass from the feature maps where only positive gradients are retained, as discussed in Sect. 13.2.1. Saliency maps generated by this approach highlight certain areas of the input image that strongly activate the neurons and excite the feature maps at any Conv layer in the network. Instead of only visualizing neurons in the Conv layers and using features learned from multiple images by DeconvNet, Simonyan et al. [4] proposed the image-specific class saliency visualization, which uses the derivative of a class score with respect to one single image. It can be applied to any layer in the network and is most commonly performed in the FC-layer. The derivative herein is a gradient matrix with the same size as the input image and can be easily computed by a single backpropagation pass, where the gradients are zero if the activation is zero in the forward pass. It is noted that the magnitudes of the entries in the gradient matrix represent the activation condition of each pixel in an image toward the class score and the backward signal shares the same activation condition of the activation function, e.g., *ReLU* (Eq. 4.1). Based on the above methods, Springenberg et al. [1] proposed the GBP, which guides the backward signal from the deep layers (last several layers) to the input image by zeroing the entries of the signal having either negative values of the gradient in the backward pass or zero activation in the forward pass. The results produced by GBP show clearer visualizations than the above two methods. However, neither DeconvNet nor GBP is a class-discriminative method.

From the perspective of feature reconstruction, Zhou et al. [2] developed a new approach to highlight the class-discriminative regions, namely, CAM. In this approach, GAP is applied on the Conv feature maps and the output of the GAP is followed by a new single FC-layer (whose outputs are class scores), replacing the original FC-layers. Through the weighted sum of the feature maps and the FC-layer weights for one specific class, CAM generates a class discriminative saliency map. After up-sampling to the image size, the saliency map can identify the most discriminative area of the specific class in a heat map format. However, since CAM alters the raw pre-trained model, it may lead to losing some information in the original FC-layers, which leads to recognition performance deterioration in the modified network. Moreover, to determine the updated weights in the added single FC-layer requires running forward propagation once again, which is computationally more demanding than other visualization methods. To address these drawbacks, Selvaraju et al. [3] extended CAM into a more generic approach, namely, Grad-CAM. In this approach, instead of modifying the model, Grad-CAM computes the weights of the feature maps by gradients of the class scores with respect to the feature maps in the particular layer of interest. Grad-CAM can produce better interpretable saliency maps with clearer boundaries and more precise object locations than CAM. Furthermore, by combining both Grad-CAM and GBP, a more fine-grained high-resolution visualization can be generated, which is known as Guided Grad-CAM. Compared to GBP, CAM, Grad-CAM, and Guided Grad-CAM are class-discriminative. Currently, Grad-CAM is the most popular and widely accepted approach in DL model interpretation [9]. There are similar limitations of Grad-CAM to those of CAM, e.g., the most active area cannot cover the entire target and Grad-CAM performs less satisfactorily if there exist multiple objects of the same class. To overcome these limitations and further enhance the interpretable performance, Chattopadhay et al. [6] modified the formulation of weights computation and formed Grad-CAM++. Their results indicate that the Grad-CAM++ algorithm has a more complete coverage toward single or multiple targets. Visual interpretation methods other than gradient-based or feature reconstruction methods include Local Interpretable Model-Agnostic Explanation (LIME) [10], DeepLift [11], etc. More information about these methods can be found in the literature review in [6].

The methods discussed above are dedicated to generating good visualizations with reasonable explanations to a well-trained DL discriminative classifier, where the highlighted pixels or the active regions in the saliency map focus on the object consistent with its class label, which is analogous to human judgment. Therefore, these interpretation methods provide an opportunity to visualize and understand the DL model, which can partially resolve the "black box" issue. This is expected to inspire and expand the practical engineering applications, e.g., vision-based SHM, through better exploration and understanding of the trained DL models. In the vision-based SHM, there are no relevant studies specifically focusing on visually understanding and explaining the working principle of the DL models in damage recognition. Therefore, a more comprehensive investigation in this direction should be pursued.

13.2   Visual Interpretation Methods of Deep CNNs

In this section, several classic visual interpretation methods, mentioned above, are introduced in detail. To further explore the high potential of XAI technologies in vision-based SHM in a more comprehensive manner, a systematic framework with a human-in-the-loop (HiL) mechanism for diagnoses and interpretation, namely, Σ-Box, or SIGMA-Box, which stands for "Structural Image Guided Map Analysis Box", is subsequently presented in Sect. 13.3.

Firstly, some basics and mathematical notations are introduced. In the *forward pass* of a common deep CNN, F^l denotes the output from the l-th layer before the activation function, i.e., the element-wise operation $ReLU(x) = \max(0, x)$ (Eq. 4.1) used herein. A^l represents the activated feature maps from F^l, i.e., $A^l = ReLU(F^l)$, where only positive entries in F^l are passed to A^l. S is the class score vector computed by the deep CNN, refer to $S(\mathbf{x})$ in Eq. 4.5, and S^c is one entry of it for a particular class $c \in \{1, 2, \ldots, C\}$, where C is the total number of classes of the task at hand. In the *backward pass*, G^l denotes the gradient of the output with respect to A^l in the l-th layer, i.e., $G^l = \partial S / \partial A^l$, which can be computed using the backpropagation algorithm [12], discussed in Sect. 4.1.2. When the signal passes back through the activation function, the *restoring gradient* to shallower layers is denoted by R^l, where only positive gradients can be passed back. The entry (i, j), corresponding to a particular spatial location in the feature map, of R^l is computed in Eq. 13.1, using the indicator function, $\mathbb{1}\{x\}$ in Eq. 13.2, also mentioned in Sects. 7.2.2 and 8.1.1 with some variations, refer to the example in Fig. 13.1.

$$R^l_{i,j} = \mathbb{1}\left\{G^l_{i,j} > 0\right\} \cdot G^l_{i,j}. \tag{13.1}$$

$$\mathbb{1}\{x\} = \begin{cases} 1 & \text{Event } x \text{ is True} \\ 0 & \text{Otherwise} \end{cases}. \tag{13.2}$$

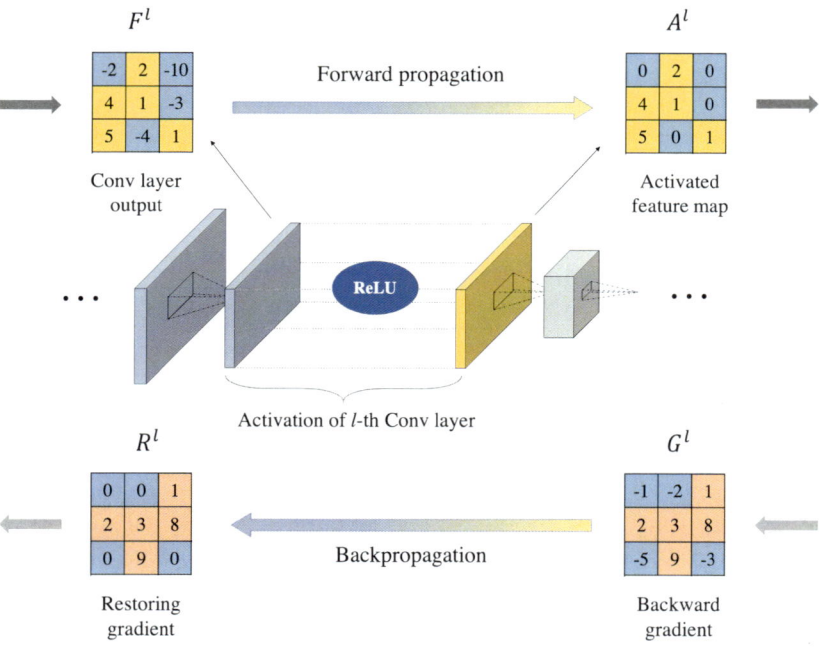

Fig. 13.1 Illustration of the forward and backward signal propagation in deep CNN

13.2.1 Guided Backpropagation

The formulation of the GBP absorbs the characteristics of both backpropagation and Decon-vNet and considers the contributions from both forward and backward passes. It is defined using $\mathbb{1}\{x\}$ in Eq. 13.2 such that while computing the restoring gradient of the l-th layer, R^l, only the entry (i, j) in the backward gradient signal with both positive gradient value $G^l_{i,j}$ and positive activation value $F^l_{i,j}$ in the forward Conv output can be passed back to the shallower layers, refer to Eq. 13.3 and Fig. 13.2 for an illustrative example. Note that the obtained gradient in this way is referred to here as the *guided restoring gradient* and is denoted by GR^l with its (i, j) entry expressed as $GR^l_{i,j}$, Eq. 13.3. This is a reasonable choice because while *ReLU* (Eq. 4.1) is applied, only positive activation contributes to the class score, S, and positive gradient contributes to the change of the parameter weights during the forward and backward passes, respectively, refer to Figs. 13.1 and 13.2. By zeroing out unrelated information, the GBP method can generate clearer saliency maps [1]. It should be noted that neurons are the detectors of particular image features. Moreover, the saliency maps aim toward visualizing what image features the neuron detects and not what features it does not detect. This is achieved by preventing the backward flow of non-positive

Fig. 13.2 Example of the formulation of guided restoring gradient in GBP

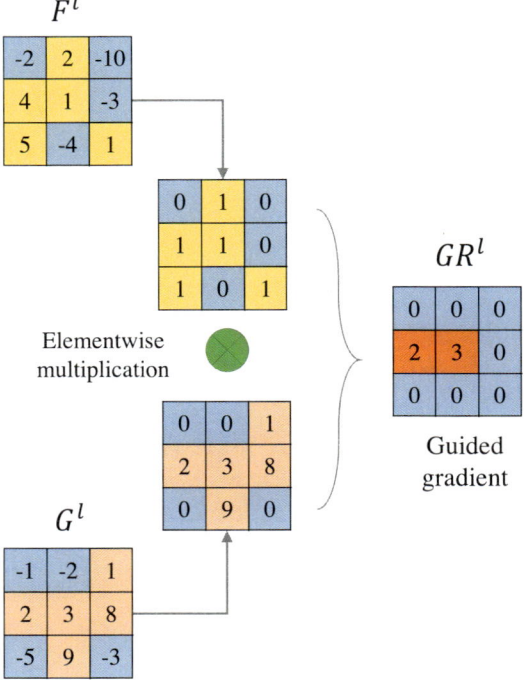

gradients, corresponding to the neurons which decrease the activation of the higher layer that the saliency maps aim to visualize.

$$GR^l_{i,j} = \mathbb{1}\left\{F^l_{i,j} > 0\right\} \cdot \mathbb{1}\left\{G^l_{i,j} > 0\right\} \cdot G^l_{i,j}. \tag{13.3}$$

13.2.2 Class Activation Map

Instead of visualizing the back-passing gradient, as conducted in GBP, CAM presents the interpretation by a weighted sum of feature maps from one selected layer in the network, Fig. 13.3. As mentioned above, performing CAM needs to modify the original model by replacing the subsequent layers of the feature maps with GAP and one FC-layer. Suppose the l-th layer is intended to be interpreted by using feature maps. Firstly, GAP projects the N activated feature maps of A^l to an N-dimensional vector, which acts as a dimension reduction. Subsequently, this vector is fully connected to the output layer which generates the class score vector, S, with a weight matrix, W. Suppose S has C classes, i.e., the size of W is $N \times C$, and the $N \times 1$ weight vector with respect to the particular class c is denoted by W^c. Thus, the new class score, S^c, for the particular class, c, can be computed as a weighted sum of the feature vector and W^c, as follows:

$$S^c = \sum_{n=1}^{N} W^c_n \sum_{i,j} A^l_{i,j;n}, \tag{13.4}$$

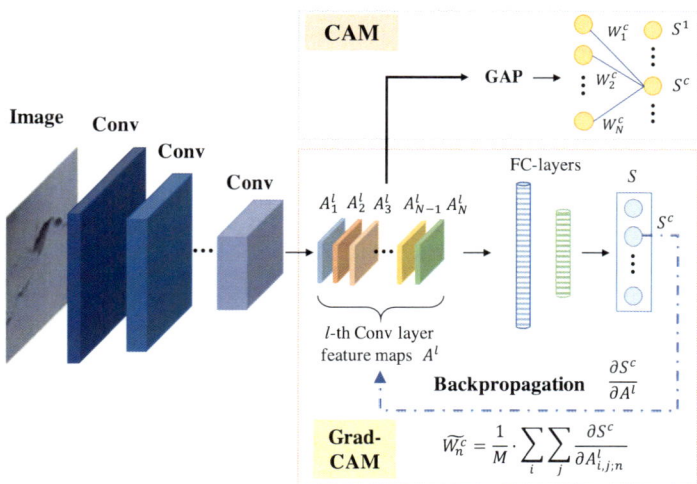

Fig. 13.3 Illustration of generating a saliency map of CAM and Grad-CAM

where for $n \in \{1, 2, ..., N\}$, W_n^c represents the weight of the n-th feature map of A^l with respect to class c, as a scalar, $A_{i,j;n}^l$ is the entry having the spatial location (i, j) in the n-th feature map of A^l, which is collectively referred to as A_n^l. $\sum_{i,j} A_{i,j;n}^l$ has a scalar output and it is a shorthand notation of $\sum_i \sum_j A_{i,j;n}^l$ representing the GAP operation over a single feature map while ignoring some constant multipliers due to the averaging calculations.

Finally, the modified DL model is retrained for one epoch and then the weight matrix, W, is obtained. Furthermore, by summing over the product of the W_n^c and A_n^l among the N feature maps, the saliency map O_{map}^c is generated, Eq. 13.5, indicating the activation of features toward class c. Herein, the "·" operation denotes a scalar multiplication, i.e., multiplying the scalar W_n^c with each entry in the matrix A_n^l. It is noted that O_{map}^c in Eq. 13.5 has the same size as the feature map. Furthermore, by up-sampling (usually using a bilinear interpolation [13]) O_{map}^c, a pixel scale saliency map is generated with the same size as the input image.

$$O_{\text{map}}^c = \sum_{n=1}^{N} W_n^c \cdot A_n^l. \tag{13.5}$$

13.2.3 Gradient-Weighted CAM

In order to avoid the architectural modification in CAM, Grad-CAM is proposed [3], refer to Fig. 13.3. Improved from CAM, Grad-CAM is a more versatile version that can produce visual explanations for any arbitrary deep CNN [3] and the weights $\widetilde{W}_n^c$ are calculated by using backward gradient signal. Suppose it is intended to interpret the model by using feature maps in the l-th layer, the weight of the c class-specific n-th feature map of A^l, i.e., $\widetilde{W}_n^c$, is computed by the partial derivatives of the class score, i.e., S^c, with respect to the different entries having the spatial location (i, j) in this n-th activated feature map of A^l, i.e., $A_{i,j;n}^l$. This weight also uses the total number of entries in A^l as a constant multiplier, M, as follows:

$$\widetilde{W}_n^c = \frac{1}{M} \sum_{i,j} \frac{\partial S^c}{\partial A_{i,j;n}^l}, \tag{13.6}$$

where in CV applications using images, M is actually the number of pixels in the feature map, i.e., $M = \sum_i \sum_j 1$. Other notations are defined in Sect. 13.2.2.

In practice, to obtain $\partial S^c / \partial A_{i,j;n}^l$, typically arranged in the form of a weight matrix $[\widetilde{W}]_{N \times C}$, the deep CNN model is usually retrained for one epoch and then the backward signal is recorded, making use of the backpropagation algorithm [12]. After computing $[\widetilde{W}]$, by summing over the product of $\widetilde{W}_n^c$ and A_n^l for all the N feature maps and then being activated by *ReLU* to remove the negative gradients, as discussed above, the saliency map S_{map}^c is generated, Eq. 13.7, indicating the activation of the features toward class c. In

Fig. 13.4 Illustration of
generating a saliency map
using the Guided Grad-CAM

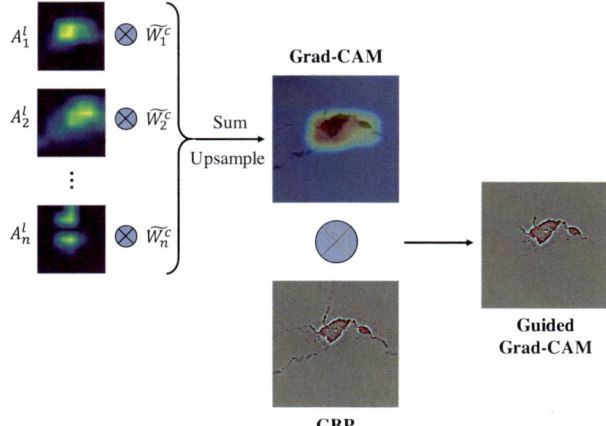

addition, *ReLU* is applied to highlight the pixels that have positive influences on the score
of the considered class c.

$$O^c_{\text{map}} = ReLU \left(\sum_{n=1}^{N} \widetilde{W}^c_n \cdot A^l_n \right). \tag{13.7}$$

Similar to CAM, a saliency map, O^c_{map}, is generated, which is further up-sampled (e.g.,
using a bilinear interpolation [13]) to have the same size as the input image.

13.2.4 Guided Grad-CAM

Further elementwise multiplication of the O^c_{map} with the saliency map generated by GBP, one
obtains a more fine-grained visualization result, which is known as the Guided Grad-CAM,
Fig. 13.4. As a fusion of GBP and Grad-CAM, the saliency map in the Guided Grad-CAM
is both class-discriminative and in high-resolution [3].

13.3   Structural Image Guided Map Analysis Box (Σ-Box)

Based on previous studies for vision-based applications [14–16], feature reconstruction
methods (e.g., Grad-CAM and Guided Grad-CAM) are more reliable, interpretable, scalable,
and have less computational costs than other methods, e.g., LIME and DeepLift, which are
not covered in this book. However, an application for the use of LIME for visualization
& explanation of deep TL results for image-based structural damage recognition can be
found in [17]. Therefore, Grad-CAM and Guided Grad-CAM are adopted as backbone
methods along with the supplement of GBP visualization, which contributes to evaluating
the rationality of each pixel in a general and expedited manner. However, these methods

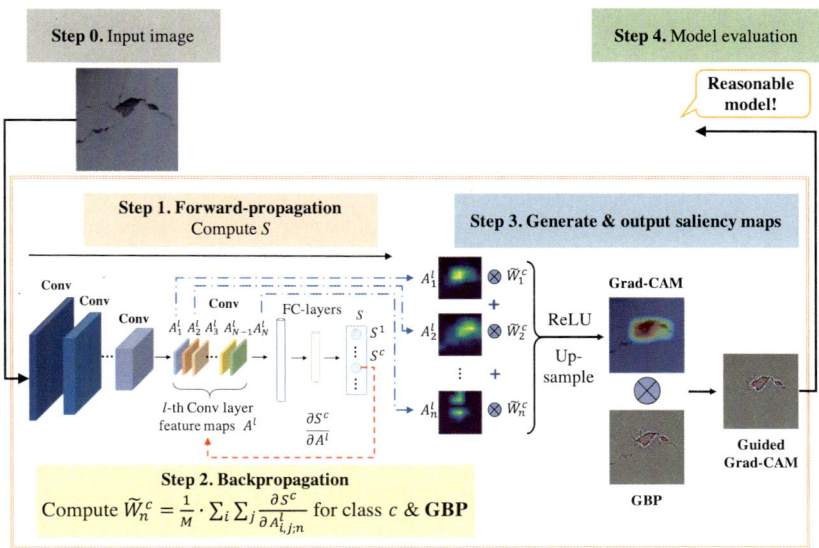

Fig. 13.5 The procedure of Σ-Box

were only performed separately in the past works, where a comprehensive investigation of them collectively is lacking.

A systematic and HiL DL model interpretation and diagnoses framework is proposed, namely, **S**tructural **I**mage **G**uided **M**ap **A**nalysis Box (SIGMA-Box or Σ-Box), which is illustrated in Fig. 13.5. In Σ-Box, different characteristics of each of these methods (GBP, Grad-CAM, and Guided Grad-CAM) are utilized to complement each other for a more comprehensive judgment basis suitable for human experts. The term "Box" originates from the objective of packaging and integrating several visual interpretation methods together with Structural Engineering knowledge. It also combats the commonly used "black box" view of deep CNNs with the goal of circumventing the negative perception associated with this view.

13.3.1   Human Interactive Evaluation

With a specified CNN model and an image containing structural damage features, the Σ-Box outputs saliency maps for engineers to visually evaluate the model. The saliency maps highlight the important pixels, where Grad-CAM and Guided Grad-CAM are class-discriminative, which can link those pixels directly to the damage-related predictions. As one step of the HiL paradigm, with these saliency maps, structural engineers can utilize domain knowledge to judge the rationale of the model. As mentioned in Sect. 13.1, GBP is class-indiscriminative and structural engineers should evaluate the model's capability and rationality via finding basic visual features consistent with the visual tools used by the human expert, e.g., edges of the crack and shape & texture of the spalled area. Conversely, Grad-CAM and Guided Grad-CAM are class-discriminative and engineers can examine the model's discriminative ability with visual results to determine whether or not the focus of the most active area of the image (i.e., the saliency map) concentrates on the right damage location.

In Σ-Box, one uniform criterion is defined for evaluation as follows: *For an acceptable model, the highlighted pixels in the saliency maps should provide reasonable visual interpretations, i.e., present correct damage features and point to correct damage objects & locations consistent with the pre-defined damage definition.* The damage definition can be based on past experience from post-disaster reconnaissance, refer to the discussion about the StEER Network in Sect. 1.3.1, and the seismic design guidelines, e.g., [18, 19]. With the application of this criterion, there are four possible outcomes: (i) correct prediction & reasonable visual interpretation, *CorRes,* (ii) *correct prediction & unreasonable interpretation,* CorUnr, (iii) incorrect prediction & reasonable interpretation, *IncRes, and (iv) incorrect prediction & unreasonable interpretation,* IncUnr. For example, *IncRes* means the prediction is wrong but the focus & active region(s) of the saliency map point to the damage location(s).

13.3.2 Outline of the Σ-Box

The Σ-Box packages three visual interpretation methods followed by a human interactive evaluation process and its outline consists of the steps shown in Fig. 13.5 and discussed below. Given (1) a structural image containing damage patterns, (2) a well-trained deep CNN, (3) a target class $c \in \{1, 2, \ldots, C\}$, and (4) a target depth l, i.e., the location of the l-th layer, the Σ-Box performs the following steps:

Step 0   Input the image into the deep CNN.

Step 1   Feedforward signals until obtaining the class score vector, S.

Step 2   Compute the class-specific weight, $\widetilde{W}_n^c$, for class, c, from the backward gradient stopping at the l-th layer. Meanwhile, based on the backward signal, generate the saliency maps of GBP.

Step 3   Generate the saliency maps of the Grad-CAM from the corresponding feature maps and weights. Subsequently, perform element-wise multiplications of the GBP and Grad-CAM saliency maps to generate the final saliency maps of the Guided Grad-CAM.

Step 4   Interpret and evaluate the deep CNN's rationality by the judgment of human experts based on the saliency maps of GBP, Grad-CAM & Guided Grad-CAM.

In the practical applications of SHM, structural engineers can adopt the Σ-Box to examine their well-trained deep CNNs and explain how the model makes the damage prediction (e.g., cracking and/or spalling) via the generated saliency maps. For example, in Fig. 13.5, the saliency maps of GBP, Grad-CAM, and Guided Grad-CAM highlight the important pixels for the predictions, where Grad-CAM and Guided Grad-CAM are related to spalling and they complement each other to provide complete information. In addition, the Σ-Box supports exploring the model's interpretable performance for any depth in the network via backpropagating signals toward the target depth l, i.e., the location of the l-th layer. Thus, it helps the engineers to understand the learning path of the model. Inherited from the packaged visual methods, the Σ-Box is robust for many types of CNNs.

In summary, the Σ-Box forms a systematic and HiL framework, where engineers input structural images and judge the model performance via the outputs of the saliency maps. The domain knowledge of Structural Engineering is exploited to compare the highlighted areas with human expert judgment. By repeating the evaluation for multiple images, a comprehensive understanding of the trained deep CNN is obtained, giving confidence to the engineers to use CNN in practical SHM projects.

13.4 Interpreting CNN in ϕ-Net Tasks

The established Σ-Box can explore the trained deep CNN in the classification of structural images for SHM. Five key questions, covering basic demands for resolving the above mentioned "black-box" model issue, are enlisted as follows:

- Can deep CNNs learn well during the basic SHM vision tasks?
- How does the CNN model learn during training?
- What are the contributions of the learned features from different depths (i.e., number of layers) of the deep CNN model?
- What are the influences of TL and DA on the performance of deep CNNs?
- Can deep CNNs learn well during more abstract SHM tasks?

In this section, adopting the Σ-Box, using deep CNN models trained in the ϕ-Net experiments (Chap. 6) as examples, is explored to answer the five questions above. The objectives and contributions of this exploration are discussed in the following list.

1. *Interpretation of spalling detection*: The Σ-Box is applied to explore the use of the trained deep CNN in the spalling detection (Task 3 in ϕ-Net, Sect. 5.2.2.3), which is one of the basic SHM vision tasks. Using the well-trained model in the ϕ-Net benchmark experiments in Chap. 6, the saliency maps for spalling are produced and analyzed. This provides a general test on whether the Σ-Box can interpret basic SHM vision tasks.
2. *Interpretation of the learning process*: During training, under ideal situations, a deep CNN model gradually converges to local minima with decreasing loss (as an objective function) making use of its gradient to descend toward the minimum loss to update the

model parameters and eventually find their optimal values, as discussed in Sect. 4.1.2. To better understand the learning process of the model used in (1) beyond numerical analysis, the saliency maps generated by the CNN at several key epochs during training are discussed.

3. *Exploration at different network depths*: It is found that the deep layers in a deep CNN capture more high-level semantic information than the shallow ones [3, 5]. Thus, visual explanations of the activated feature maps with respect to "spalling" (SP) extracted at different network depths, i.e., from shallow to deep, are studied. This verifies the findings in Sects. 6.3 and 6.5 that deep layers are more task-dependent and contain essential information for damage recognition.

4. *Influence of the training techniques*: During training, several strategies may help deep CNNs to achieve better performance with limited data, e.g., TL and DA. For TL, through fine-tuning model parameters retrained from the source domain with a large dataset, the model can adapt to the target domain with a small dataset. For DA, the conventional way is to perform certain transformations on raw images, e.g., translation, flip & scale, and then augment the raw dataset with the processed images. Results in Chap. 6 demonstrated that TL significantly enhances the performance of spalling detection. On the contrary, also results from Chap. 6 indicated the ineffectiveness of DA where the accuracy did not change if solely using DA. Moreover, using TL and DA simultaneously may even render worse results. Thus, comparisons of the saliency maps between the baseline model trained from scratch (BL), BL using TL (BL-TL), and BL using both TL & DA (BL-TL-DA) are conducted and discussed.

5. *Interpretation of damage type classification*: The above investigations are within the scope of concrete spalling detection, which is a basic and non-expert task where identifying the occurrence of spalling does not require much professional knowledge. As an extension, a similar exploration is conducted on a more complex task with higher semantic abstraction, namely, damage type classification (Task 8 in ϕ-Net, Sect. 5.2.2.8). This involves domain knowledge, e.g., the definitions of flexural-type and shear-type damages with their corresponding visual patterns caused by ductile and brittle failure modes of RC structural components, as part of the seismic design guidelines, see for example [18]. Those patterns are not so visually obvious and sometimes even experts may find it challenging to make accurate visual decisions on the respective type during reconnaissance missions or even laboratory experiments. In addition, the label definitions of the damage types are more abstract and may have different semantics, e.g., the occurrence of the damage may lead to different damage type labels due to different damage locations, cracking angles, etc.

13.4.1 Dataset Preparation

All labeled image data used in the experiments are from the ϕ-Net subsets of the spalling detection task and damage type task. From the definitions in Sect. 5.2.2, the scene level describes the scale of the target objects in the images, which is usually reflected by the distance to the target. The pixel and object levels represent images taken from the respective close-range and mid-range distances from the structural component. On the other hand, the structural level is for images containing most parts of the structure or identifying its outline. Most of the images fall within the scale of *pixel level* or *object level* and few of them are in the *structural level*.

Images having damage patterns are the only ones used with the Σ-Box. This is because labels representing no damage patterns, e.g., "non-spalling" (NSP) and "undamaged state" (UD), are more abstract and recognizing them in a classification task is an *occurrence detection problem*. For example, in NSP images, the active regions in the saliency maps (in the absence of spalling) can point to anywhere in the images, which is *indeterminate* and does not provide useful information for interpretation. On the contrary, if the saliency maps are generated for the SP images, the active regions should only point to the corresponding spalled areas, which is *determinate*. As a result, images with NSP and UD labels are not considered herein. In the experiments, most of the shown results are *CorRes & IncRes*. However, *CorUnr & IncUnr* results are also shown for model diagnoses in some cases. For investigating the model's generalization performance, the saliency maps are mainly generated based on the test dataset of the labeled images.

13.4.2 Interpretation of Spalling Detection

In this experiment, the used deep CNN is the best model trained in the ϕ-Net benchmarking spalling detection task in Sect. 6.3.5. It is a variant version of the VGG-16 model pre-trained from ImageNet [21] using TL (denoted, for short, as BL-TL). The values of training and test accuracy (Eq. 3.12) are 99.2% and 81.6%, respectively. As a binary classification, the number of correct predictions for SP and NSP is denoted as TP and TN, respectively, and the number of incorrect predictions for SP and NSP as FP and FN, respectively. Accordingly, the TPR and TNR, Eqs. 3.14 and 3.15, respectively, are computed in Table 13.1 to analyze the model bias in the test dataset. In addition, the saliency maps are sourced from the feature maps of the BL-TL model from the last Conv layer.

The saliency maps of GBP, Grad-CAM & Guided Grad-CAM are generated via the Σ-Box for all test SP images. It is observed that most results are *CorRes* and there are very limited *CorUnr* and few *IncUnr* results. Examples of *CorRes*, 16 cases, are presented in Fig. 13.6 and examples of *IncRes* and *IncUnr*, 8 cases, are shown in Fig. 13.7, where the predicted probability of the class at hand, i.e., SP in this case, is listed at the top of the

Table 13.1 Test accuracy, TPR & TNR (%) of the spalling detection task

Model	BL	BL-TL	BL-TL-DA
Accuracy	63.0	81.6	81.5
TPR	0.0	85.5	74.2
TNR	100.0	79.3	86.0

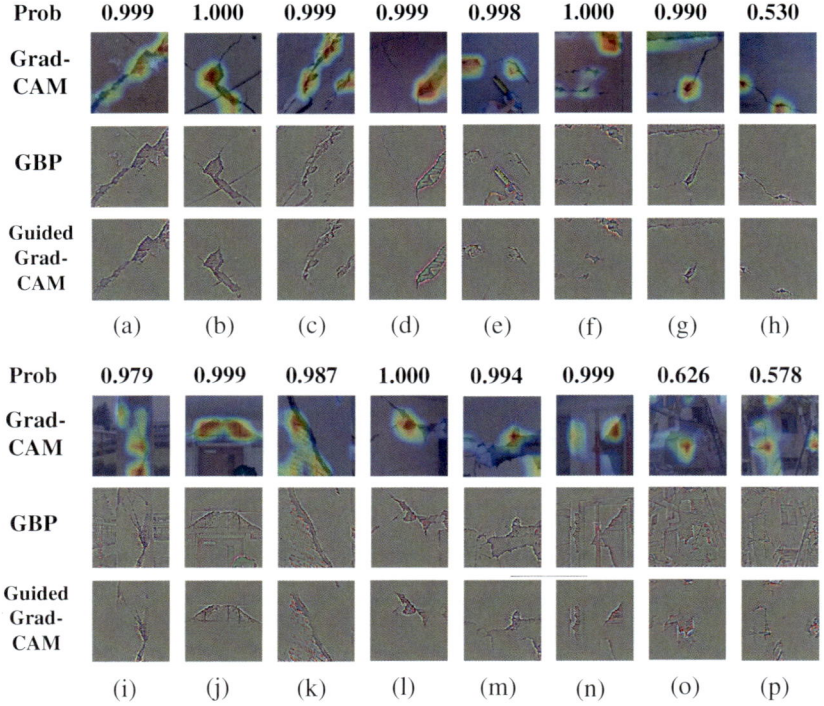

Fig. 13.6 Saliency maps in predicting the SP (*CorRes*) using BL-TL on test dataset

set of 3 maps (Grad-CAM, GBP & Guided Grad-CAM) for each case in both figures and also subsequent figures of this chapter. In general, the saliency maps furnish good visual explanations in most cases for the basic spalling detection. For the GBP, as it is a class-indiscriminative method, the saliency maps capture all pixels contributing to predictions including cracks, spalled areas, and objects unrelated to damage, e.g., Fig. 13.6e shows a human hand with a measuring tap for GBP, which is not highlighted in Grad-CAM and completely removed in Guided Grad-CAM. The findings reflect the BL-TL capability in identifying the basic visual features, i.e., line-like edges and texture difference between two adjacent regions. The saliency maps for Grad-CAM & Guided Grad-CAM ignore the

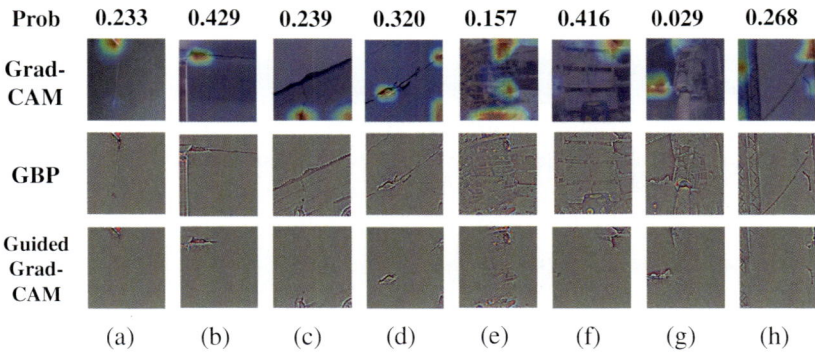

Fig. 13.7 Saliency maps in predicting SP (*IncRes* & *IncUnr*) using BL-TL on test dataset

irrelevant information and only highlight the regions contributing to the SP class. Most results correspond to the *CorRes* outcome, where the active regions are very close to human judgment.

The BL-TL model is prone to predict the image as SP once identifying the ellipse-like spots within the entire spalled area, especially in Fig. 13.6a–h. For images with irregular spalling shapes, the Grad-CAM saliency maps still indicate ellipse-like active areas even with limited coverage. For example, in Fig. 13.6j, the actual spalled shape is close to a trapezoid and the model recognizes it as SP with two elliptical regions near the two hypotenuses of the trapezoid. This observation is common among most SP training images which indicates that the model is more sensitive to the ellipse-like visual patterns. It is expected that the BL-TL model can achieve good and explainable recognition on other unseen images with regular ellipse-like spalling patterns. Moreover, BL-TL is capable of identifying small spalled areas, but it loses confidence (lower probability) in decision-making. In Fig. 13.6h, the probability of SP is low (0.53) increasing the risk of missed detection.

Compared to the coarse heatmaps generated by Grad-CAM, the saliency maps by Guided Grad-CAM are high-resolution and represent pixel-scale visualization, which clearly highlight the pixels that positively activate the prediction results. Beyond the precise damage location, the results in the Guided Grad-CAM also illustrate the boundary (i.e., edges) of the spalled area along with nearby cracks. Moreover, the texture of the spalling is highlighted in some cases, e.g., exposed reinforcing bars in Fig. 13.6j and masonry head and bed mortar joints in Fig. 13.6k, providing the inference bases for the trained model.

The interpretable quality of Grad-CAM and Guided Grad-CAM is related to the scale of the image, model recognition ability, and preference. It is observed that most pixel level (close-up) images have better visualization results, Fig. 13.6a–m, where the heatmaps almost cover the entire spalled area and also accurately capture multiple spalled areas. Although the heatmaps still can locate the correct damage areas in images with a border view, i.e., object level images, some results only highlight partial damage areas, Fig. 13.6n, o and p. From previous studies, this observation can be partially explained by the limitation of Grad-CAM, where its heatmaps often do not cover the entire object [6]. Other reasons are related to the recognition ability and prediction preference of the trained deep CNN. The receptive field in a deep CNN refers to the part of the image or feature map visible to one filter at a time. The BL-TL used herein requires resizing the images to the fixed 224×224 resolutions as inputs and this results in the same size of the respective field of each Conv layer for images corresponding to both pixel and object levels. Under the same size of the respective field, it is easier to recognize the SP class with visual patterns on the pixel level, which describes these patterns with less noise and more distinctive features compared to those on the object level. Therefore, larger areas in the feature maps are more likely to be activated in a close-up image. However, incomplete active regions do not negatively affect the rationality of the deep CNN model in any significant manner because it is still reasonable for the deep CNN to predict the SP class by only finding a partial spalled area instead of the entire region, which is similar to the usual human behavior in making classification decisions.

The results of *IncRes*, Fig. 13.7a–e, and *IncUnr*, Fig. 13.7f–h, are studied for model diagnoses. Figure 13.7a–d present similar small spalling patterns to Fig. 13.6h. However, here, the model wrongly classifies the images as NSP. This is expected to take place analogous to the observation in Fig. 13.6h, where the probability of SP class is low (< 0.5, i.e., worse than the random guess probability). Therefore, such model limitations result in an increasing number of the FN. Figure 13.7e describes a heavily spalled column, where the visual pattern is much different from most images and there is no clear boundary between the intact and the spalled parts. From the saliency maps of GBP and Guided Grad-CAM in Fig. 13.7a–e, the BL-TL captures some key features, e.g., texture and location of the spalled area. However, the model makes the wrong predictions. This indicates that the model is capable of finding the spalling-related features, but it encounters the problem of finding the correct relationship between these features and the label due to only learning from very few of these types of images leading to a lack of generalization in the prediction. The result of *IncUnr*, Fig. 13.7f, is mainly due to the image having a much broader scope (object level or even close to the structural level) causing sharp differences between most SP images in the training and the test datasets. Thus, the model is incapable of effectively generalizing to such types of images. Similarly, Fig. 13.7g is also a rare heavily spalled case and Fig. 13.7h contains some jammed in objects, e.g., wire mesh and electric cables, which make the model incapable of correctly focusing on the target and leading to incorrect predictions.

The above observations improve the understanding of the trained BL-TL model in predicting the SP class. Conclusions related to the model diagnoses are as follows:

- The current model has both accurate and explainable recognition ability for the pixel level images, but its performance degrades with the increasing distance from the target object, which indicates that the model cannot generalize well to all scene levels.
- The model is still less accurate in identifying small and uncommon spalling patterns and for images containing high noise (e.g., unrelated objects to the spalling detection task).

The above shortcomings can be alleviated by retraining the model with more images. A more straightforward approach to improve the model generality is to separate the data having more fine-grained scene levels to learn the spalling patterns separately based on these different levels. For example, by splitting the SP image data into pixel and object levels, two separate deep CNN models can be trained for pixel level (close-up) and object level (mid-range) SP images. This is analogous to the discussion in Sect. 5.5 regarding the extensions of the ϕ-Net dataset hierarchy, Fig. 5.13, particularly going from "depth-2" to "depth-3", as illustrated in Fig. 5.14.

13.4.3 Interpretation of the Learning Process

The deep CNN usually needs multiple training epochs and the accuracy increases along with the model parameter updating. Continuing with the BL-TL model, a few featured epochs (corresponding to when the test accuracy increased during the training process) are investigated. According to the training history in Fig. 6.13c, the selected epochs are 1, 2, 3, 5, 8, 11 & 14 (best test accuracy) as listed in Table 13.2 where epoch 0 corresponds to direct prediction based on ImageNet [21] pre-trained weights without fine-tuning with the ϕ-Net dataset.

The saliency maps of the BL-TL model at different training epochs are studied and two examples are presented. Figure 13.8a is *CorRes* and has an almost complete heatmap coverage that is converging toward epoch 14, achieving the best test accuracy. The shown rows of the saliency maps in Fig. 13.8 illustrate the learning process of the model. Before fine-tuning the model (epoch 0), only using pre-trained weights from ImageNet [21] to capture the image features, the model is totally biased and predicts all data as NSP. The active regions of Grad-CAM and Guided Grad-CAM with respect to SP are deviated from the damage areas. On the contrary, as class-indiscriminative, the saliency maps of GBP highlight all important pixels that the model captures, where the edges of the spalled area and the cracks are well-captured. However, these GBP maps also highlight other edges, e.g., the boundary of the column and wire meshes and even the two human inspectors in the photographs, Fig. 13.8b. This observation indicates that the ImageNet [21] pre-trained model is good at detecting line-like edges, which is the most dominating characteristic of cracking and spalling features. Transferring knowledge to the target domain of the SP

Table 13.2 Test accuracy, TPR & TNR (%) of the BL-TL at different training epochs

Epoch	0	1	2	3	5	8	11	14
Accuracy	63.0	68.5	78.6	79.1	80.0	80.6	81.5	81.6
TPR	0.0	87.7	84.8	81.9	88.4	74.2	78.4	85.5
TNR	100.0	58.1	74.6	77.2	75.1	84.6	83.3	79.3

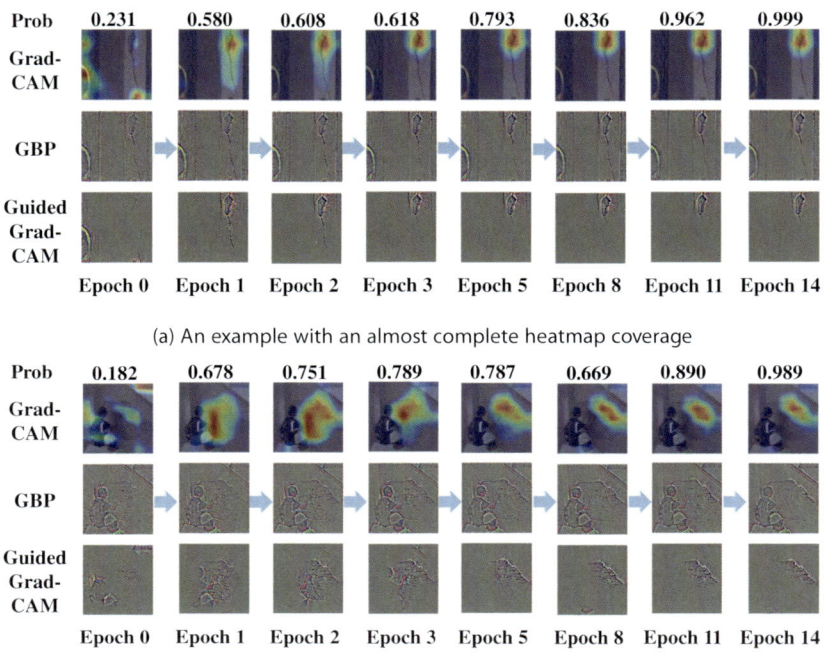

(a) An example with an almost complete heatmap coverage

(b) An example with incomplete heatmap coverage

Fig. 13.8 Two test examples of *CorRes* results for different training epochs

images using the ϕ-Net dataset, the model starts to discriminate spalling-related edges and ignore the irrelevant ones. From epochs 1 to 14, it is observed that the active regions are approaching the correct spalled locations and shapes with accurate coverage. Moreover, the probability of predicting SP is increasing. This indicates the gaining and accumulation of the discriminative abilities of the deep CNN model.

The deep CNN model is learning rapidly based on TL and it obtains a significant performance enhancement in detecting SP (TPR = 87.7%) with only one epoch training, refer to Table 13.2. Moreover, the model is becoming "smarter" linking the predictions with the visual patterns of cracking and spalling, where both are highlighted in the Grad-CAM results. As the training process continues, the model learns a better relationship between the label SP and the visual patterns with increasing confidence (i.e., higher probability of SP) by ignoring the occurrence of cracks and only focusing on the spalled areas. Unlike Grad-CAM and Guided Grad-CAM, the GBP results are less variant due to its class-indiscriminative property making its saliency maps always maintain focus on the edge features, e.g., shape and boundary of cracking and spalling regions.

Figure 13.8b illustrates the learning process of a *CorRes* example with an incomplete heatmap coverage. This example further demonstrates the inference that the incompleteness of the saliency maps is mainly due to the recognition ability and the preference of the model. The active regions of both Grad-CAM and Guided Grad-CAM are large at epochs 1, 2, 3 & 5, but they gradually shrink toward the top right corner until achieving the best accuracy at epoch 14. From the saliency maps of the Guided Grad-CAM at epochs 1 to 5, the edges of irrelevant objects to the spalling detection task, e.g., the hard hat of one of the inspectors and the arm of the other inspector, also contribute to the prediction and the model tends to eventually ignore these objects by shrinking the discriminative regions. As concluded in Sect. 13.4.2, the BL-TL model is active in identifying an ellipse-like shape of the feature, which is *convex* shape, i.e., curving outward, and easy to learn. On the contrary, in Fig. 13.8b, the complete spalled area without the human arm forms an almost *concave* shape, i.e., curving inward, especially its top boundary near the right edge of the image. Even though such concave shape can be represented by activating multiple elliptical regions, it is usually harder to learn because the model is trained by the gradient descent method (a *greedy* optimization approach for locally optimal choices, refer to Sect. 4.1.3), where gradients are along the steepest descent direction corresponding to the largest reduction of the loss function [12]. In this manner, compared to bypassing the pixel region of the human arm to form multiple ellipse-like active regions, it is more efficient for the deep CNN model during training to move and shrink the active regions toward the upper corner of the spalled area with only one ellipse-like region. To support these interpretations, the comparison of epoch 3 to epoch 5 shows that the model mostly ignores the influence of irrelevant objects, i.e., the pixel region of the human arm disappears in both Grad-CAM and Guided Grad-CAM. Therefore, the TPR of the model increases from 81.9% to 88.4%. Even though the confidence of the model occasionally drops where, e.g., going from epoch 5 to epoch 8, the probability of SP decreases from 0.787 to 0.669 and TPR decreases from 88.4% to 74.2%, refer to Fig. 13.8b and Table 13.2, the model continues to learn the trend, where the covered pixels in the regions eventually lead to the most discriminative prediction with higher confidence and TPR at the later epochs, e.g., 11 and 14.

13.4.4 Exploration at Different Network Depths

According to Sect. 6.3, the BL-TL is composed of five Conv blocks followed by multiple FC-layers. In this experiment, the saliency maps are generated from the final layer (after max pooling) of each Conv block, from shallow to deep, which represents the learning performance of the network at different depths. The results of a *CorRes* example from five different depths are presented in Fig. 13.9. The GBP, a class-indiscriminative method, plots all important pixels contributing to predicting both SP and NSP classes. In the shallow layers, i.e., Conv block #1, the saliency maps are not so informative where most pixels are highlighted and resemble noise. However, starting from block #2, the saliency maps show certain edge features. When increasing the depth to block #3, the edge features of the cracking and spalling become apparent with decreasing noisy pixels. For deeper Conv blocks #4 & #5, the edge features are more evident. This observation shows that, even

Fig. 13.9 Saliency maps of test data from different network depths

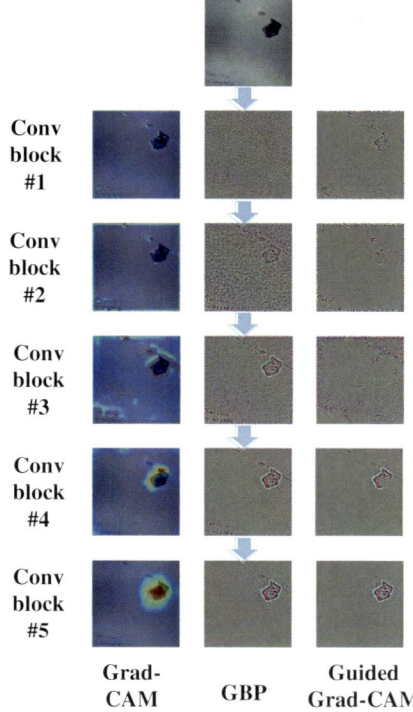

Conv block #1

Conv block #2

Conv block #3

Conv block #4

Conv block #5

Grad-CAM   GBP   Guided Grad-CAM

in shallow layers, the model is capable of learning and identifying edges of cracking and spalling, confirming the feasibility of using a relatively shallow network to identify the concrete cracking and spalling from images for the purpose of practical applications of vision-based SHM. From the results of Grad-CAM and Guided Grad-CAM, shallow layers are not sensitive to discriminate the SP class, where the active regions are small with a limited number of activated pixels. With increasing depth toward Conv blocks #4 & #5, the active regions start to focus on the correct damage locations. This observation validates the characteristics of the deep CNN that deep layers are more task-dependent and explains the findings that retraining the last two Conv blocks of the deep CNN leads to significant accuracy enhancements, as discussed in Sect. 6.3.

13.4.5 Influence of the Training Techniques

The above experiments have explored visual interpretation analyses of the saliency maps of BL-TL, which reveal the good performance of the deep CNN using TL. To investigate the influence of the TL and DA, visual results of the BL and BL-TL-DA models for spalling detection are produced for comparison in this section. BL is trained from scratch, where the model's parameters are initialized from a random space. As reported in Sect. 6.3.5, due to the poor initialization, BL's accuracy curve is flat and almost unchangeable during the training process. BL's best classification performance is listed in Table 13.1, where all data are predicted as NSP, indicating a severely biased model. While solely using DA in training the BL, the model remained untrainable with poor accuracy. Therefore, the results of the BL-DA model are not shown here. Comparing the BL-TL-DA model to the BL-TL model in this experiment, Table 13.1, shows that the model gained performance improvement in identifying NSP images (higher TNR from 79.3% for BL-TA to 86.0% for BL-TA-DA) with the trade-off in the decrease of the accuracy to detect SP as reflected in lower TPR (from 85.5% for BL-TA to 74.2% for BL-TA-DA).

Several examples of the BL's saliency maps are shown in Fig. 13.10, and all results are *IncUnr*. The saliency maps of GBP in this case show several dispersed points. Regardless of any discriminative properties, the BL model cannot capture any basic visual features of damage, e.g., edge or texture, and simply randomly treats each pixel as important, i.e., a scenario of *no learning*. When investigating the discriminative ability, the important regions of Grad-CAM and Guided Grad-CAM for all images are similar but focus on the margins and corner areas. Therefore, they do not show any clear relationships toward either cracking or spalling. In conclusion, by visualization, it is easier for the structural engineers to understand and diagnose the images than the BL model because: (1) the BL model cannot capture the damage features, and (2) the BL model lacks the discriminative ability toward the SP label. Combining the flat training history curve, Fig. 6.13c, and the non-informative saliency maps, the poor performance of BL has one possible explanation related to the poor parameter initialization. On the contrary, while adopting TL, the model parameters are pre-trained from ImageNet [21] instead of from a random space. The abundant information obtained from this pre-training empowers the model with some sense of edge and texture of the natural objects. This has been validated from the results of epoch 0 in Fig. 13.8, where the saliency maps of GBP are able to capture the edges, shapes, and textures of cracking and spalling. From this basic feature capturing, the BL-TL model starts, as early as in epoch 1, to gain accurate discriminative ability along with parameter fine-tuning during the training process.

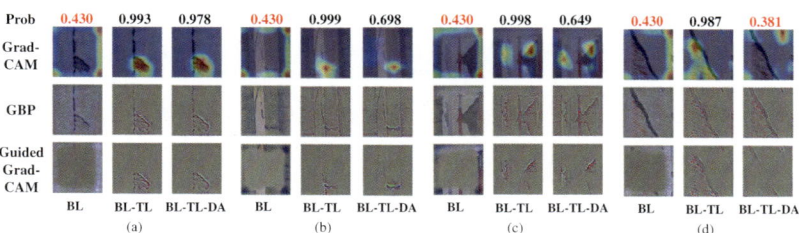

Fig. 13.10 Saliency maps of 4 images in test dataset considering 3 different training techniques

Comparing BL-TL to BL-TL-DA, the negligible 0.1% difference in the accuracy (81.6% for BL-TL vs. 81.5% for BL-TL-DA, Table 13.1) is deceptive, concealing the performance of TPR and TNR. Even though the BL-TL-DA model gains the ability to detect NSP (higher TNR), Table 13.1, its performance deteriorates in detecting SP (lower TPR) and increasing FN, which brings more risks to practical SHM applications. Figure 13.10a is an easy example for capturing the spalling feature, where the performance of BL-TL-DA is similar to that of BL-TL. On the other hand, Fig. 13.10b represents an object level SP image, which is harder to classify for SP/NSP than a pixel level image, as discussed at the end of Sect. 13.4.2. It is observed that the active regions in Grad-CAM not only have the spalled area but also contain the vertical wide crack connected to the spalled region, where both parts contribute to the final classification decision. With the use of DA, the probability of SP decreases significantly from 0.999 (BL-TL) to 0.698 (BL-TL-DA), indicating that BL-TL-DA can be less discriminative for SP. Figure 13.10c and d are two other relatively difficult examples for SP/NSP classification. For Fig. 13.10c, Guided Grad-CAM and Grad-CAM of BL-TL-DA produce similar visualization to those of BL-TL, which indicates a similar incompleteness visualization in some mid-range or far-distance images. However, the probability of SP with DA significantly drops from 0.998 (BL-TL) to 0.649 (BL-TL-DA) with a higher risk for wrong prediction, another indication of the worse discriminative ability with the addition of data using DA. Figure 13.10d represents an example where the BL-TL-DA fails in prediction and shows unreasonable saliency maps but BL-TL performs almost perfectly on both fronts. This analysis via the saliency maps in Fig. 13.10 further provides the support that BL-TL-DA loses the generalization for images from the object (i.e., mid-range) and structural (i.e., far-distance) levels and makes wrong predictions. The worse performance of BL-TL-DA from the saliency maps visualization interprets well the 11.3% TPR downstream performance compared to BL-TL in Table 13.1, where the TPR of BL-TL and BL-TL-DA are 85.5% and 74.2%, respectively.

In summary, even though BL, BL-TL, and BL-TL-DA share the same network architecture, due to different training techniques and initial conditions, they produce very different models. Moreover, the numerical values, i.e., accuracy, TPR, and TNR, can not fully reflect the intrinsic nature of these models, where the accuracy itself can be deceptive. Instead, a better understanding and comprehensive model evaluation is effectively achieved via a complementary saliency maps analysis, which is useful for *model selection*, e.g., concluding that BL-TL is better than BL-TL-DA and BL is totally unacceptable based on the discussed computer experiments.

13.4.6 Interpretation of Damage Type Classification

In this experiment, the Σ-Box is applied to a more complex and abstract damage type classification (Task 8 in ϕ-Net) with more fine-grained labels: undamaged (UD), flexural (FLEX), shear (SHEAR), and combined (COMB). It is noted that in Sects. 5.2.2.2 and

Fig. 13.11 CM of BL-TL for the damage type classification task (test accuracy)

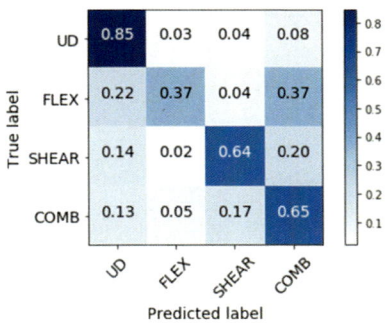

5.2.2.8, these four labels are referred to as "US" for *undamaged state*, "FD" for *flexural damage*, "SD" for *shear damage*, and "CD" for *combined damage*. From Sect. 5.2.2, FLEX, a ductile mode of damage, mostly occurs in the horizontal or the vertical direction at the restrained ends against rotations of structural components oriented vertically or horizontally, respectively, SHEAR, a brittle mode of damage, is manifested in the form of diagonal, "X" or "V" shaped crack patterns, and COMB is for complex distribution of cracks or for cracks accompanied by heavy spalling.

Analogous to explaining deep CNNs in the spalling detection, similar work using BL-TL is repeated for the damage type task. According to the benchmark experiments in Sect. 6.3.5, the values of training and test accuracy of BL-TL are 98.5% and 70.7%, respectively. The CM in Fig. 13.11 shows that the test accuracy for FLEX is very low (37%) and those for SHEAR (64%) and COMB (65%) are not high enough for practical usage, rendering the model unreliable. Besides intuitive reasoning to interpret these results, as discussed in Sect. 6.3.5, the saliency maps provide additional information and insights for informed model diagnoses. The saliency maps from the training data for this damage type classification task are discussed to complement those from the test dataset.

The saliency maps in predicting FLEX from images are plotted in Fig. 13.12. Figure 13.12a–f are from the test dataset, where Fig. 13.12a–d are *CorRes* and Fig. 13.12e and f are *IncUnr*. Figure 13.12g and h are from the training dataset, where Fig. 13.12g is *CorRes* and Fig. 13.12h is *CorUnr*. Unlike spalling detection, due to only 37% accuracy in detecting FLEX images (Fig. 13.11), most visual results are *IncUnr* and point to irrelevant pixel patches in this case. This indicates that the BL-TL model did not learn well about the FLEX class supporting this low value of accuracy. The saliency maps in Fig. 13.12a–c illustrate that the model is capable of capturing thin and horizontal cracks, representing the most successful prediction cases of FLEX. However, an inaccurate case is shown in Fig. 13.12e. Figure 13.12d–f are examples of FLEX concentrating at the ends of the component as plastic hinges. Figure 13.12d roughly captures the boundary of the spalled region, but Fig. 13.12f is incorrect. Figure 13.12g indicates that the BL-TL model learns well to recognize horizontal cracks during the training providing the foundation for further generalization to unseen test data. On the contrary, Fig. 13.12h shows the unreasonable active regions toward prediction

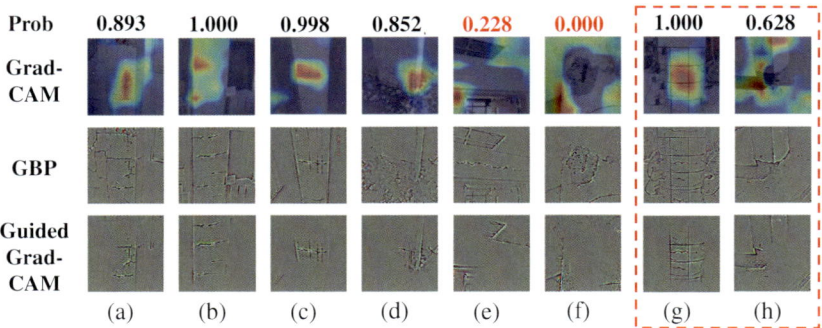

Fig. 13.12 Saliency maps in predicting FLEX (test (a–f) and training (g & h) cases)

even though the model makes the correct prediction with an acceptable probability. It is therefore inferred that the deep CNN model did not fully understand the semantics of the FLEX class even during training and, of course, did not perform well on the test dataset, especially for the case of damage occurring at the ends of the structural components. From the perspective of GBP, the edges of the cracks and spalled areas are sharp and easy to highlight. Thus, it is evident that the model is still able to identify some basic damage features of FLEX.

The saliency maps in predicting the SHEAR class from images are plotted in Fig. 13.13. Figure 13.13a–f are from test dataset, where Fig. 13.13a–e are *CorRes* and Fig. 13.13f is *IncUnr*. Figure 13.13g–h are from the training dataset and both are *CorRes*. Compared to the FLEX class, the SHEAR class accuracy is much higher at 64%, Fig. 13.11. As a result, there are more *CorRes* and very few *IncUnr* results, where wrong predictions usually correspond to *IncRes*, such as Fig. 13.13f. Consistent with the definition of SHEAR, the BL-TL model is capable of detecting thin diagonal cracks in multiple directions (Fig. 13.13a, b and e), "V" shape cracks (Fig. 13.13c) and "X" shape cracks (Fig. 13.13d) with a high detection accuracy and excellent saliency map visualization. From the CM in Fig. 13.11, 20% SHEAR images are misclassified as COMB. The most common characteristics of these images belong to the object level or have complex visual patterns, e.g., wide cracks or large spalled areas leading to classification failure cases like Fig. 13.13f. Further investigations of the training data, Fig. 13.13g–h represent object level images with complex visual patterns similar to Fig. 13.13f. However, during training, the model not only can recognize and locate damage but also it can correctly relate the SHEAR label to diagonal cracks, consistent with the SHEAR definition, instead of other irregular and heavily spalled areas with increasing confidence. This is reflected by the probabilities of the SHEAR class, i.e., 0.65 in Fig. 13.13g and 0.99 in Fig. 13.13h. It is concluded that the model is able to recognize key features of SHEAR under several typical vision patterns, especially for pixel level images. This is a promising conclusion for the application of vision-based SHM for RC structures where shear

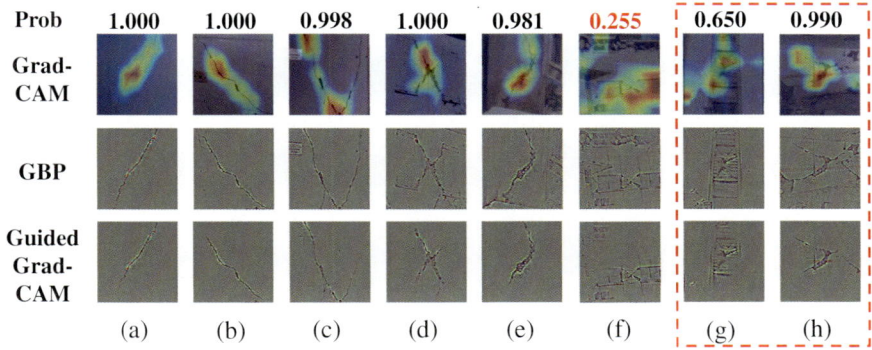

Fig. 13.13 Saliency maps in predicting SHEAR (test (a–f) and training (g & h) cases)

damage is one of the most critical modes of brittle failures leading to sudden or progressive collapse [20].

The saliency maps in predicting the COMB class from images are plotted in Fig. 13.14. Figure 13.14a–f are from the test dataset, where Fig. 13.14a–d are *CorRes*, Fig. 13.14e is IncRes, and Fig. 13.14f is *IncUnr*. Figure 13.14g–h are from the training dataset and both are *CorRes*. Figure 13.14a–b illustrate that the model made the decisions based on the simultaneous occurrence of horizontal and inclined cracks along with spalling patterns. This is consistent with the COMB definition stated at the end of Sect. 5.2.2.8 (item #3). Figure 13.14c–e show that the active areas of the saliency maps by both Grad-CAM and Guided Grad-CAM focus on the heavily spalled areas, where the damage is mainly caused by complex mixture effects, consistent with the COMB definition, mentioned above. The *IncRes* case (Fig. 13.14e) represents the poor discriminative result of the model toward damage in distant images. Even in the training process, the confidence in learning from this type of distant images is low, e.g., Fig. 13.14h. Unlike the SHEAR cases, there still exists a certain number of *IncUnr* results like Fig. 13.14f, where the raw image describes a heavily damaged structural component with significantly spalled concrete and buckled reinforcement. In this case, due to irregular visual patterns and sharp differences between these patterns, the difficulty in accurately extracting the COMB features increases. From Fig. 13.14g, the model is actually capable of predicting the correct result with an incomplete heatmap during training, where the key discriminative visual pattern is the occurrence of exposed buckled reinforcing bars. Therefore, low test accuracy of 65% for the COMB class, Fig. 13.11 and poor model prediction, as in Fig. 13.14f, are expected to be improved if more images of the type in Fig. 13.14g are collected and used for training.

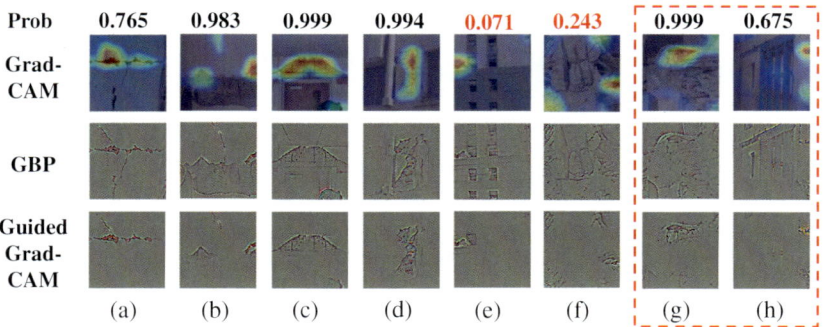

Fig. 13.14 Saliency maps in predicting COMB (test (a–f) and training (g & h) cases)

13.5   Summary and Closure

In this chapter, for further consideration of resilient AI disaster prevention systems, one of the key factors, namely, the "black box" issue of applying deep CNNs in vision-based SHM, is explored and discussed. Accordingly, an integrated HiL XAI framework, named Σ-Box, is proposed to visualize, understand, and validate the performance of the deep CNN in vision-based SHM tasks.

Through case studies, the Σ-Box is applied to the well-trained deep CNN models obtained from the ϕ-Net benchmark experiments and their predictive performances are explored via the saliency maps. Beyond the deep CNN performance in the non-expert spalling detection task, other factors in the CNNs training are investigated, including learning process, learned features from different network depths, training techniques using TL & DA, and level of semantic abstraction represented by the damage type classification. Several key observations and conclusions are obtained and summarized in the following points.

- The saliency maps of Grad-CAM and Guided Grad-CAM perform well in interpreting the deep CNN models for spalling detection. The visual interpretation qualities are related to the scene level of the images and both recognition ability and preference of the deep CNN model. In this case study, the best interpretation results are obtained in the pixel level SP images.
- While identifying spalling, highlighted ellipse-like regions are usually used as evidence of the model prediction. The deep CNN model is also capable of finding some small spalled areas. However, its performance degrades with the increase of image complexity and existence of other small or uncommon damage patterns.
- Benefiting from TL, the BL-TL model learned rapidly in the first few epochs, i.e., 79% accuracy, 82% TPR, and 77% TNR at epoch 3. The active regions in the saliency maps of Grad-CAM and Guided Grad-CAM changed to accommodate the increasing discriminative ability of the model, which may sometimes end up with incomplete coverage

of the damaged area. Regardless of the class-discriminative property, the results of the GBP indicate that the model is able to capture line-like edge features, whose property is inherited from the pre-trained model with ImageNet [21]. These features are suitable for making judgments on the occurrence of the damage (cracking and spalling in this case study) from a human expert perspective, i.e., effectively integrating the HiL as part of the Σ-Box framework.

- Through investigating the deep CNN's features at different depths, the dominating influence of deep layers in spalling identification is clearly illustrated. From the saliency maps of the GBP, the model has some sense of the damage features from the shallow and intermediate layers, and the discriminative ability is achieved in the deep layers, e.g., layers in Conv block #5.

- Different training techniques produce different deep CNN models even though they share the same network architectures. The saliency maps of the BL model consist of several dispersed points indicating untrainability, which implies that the BL is incapable of extracting damage features and lacks discriminative ability. Such shortcomings can be significantly alleviated by better initialization methods such as using TL, i.e., BL-TL and BL-TL-DA models. Therefore, the saliency map results provide useful information for the structural engineers to consider the physical meaning of the obtained model predictions for reliable model selection.

- For the more abstract damage type classification, it is concluded that with the limited training data, the current model is still not able to recognize well the abstract semantics of the damage type, especially for the flexural class. However, some saliency maps of Grad-CAM and Guided Grad-CAM show that the model has some sense of recognizing specific damage patterns for the shear and combined damage types.

- From the perspective of the model performance, the visual explanations provided by the saliency maps give more intuitions to improve the model, e.g., enlarge the training dataset, and further split data into more fine-grained scene (pixel, object & structure) levels to conduct classification within each level. It should be noted that besides cracks, other line-like edges were highlighted in the saliency maps of the current model, e.g., outlines of the structural components and grid lines in the test specimens. Therefore, not only introducing more data but also more variety of data including noise are expected to improve the robustness of the DL model.

In closing, the Σ-Box, as a representative of other fundamental and applied advances discussed in this book, is intended to encourage the Structural Engineering community to confidently adopt these AI technological advancements for better monitoring and mainly for effective rapid assessment of the built environment under service conditions, as part of the routine maintenance, or following extreme events like major earthquakes, as part of the reconnaissance efforts. This is expected to boost the decision-making process for repair and operation of the structural systems and to eventually reduce the *functional recovery time* following shocks to these systems caused by extreme events. Ultimately, this will lead to enhancing the *resiliency of the built environment*.

Fig. 13.15 Sample signals in forward and backward propagations

(a) Conv output, F^l (b) Backward gradient, G^l

13.6 Exercises

1. Discuss the potential risks of implementing AI as a "black box" in SHM.
2. Conduct a literature review and create a timeline plot to depict the history of the visual interpretation methods in DL.
3. In forward propagation, suppose the Conv output, F^l, from the l-th layer before activation is as shown in Fig. 13.15a, compute the activated feature map using the *ReLU* activation function (Eq. 4.1).
4. In backpropagation, if the backward gradient, G^l, at the l-th layer is as shown in Fig. 13.15b, compute the restoring gradient, R^l.
5. While adopting GBP for visualization, compute the guided restoring gradient, GR^l, based on the above F^l and G^l in Fig. 13.15.

References

1. J.T. Springenberg et al., Striving for simplicity: the all convolutional net (2014). arXiv:1412.6806
2. B. Zhou et al., Learning deep features for discriminative localization, in *Proceedings of the IEEE Conference on Computer Vision and Pattern Recognition* (2016), pp. 2921–2929
3. R.R. Selvaraju et al., Grad-cam: visual explanations from deep networks via gradient-based localization, in *Proceedings of the IEEE International Conference on Computer Vision* (2017), pp. 618–626
4. K. Simonyan, A. Vedaldi, A. Zisserman, Deep inside convolutional networks: visualising image classification models and saliency maps (2013). arXiv:1312.6034
5. M.D. Zeiler, R. Fergus, Visualizing and understanding convolutional networks, in *European Conference on Computer Vision* (Springer, 2014), pp. 818–833
6. A. Chattopadhay et al., Grad-cam++: generalized gradient-based visual explanations for deep convolutional networks, in *2018 IEEE Winter Conference on Applications of Computer Vision (WACV)* (IEEE, 2018), pp. 839–847
7. M.I. Razzak, S. Naz, A. Zaib, *Classification in BioApps* (Springer, New York, 2018)
8. J. Kim, J. Canny, Interpretable learning for self-driving cars by visualizing causal attention, in *Proceedings of the IEEE International Conference on Computer Vision* (2018), pp. 2942–2950

9. K. Li et al., Tell me where to look: guided attention inference network, in *Proceedings of the IEEE Conference on Computer Vision and Pattern Recognition* (2018), pp. 9215–9223

10. M.T. Ribeiro, S. Singh, C. Guestrin, Model-agnostic interpretability of machine learning (2016). arXiv:1606.05386

11. A. Shrikumar, P. Greenside, A. Kundaje, Learning important features through propagating activation differences, in *Proceedings of the 34th International Conference on Machine Learning*, vol. 70 (JMLR. org., 2017), pp. 3145–3153

12. I. Goodfellow, Y. Bengio, A. Courville, *Deep Learning* (MIT Press, Cambridge, 2016)

13. E.J. Kirkland, E.J. Kirkland, Bilinear interpolation, in *Advanced Computing in Electron Microscopy* (2010), pp. 261–263

14. P.P. Angelov et al., Explainable artificial intelligence: an analytical review. Wiley Interdiscip. Rev. Data Min. Knowl. Discov. **11**(5), e1424 (2021)

15. D. Cian, J. van Gemert, A. Lengyel, Evaluating the performance of the LIME and Grad-CAM explanation methods on a LEGO multi-label image classification task (2020). arXiv:2008.01584

16. I. Kakogeorgiou, K. Karantzalos, Evaluating explainable artificial intelligence methods for multi-label deep learning classification tasks in remote sensing. Int. J. Appl. Earth Obs. Geoinformation **103**, 102520 (2021)

17. Y.Q. Gao et al., Deep residual network with transfer learning for image based structural damage recognition, in *Eleventh US National Conference on Earthquake Engineering, Integrating Science, Engineering & Policy* (2018)

18. J. Moehle, *Seismic Design of Reinforced Concrete Buildings* (McGraw Hill Professional, 2014)

19. Y. Gao, K.M. Mosalam, PEER Hub ImageNet: a large-scale multiattribute benchmark data set of structural images. J. Struct. Eng. **146**(10), 04020198 (2020)

20. M. Talaat, K.M. Mosalam, Modeling progressive collapse in reinforced concrete buildings using direct element removal. Earthq. Eng. & Struct. Dyn. **38**(5), 609–634 (2009). https://doi.org/10.1002/eqe.898. eprint: https://onlinelibrary.wiley.com/doi/pdf/10.1002/eqe.898. https://onlinelibrary.wiley.com/doi/abs/10.1002/eqe.898

21. J. Deng et al., Imagenet: a large-scale hierarchical image database, in *2009 IEEE Conference on Computer Vision and Pattern Recognition* (2009), pp. 248–255

Correction to: Artificial Intelligence in Vision-Based Structural Health Monitoring

Correction to:
K. M. Mosalam and Y. Gao, *Artificial Intelligence in Vision-Based Structural Health Monitoring*, **Synthesis Lectures on Mechanical Engineering,**
https://doi.org/10.1007/978-3-031-52407-3

The original version of this book was inadvertently published with incorrect fonts in the figures for the following chapters 4 (Figs. 4.8, 4.9, 4.10) and 12 (Fig. 12.1). The same has been corrected.

The correction chapters and the book have been updated with the changes.

The updated version of these chapters can be found at
https://doi.org/10.1007/978-3-031-52407-3_4
https://doi.org/10.1007/978-3-031-52407-3_12

Index

K. M. Mosalam and Y. Gao, *Artificial Intelligence in Vision-Based Structural Health Monitoring*, Synthesis Lectures on Mechanical Engineering,
https://doi.org/10.1007/978-3-031-52407-3